The
CLASSICS
of
SCIENCE

The CLASSICS *of* SCIENCE

A Study of Twelve Enduring Scientific Works

DEREK GJERTSEN

LILIAN BARBER PRESS, INC.
NEW YORK

First published in the United States of America 1984 by

LILIAN BARBER PRESS, INC.
Box 232
New York, NY 10163

Library of Congress Cataloging in Publication Data

Gjertsen, Derek.
The classics of science.

Bibliography: p.
Includes index.
1. Science – History – Sources. I. Title.
Q125.G54 509 83-27539
ISBN 0-936508-09-4

Manufactured in the United States of America

CONTENTS

Preface and Acknowledgements

Scholars of early Islam used to distinguish between those who, like themselves, Jews and Christians, were *ahl al-kitab*, people of a book, and those who lacked a religion revealed in a sacred text. The former they accorded certain legal privileges and rights, the latter could be enslaved and were thought little of. Something comparable can be seen in science where its practitioners much more than most other intellectuals are, or certainly were, people of not one book but, as *ahl al-kutub,* of several. Without too much exaggeration it could thus be claimed that for many centuries an astronomer was not particularly someone who spent his nights observing the heavens but rather someone who spent his days mastering the intricacies of a text like *The Almagest* of Ptolemy.

And so on throughout much of its history science, or different parts of it, has been dominated by some particular book. Some of these books have been innovational and like Darwin's *Origin of Species* have presented a new theory, broad and deep enough to allow scientists from many disciplines to pursue and to continue to pursue its various implications. Other works have not so much introduced new theories but like *The Elements* of Euclid collected and arranged the labors of others in so supreme and authoritative a manner as to permit no competition. It would be absurd to claim that there is *no* more to the history of science than the publication of a dozen or so works of this kind; it is equally true that much of the development of science can be seen in such books. Science, at least in its history, is the most literary of disciplines.

There already exist many excellent studies of most of the works discussed here, and for those interested in the details of the dynamics of Newton, Euclid's treatment of irrational magnitudes or other specialized topics it is to these excellent monographs they should turn. Instead, I have tried to do something different by simply telling the story of twelve of the most important classics of science. This has inevitably involved giving some account of the contents of each volume, particularly when they may be unfamiliar to most readers, but equal space has also been given to how the books came to be written, their reception and their subsequent publishing history. The assumption behind this approach is that virtually nothing can be too trivial to record about works of such importance in the intellectual history of Man. No attempt has been made, however, to compete with the works of the professional bibliographers although their results are incorporated in virtually every page of this book. Few are really interested in knowing that in the first edition of *Origin* the word "species" is misspelt on line eleven of page twenty; consequently such details have been ignored, and throughout the book it is the publication and not the printing of the classic works of science that has attracted me.

There is thus little talk of type faces and paper sizes although the various editions of the works discussed are comprehensively and, I trust, accurately recorded. The aim has been instead to concentrate on the dramatic events that regularly have accompanied the publication of such works. And dramatic indeed such events have often been. Works which have shattered the accepted views of the bulk of mankind have seldom emerged with any less controversy and publicity than that which has greeted in more recent times such familiar books as *Lady Chatterly's Lover* and *The Pentagon Papers.*

Anyone who attempts to write about important works soon becomes aware of how much he depends on the labors of the numerous scholars who, over many decades, have edited texts, sought out manuscripts, produced translations, collected letters and engaged in that minutiae of exact scholarship that allows synoptic works like the present to be compiled. Where a specific source has been used I have endeavored to make the obligation clear at the appropriate place; there are some debts, however, which are so basic and persistent that something more than an acknowledgement is called for. I would therefore like to express my especial indebtedness to the following works for valuable insights and much essential information:

On Hippocrates and ancient science in general: G. Sarton, *A History of Science*; G. E. R. Lloyd, *Hippocratic Writings*; on Euclid, the quite invaluable three-volume edition of *The Elements* by T. Heath and his equally superb two-volume work on *Greek Mathematics*; on Ptolemy, everyone is deeply indebted to the two works of O. Neugebauer, *The Exact Sciences in Antiquity* and *A History of Ancient Mathematical Astronomy* as well as the important commentary, *A Survey of the Almagest* by O. Pedersen; on Copernicus, there is the edition of *De revolutionibus* by E. Rosen as well as his *Three Copernican Treatises* while no one can approach Copernicus anymore without first reading Arthur Koestler's *The Sleepwalkers*; students of Vesalius have been admirably served by the brilliant biography of C. D. O'Malley, *Andreas Vesalius of Brussels* and the *A Bio-bibliography of Vesalius* of H. Cushing; on Galileo, the debt to Koestler's *Sleepwalkers* must once more be mentioned together with G. de Santillana's *The Crime of Galileo* and S. Drake's *Galileo at Work*; on Harvey, there is the great *The Life of William Harvey* by Geoffrey Keynes together with his *A Bibliography of the Writings of William Harvey*; the chapter on Newton is particularly indebted to the work of I. B. Cohen, his *Introduction to Newton's Principia* and the two-volume variorum edition of *Principia* he produced with A. Koyré. Mention must also be made of the biographical work of R. Westfall, *Never at Rest,* the bibliography of Newton by P. and R. Wallis, *Newton and Newtoniana,* and the various writings of Frank Manuel: *I. Newton Historian, The Religion of Newton* and *A Portrait of I. Newton*; on Linnaeus it is the writing of W. T. Stearn I have found most helpful with his introduction to *Species plantarum* and his appendix to Wilfred Blunt's biography of Linnaeus, *The Compleat Naturalist,* being the most useful; on Dalton I must mention M. P. Crosland's *Historical Studies in the Language of Chemistry*, the magisterial tomes of J. R. Partington's *History of Chemistry* and the bibliography by A. L. Smith, *John Dalton 1766-1844*; with Lyell I have found the biography of L. Wilson, *Lyell: The Years to 1841* the most useful while mention must also be made of M. Rudwick's *The Meaning of Fossils* and C. Gillespie's *Genesis and Geology*; finally, amongst the massive literature on Darwin I would like to express my indebtedness particularly to P. Vorzimmer's *Charles Darwin: The Years of Controversy*, the variorum edition of *Origin* produced by M. Peckham and the bibliography of R. Freeman, *The Works of Charles Darwin*.

There are also obligations of a more personal kind. Above all I owe a special debt to Richard Greenfield whose patience and friendship have done much to make this book possible. Even more patient have been an uncomplaining family waiting for the completion of a task which at one time must have seemed unending; help though from my wife Kirsty and daughter Polly of a more practical kind eased the burden at one point. I wish, however, to dedicate the book to an old friend and one time colleague Edmund Collins in memory of the many hours of conversation spent with him in the once happy land of Ghana during the period 1960-78.

CHAPTER 1

Are There Classics of Science?

No man ever read a book of science from pure inclination.

– Dr. Johnson

Who lasts a century can have no flaw
I hold the Wit a Classic, good in law

– Pope

Madoc will be read, when Virgil and Homer are forgotten.

– Richard Porson[1]

Qu'est-ce qu'une classique?

– Sainte-Beuve

A classic then is a book that is read a long time after it was written . . . without institutional constraint . . . by the competent.

– Frank Kermode

Are there classics of science in the same way that there are classics of literature and philosophy? Are there works of science that are read in the same way and as frequently as the dialogues of Plato, the plays of Sophocles, the verse of Goethe, the essays of Montaigne, or the novels of Tolstoy?[2] An initial response is likely to be that there are no such works or, if there are any, they must be so few as to be incapable of forming a significant class. Thus J. W. Burrow in his recent introduction to the Penguin edition of *Origin of Species* (1968) has noted that, unlike works of literature and philosophy, texts of science endure as "contributions" rather than "books." Nor are reasons hard to find for this widely recognized state of affairs. To most readers, even scientists themselves, the works of the past are likely to be expressed in a vocabulary that is both technical and unfamiliar. But, in addition, "however successful the theory they propound, there is always a sense in which they become superseded; at best they are overlaid by the subsequent work they inspire, which thereupon becomes the most authoritative work on the subject and is usually all that is referred to by the students " (Darwin 1968: 11). There are exceptions, Burrow declares, but they are extremely few. *Origin* is clearly exempt from such comments as is, perhaps, "Galileo's dialogues on motion."[3] A pretty slim subject for a book!

There is much truth in Burrow's judgment. Newton's *Principia* and Ptolemy's *Almagest* are, because of their rich mathematical content, fully accessible to few while Vesalius's *De fabrica*, although published over four hundred years ago, can still only be read in its original Latin. It is equally true, as Burrow pointed out, that great scientific works of the past have been superseded and regularly continue to be superseded in ways unknown in literature and philosophy. Today's consultants interested in cardiac anatomy and physiology simply do not consult textbooks fifty or even twenty-five years old, still less are they likely to refer to Harvey's *De motu cordis* published in 1628.

In contrast, classic works of literature and philosophy seem to escape this fate. Some may date badly and if not of the very highest class drop out of sight altogether or become that curiosity, an "obscure" or "little-known" classic.[4] The success of *West Side Story* or *My Fair Lady* has not meant that *Romeo and Juliet* or *Pygmalion* have in any way become out of date, inadequate or no longer of interest; in fact a quite opposite reaction is usually noticeable in a situation of this kind and the success of the modern version tends to send many back to its classic source. It is well known that the surest way to get a work like Jane Austen's *Pride and Prejudice* back onto the bestseller list is to commission a television adaptation. In contrast, the quickest way to fill the second-hand bookshops with unwanted textbooks is to publish a new edition.

It would be absurd to deny that scientific works, however significant, do date and are superseded by later texts. At the same time it would be imprudent to suppose the process to be rapid and universal. Books on the microchip, interferon, superconductors, quarks or quasars are likely to date as rapidly as those on the 1982 World Cup or the Pope's visit to Britain. Yet there are other books whose history must be measured in centuries rather than decades. *The Elements* of Euclid, for example, was still a working text in schools and universities in the late 19th century when it was well over 2,000 years old. *The Almagest* lasted less long, only until the 16th century, when it was a mere 1,500 years old. It was not just that astronomers used the constructions of Ptolemy or adopted his geocentric framework but that the actual text was pored over in the way of any advanced textbook. It would seem, therefore, a little strict to admit as classics works of fiction that have been available for little more than a few decades while objecting to the use of the term to describe works like *The Almagest* and *The Elements* which survived in regular use for many centuries.[5]

At the same time it should also be recognized that while such works may no longer be used as working texts they continue to arouse interest in a wide variety of readers. To meet this demand modern editions are regularly published, some specialized and expensive, some facsimiles but others cheap and readily available. Thus the main elements of the *Hippocratic Corpus* can be found in Penguin paperback, as can Darwin's *Origin;* Newton's *Optics,* together with Euclid's *The Elements* are all available in Dover paperback while *De motu cordis* can be had in an Everyman edition and Galileo's *Siderius nuncius* in a Mentor edition. Admittedly, there are exceptions: it is not easy to get hold of a copy of Lyell's *Principles* and to obtain a copy of Copernicus's *De revolutionibus* one needs access to a large library or a substantial book budget. Still, the fact that most science classics, if such they be, can be purchased easily speaks of the continuing interest in them.

A further mode of access to the classic works of science is through currently available anthologies and collections of readings.[6] Nor is there anything particularly new about attempts to make the contents of classic works available to the wider public in either summary or excerpted form. Both Ptolemy and Vesalius produced companion volumes to their major works by which they hoped to make them available in cheaper and more accessible forms.

At the same time it must be conceded that the fate of the so-called classic of science is often one of obscurity. Even such basic works as *The Almagest* and *De revolutionibus* can remain out of print and untranslated into modern languages for centuries. Consider the fate of a few:

The Almagest: Written in the second century A.D., it was printed in 1515 but remained out of print between 1549 and 1813 when it first appeared in a modern language (French). It was published in English only once – in 1952 – about 1,800 years after it was written.

De revolutionibus: First edition was published in 1543; subsequently it remained out of print between 1617 and 1854, when it was first translated into a modern language (Polish). It first appeared in English in 1952.

Méchanique céleste: The five volumes of Laplace's classic work were published during 1799-1825. It has never been completely translated into another language, although the first two volumes were published in a German translation in 1800-02 and Vols. 1-4 were translated into English by Nathaniel Bowditch and published in Boston in 1829-39.

Cours de chymie: Written by Nicolas Lemery and published in 1675, it sold, in Fontenelle's words, as well as "a work of gallantry and satire." Translated into Latin, English, German, Dutch, Italian and Spanish, it is known in some forty editions. But since 1756 it has not been reprinted and few, if any, outside the ranks of historians of chemistry have read it.

The fate of Lemery and his *Cours* is perhaps the more typical fate awaiting major works of science. However popular they proved in their day, and their day may well have been a long one, once out of date they disappear from view completely. Known as a name, an entry in a bibliography or in a footnote, they survive only in encyclopedias and large reference works and are as much read as the sermons of a previous century. What follows from this, however, is not that there are no classics of science but that the criteria by which they are judged cannot be the same as that applied to works of literature and philosophy. How should they be judged?

One of the most successful books of modern times has been, surprisingly, *Origin of Species.* Translated into twenty-eight languages and gone through 326 editions in 116 years, it established a record equalled by few other works. Yet its classic status depends on more than this if only because books of a relatively trivial kind can be produced with no less impressive records. From the medieval period, for example, *Tractatus de sphaera* can be noted. Written by John Holywood (1200-1250), better known in the latinized form of Sacrobosco, it is a brief and simple account of the various spheres and circles of traditional cosmology. It exists in numerous manuscripts and was first printed in 1472 in Ferrara with a further twenty-four editions to appear before 1500. In the 17th century well over two hundred further editions were printed, in Latin or translation.[7] Just about every press in Europe must have produced a Sacrobosco at one time or another in marked contrast to the care which was taken to avoid any contact with the works of Copernicus.

In modern times, excluding textbooks and telejournalism, there is *Weltrasel* (1899), translated as *The Riddle of the Universe,* by Ernst Haeckel (1834-1919). Its statistics are impressive: 100,000 copies were sold in its first year; ten editions had appeared by 1919; translated into twenty-five languages by 1933 it went on to sell half-a-million copies in Germany alone. Yet, according to one biology historian, the work is "wholly valueless" from a scientific point of view. Whatever its scientific value may or may not have been, it is difficult today to appreciate what the half-million readers can have seen in it. Without illustrations in black and white, let alone the glossy color plates demanded by today's mass audiences, Haeckel's work is a densely argued defense of mechanism and materialism replete with the neologisms he so much favored.[8] As volume No. 3 in the Thinker's Library, *The Riddle of the Universe* is found in almost every secondhand bookshop in Britain for only a few pence, evidence of how it once caught the intellectual mood of the times. Also evident in some profusion are such works as *The Mysterious Universe* (1930) by James Jeans (1877-1946) and *The Expanding Universe* (1933) by A. Eddington (1882-1944) which so appealed to the interwar generation, while the discards of today's generation, such works as Monod's *Chance and Necessity* (1971) and Dawkin's *The Selfish Gene* (1976), are already beginning to compete with them for space. These are works that tend to date and become virtually unreadable to a later generation; the classics, in contrast, though they may remain as difficult to read as they were the day they were first published, never become unreadable.

If it is not the popularity and sales of a work which make it a classic, what criteria can be appealed to? The simple answer is that classic works are ones which transform science, or, in more fashionable language, which produce a major intellectual revolution. Some years ago T. S. Kuhn published *The Copernican Revolution* (1957) and his approach has been followed more recently by I. B. Cohen in *The Newtonian Revolution* (1980) and by Michael Ruse in *The Darwinian Revolution* (1979). The volumes are – once the editorial and background material is removed – little more than studies of *De revolutionibus, Principia* and *Origin of Species* of Copernicus, Newton and Darwin respectively. The revolution was the book and study of the revolution is little more than study of the book, its sources, its content, its reception and its publishing history. Without *Principia* there would have been no Newtonian revolution; equally, without *De revolutionibus* and *Origin* there could have been neither Copernican nor Darwinian revolution. This is not meant in the trivial sense that if someone other than Darwin had written *Origin,* then the revolution clearly would not have been a Darwinian one but, rather, in the sense that the revolution associated with Darwin and *Origin* could only take place through the publication of a comparable work, whoever its author might be.

Having identified classic works of science as the mediums of intellectual revolutions, a number of general explanatory comments are called for. In the first place it is important to realize that we are discussing classic works of *science* and not some sub-discipline such as organic chemistry, bacteriology, topology or the anatomy of the heart. In each of these areas there are, no doubt, major works which have had a profound impact on their respective disciplines but have yet to influence wider areas. Thus a work which revolutionizes topology yet leaves organic chemistry and nuclear physics untouched may well be a topological classic but it is not, in the sense assigned to the term here, a classic of science. Classics such as Galileo's *Siderius nuncius* and Harvey's *De motu cordis* are thus included not because of their contributions to our knowledge of the satellite system of Jupiter or the anatomy of the heart but because through their explorations in these fields new possibilities were revealed to all scientists.

A better example, perhaps, is *Réflexions sur la puissance motrice du feu* (1824) by Sadi Carnot (1796-1832). This work begins by considering the thermal efficiency of the great new power source of Carnot's time, the steam engine, but in doing so it introduces ideas into science that proved so deep and general that they could apply to any process whether it be physical, chemical or biological. Carnot began by asking how engineers could improve the efficiency of their new machine. Could they continue to improve their design so that more and more work would be obtained from the consumption of the same amount of fuel? Or is there perhaps, Carnot asked, a limit which "nature . . . will not allow to be passed by any means whatsoever?" Carnot showed that there was such a limit and that it was a very general limit completely independent of the nature of the engine contemplated. Throughout the century the implications of Carnot's arguments became increasingly apparent to scientists in most fields. It soon became clear, for example, that chemical reactions could not be analyzed simply in terms of the affinity that various elements may or may not have for each other; there were also thermodynamic considerations to take into account. Physicists too found that many of the questions they had traditionally considered, like the radiation of heat from hot bodies or the distribution of velocities amongst molecules, made little sense until the factors raised by Carnot were first considered. And so the process has continued with, for example, the emergence of the new scientific discipline of ecology being, at one level, an exercise in thermodynamics.[9].

Or, to take a more recent example, consider the case of *Die Enstehung der Kontinente und Ozeane* (1915) by Alfred Wegener (1880-1930). First translated into English in 1924 under the title *The Origin of Continents and Oceans,* the work introduced into modern science the enormously fruitful idea of continental drift.[10] Although initially not much favored, the hypothesis was quickly shown in postwar years to be applicable to an exceptionally wide range of questions in geology, geophysics, evolutionary biology and cosmology. It would be absurd to claim that every single area of science had been deeply affected by Wegener's work for presumably the topologists and crysallographers amongst many others were little moved by thoughts of continental drift; yet the presence of exceptions does not weaken the main point that Wegener's book has had a *deep* and *novel* influence on *many, major* areas of science. If these four criteria can be satisfied in a single work then it can be counted as a science classic.

Nothing already said prevents there being classic works of, for example, psychiatry which are not classic works of science. Freud's ideas are novel and deep but they have failed to exercise much influence over "many, major" areas of science. Indeed, the central disciplines of chemistry, physics and mathematics are totally outside the scope of his work. Consequently it is irrelevant to point to the impact of his work on psychiatry, art, history, literature and psychology. A classic of modern thought, perhaps, but not of science.

It is to be hoped that the works actually selected more or less speak for themselves in this respect. The atomic doctrine of Dalton, the evolutionary ideas of Darwin, the notion of a universal attractive force proposed by Newton have clearly exercised considerable and persisting influence over most areas of science. They were all, also, original works; or, at least, as original as any work ever is in science. It requires no willful act of destruction of the continuity of history to see that physics, chemistry and biology were never really the same after the publication of *Principia* (1687), *New System* (1808) and *Origin* (1859). A word first, however, about originality. As soon as Newton, Dalton and Darwin had published their work others came along who claimed the priority for themselves. The efforts of historians of later generations have shown that in virtually all cases the ideas of classic works had been formulated well in advance by others. Dalton was not the first atomist, Darwin was not the first evolutionist and Copernicus was not the first heliocentrist. The point is true but trivial. Great intellectual works are not reducible to some simple phrase or formulae. Newton was not a great scientist because he said all bodies attract each other, nor can the insight of Darwin be determined by his claim to have been the first to state that animals have descended from a common ancestor. Such thoughts are merely the beginning of science.

What Newton and Darwin did in their classic works which was completely novel was not simply to state such views but to draw their consequences, shape them, point out the difficulties and how they could be met, deploy them in the various areas to which they apply and attempt to establish the plausibility of the theory. In brief, they have produced a *developed* theory which can be examined, evaluated, and advanced; they have not simply uttered some motto or dictum. New ideas may well be the life blood of science but developed theories are then its heart and arteries. But developed theories are seldom constructs which can be formulated with the brevity of a proverb or a mathematical formulae; hence the need for the classic work.

One consequence of this account of classic works of science is the realization that classics are relatively rare. Whereas classics of literature number in the thousands and individual authors such as Plato, Shakespeare, Goethe or Balzac could contribute a score or so each, the same cannot be said for the authors of science classics. A Newton, a Gauss or a Darwin at best produced two or three classic books; it is ludicrous to suppose any man could produce more. The effort to transform scientific thought is too considerable to be made often. Indeed, the half serious attempt to list precisely "100" classic books of science requires a considerable loosening of criteria before the full complement can be reached.

The difficulties are three. In the first place, as already mentioned, there is no shortage of classic works in astronomy, selenography, and so forth, but works with a more comprehensive effect are thinner on the ground. Again, the search was restricted to books and deliberately excluded papers. If these were allowed, then the number of 100 would be attained with ease. Finally, it could well be the case that the classic books of science are things of the past; they are certainly easier to identify in earlier centuries than today.

The reasons for this are twofold and begin with the growth of the scientific paper at the expense of the book.[11] The great revolutions in science in this century have mainly been the products of particular papers published in specialized periodicals. The names of Einstein, Rutherford, Bohr, Dirac, Watson and Crick, Planck, and Hubble are all associated with the publication of classic papers rather than books. So, too, with many 19th-century figures. Neither Faraday nor Pasteur, two of the most creative intellects of any time, prolific writers and industrious researchers both, ever produced a classic book. Rather, their work came out in a series of papers, reports, monographs, and essays. Presumably this trend will continue and when they do appear, scientific books will be textbooks, collections of previously published papers, highly specialized and bulky monographs, and works of popularization.

Another reason for the relative shortage of classic books in recent times is the greater fragmentation and stability now apparent in science. Science as a whole is less prone to revolutionary takeovers and, further, a revolution in one area – however profound – is less likely to have much of an impact on adjacent disciplines. It is today less easy for a Copernicus to come along and claim the whole system is wrong and must be turned upside down.[12] The elementary particle physicists and the molecular biologists, for example, have staked out for themselves such independent and powerful domains that it is difficult to see how they could ever be overthrown. Some critics have even claimed that so great is the capital investment in them, both intellectual and financial, that no one can afford to see them overthrown.[13]

In contrast, other areas are too fluid to allow, as yet, the emergence of classic works. In order to produce a text that completely overthrows an established intellectual orthodoxy, there must first be an agreed orthodoxy. Yet, in such areas as cosmology, astrophysics and geophysics strange and exciting results are appearing so frequently that before it has been possible to absorb and comprehend the wonders of last week another batch are emerging from the observatories of the world.

It is, therefore, to the past that the reader must go in search of the classic works of science. As most things in science have their origin in ancient Greece, it is there that the earliest specimens are likely to be found. Many failed to survive and the works of such early geniuses as Pythagoras and his followers are known only in fragments recorded by later scholars. The first recognizable scientific work of classic proportions to have survived in anything like its original form is the *Corpus* of Hippocrates, and it is consequently this work which will be examined first.

Further Reading

There is, unfortunately, no comprehensive, single volume on the history of science which can be unreservedly recommended; nor, given the magnitude of the task, is one ever likely to appear. It is a subject that thrives at the monograph level but emerges only indifferently when anything more general is attempted. Of the multi-volume efforts the best on offer at the moment and the most readily available is the Wiley (or Cambridge) History of Science series, edited by William Coleman. Volumes relevant to works discussed here are: Edward Grant, *Physical Science in the Middle Ages* (1971); A. G. Debus, *Man and Nature in the Renaissance* (1978); R. S. Westfall, *The Construction of Modern Science* (1971) and William Coleman's *Biology in the Nineteenth Century* (1971). Although initially promising, the Fontana series edited by A. R. Hall has advanced no further than Hall's own *From Galileo to Newton 1630-1720* (1963) and *The Scientific Renaissance 1450-1630* (1962) by Marie Boas. The three volumes of S. E. Toulmin and June Goodfield, available in Penguin paperback, *The Fabric of the Heavens* (1963), *The Architecture of Matter* (1965) and *The Discovery of Time* (1967), also provide an adequate background. Otherwise it is to the more specialized volumes that the inquisitive reader must go. For ancient science, and for those with a bibliographical appetite, Sarton's two-volume work, *A History of Science* (1953-59), will have most to offer. More reliable and up to date, however, are the two excellent works of G. R. Lloyd: *Early Greek Science: Thales to Aristotle* (1970) and *Greek Science After Aristotle* (1973). For medieval science A. C. Crombie's *Augustine to Galileo* (1964) remains unchallenged as does, for a later period, A. R. Hall's *The Scientific Revolution 1500-1800* (1962), although a new and much revised edition of this latter work has been promised. Still valuable, and indeed, indispensable when data on some of the less central disciplines are sought, are the two volumes of A. Wolff, *A History of Science, Technology and Philosophy in the 16th and 17th Centuries* (1935) and the corresponding volume for the 18th century (1938). As a reference work *The Dictionary of Scientific Biography* (1970-80, 16 vols.) is likely to meet most demands while the standard work on scientific literature remains J. Thornton and R. Tully's *Scientific Books, Libraries and Collectors* (1971, third edition). The role of the classics of science has, since the appearance in 1962 of Thomas Kuhn's seminal work, *The Structure of Scientific Revolutions,* been of considerable interest to philosophers of science. The issues raised by Kuhn, and many others, can be followed in Popper (1972), Feyerabend (1975), *Criticism and the Growth of Knowledge* (1970) edited by I. Lakatos and A. Musgrave and Stephen Toulmin's *Human Understanding* (1972).

Notes

1. Surprisingly, Richard Porson was Regius Professor of Greek at Cambridge from 1792 until his death in 1808 and was presumably familiar with the works of Homer and Vergil. For those more familiar with the *Iliad* than *Madoc* and without a reference work at hand, the *Oxford Companion to English Literature* notes that *Madoc* was an epic poem of Robert Southey published in 1805 which attempted to do for the conflict between the legendary 12th-century Welsh hero Madoc and the Aztecs what Homer had done for the Trojan war.

2. Although anyone who takes the trouble to examine the publishing history of many of the supposed classics of today is likely to be in for some surprises. One could be forgiven for supposing, for example, that Plato had been widely read by generations of British intellectuals but, in fact, it turns out that the first English translation of his works appeared only in 1804. There had been an earlier translation of a French version (1701), but a careful look at the history of Plato shows quite clearly that for the British, Plato only began to be read during the 19th century.

3. Burrow's reference here is obscure. He cannot mean Galileo's *Two New Sciences* which is about motion but could hardly be described as "most readable." In any case, before the present century, demand for it was negligible. Since its publication in 1638 it was twice translated into English (1665 and 1730). Nor does Galileo's *Dialogues on Two Chief World Systems* (1632) fare much better. Translated into English in 1661 in a very rare work, it remained ignored until 1953 when a new translation was published and the old one of Salusbury reissued. Readable it may well be; read in English it cannot have been for three centuries.

4. Such works have their own fascination and tempt the odd scholar to their study; hence the regular appearance of such books as *Sunk Without Trace* (1962) by R. Birley, a study of "forgotten masterpieces" in which the status of works like *Lalla Rookh* (1817) by Thomas Moore and *Festus* (1839) by P. J. Bailey is carefully considered.

5. Penguin Books, for example, together with several other publishers, issue a Modern Classics series in which such writers as Huxley, Forster, Waugh, Faulkner and Gide are represented. While not wishing to quarrel with their editorial judgment that Joyce's *Ulysses* is indeed a classic, it is difficult to see what is gained by deeming such an obscure novel as Herbert Read's *Green Child* (1935) a classic while objecting to it being applied to the writings of a Gauss, a Lavoisier or a Maxwell. The impression is sometimes gained that almost *any* serious literary work is a candidate for classic status whereas for a work of science to be so described it must rank on the same level as those of Newton and Darwin.

6. A mode of access not unknown in the literary and philosophical world. It seems perfectly proper for people to claim familiarity with the *Decline and Fall* of Gibbon, *The Prelude* of Wordsworth or even *The Sonnets* of Shakespeare when what in fact they have read are fairly sizeable extracts from a much longer work. In contrast, it is often demanded that familiarity with such a work as Newton's *Principia* can only be claimed by those rare souls who have absorbed the full implications of every

theorem, lemma and proposition. It was not a view shared by Newton. On at least two occasions, to the classicist Richard Bentley and the philosopher John Locke, he sent directives on how they could hope to attain an adequate understanding of his work even though they were not mathematicians.

7. There are few literary works of the period with such credentials. The only one I am aware of is the *Opera* of Vergil which, according to Sarton (1953-9), was published in 275 editions before 1600.

8. Anyone intrigued by such terminology as "erotic chemicotropism," "the perigenesis of the plastidule," "phyletic psychogeny," or the "histopsyche" should spend a few pennies on one of the thousands of available Haeckels.

9. Modern ecology began with the work of Charles Elton who, in *Animal Ecology* (1927), first raised the apparently simple question: why are large animals so rare? Why have life forms adopted discrete sizes? The answer came twenty years later from R. Lindeman and G. Hutchinson of Yale who saw that the problem could only be tackled "by thinking of food and bodies as calories rather than flesh" (P. Colinvaux, *Why are Big Fierce Animals So Rare* [1980], 21).

10. The subject of A. Hallam's *A Revolution in the Earth Sciences* (1973). I suspect that if it had sounded better Hallam would have preferred the title *The Wegenerian Revolution.*

11. The first purely scientific periodical was the *Philosophical Transactions* which dates from 1665; by this time seven out of the twelve works discussed in this book had appeared.

12. Attempts are still being made. In 1981 Rupert Sheldrake, a cytologist, published *A New Science of Life* in which he argued for the hypothesis of "formative causation" and the existence of "morphogenetic fields." If Sheldrake's arguments are sound then science is wrong in the way in which it describes such basic processes as the formation of molecules, the growth of cells and even the manner in which rats learn to run mazes. Sheldrake has won little support for his revolutionary claims.

13. "Now, everybody is working away at 'projects', the outcome of which must be known in advance, since otherwise the inordinate financial investment could not be justified" (see Erwin Chargaff, *Nature,* 26 April 1974, 777).

CHAPTER 2

Corpus Hippocraticum

(5th-4th Century B.C.)

In 1933 when Thomas began his studies at the Harvard Medical School, doctors were for the most part not taught to cure disease, and were not particularly expected to bring about cures. The great thing was to find out the name of the disease and then to let nature take its course. We were taught to make very accurate diagnoses, so that accurate prognosis was possible. Patients and their families could be told not only the name of their illness but also, with some reliability, how it was likely to turn out.

– From a profile of Lewis Thomas by J. Bernstein, *Experiencing Science*

A man of great eloquence, imposing presence, and unrelenting combativeness, who wielded great power in the medical world. He imagined that the cause of all diseases was inflammation, particularly in the intestines. He prescribed abundant bloodletting, leeches and severe diets. His starving patients, bled white, died like flies, but he was nevertheless made a Professor at the faculty of Paris (1831).

– A description of F. J. V. Broussais (1772-1838) by T. Zeldin, *France 1848-1945*

He knew the cause of everich maladye
Were it of hoot or cold, or moiste, or drye,
And where engendered, and of what humour;
He was a very parfit practisour.

– Chaucer, *Doctor of Physik*

There is no shortage of claims for the extraordinary nature of the *Corpus*: "It is scientific medicine, the first in Greece, if not in the world" (Sarton 1953: 344); "It is undoubtedly true that the scientific study of medicine began in the 5th century B.C." (Kirk and Raven: 89). Nor do such tributes belong only to the modern period. To Celsus, a Roman medical encyclopedist of the first century A.D., Hippocrates was "primus ex omnibus memoria dignis," the first physician worthy of being remembered (Celsus: 1-18, 12-13) while to Thomas Sydenham (1624-89), known incidentally as the "English Hippocrates," Hippocrates was the "Romulus of medicine" and "he it is whom we can never duly praise" (King 1971: 121).

Beneath the rhetoric two features in particular are worth stressing. In the first place the *Corpus* is arguably the oldest extant western scientific text. There were scientists in the Greek world before Hippocrates and they certainly produced a fair number of texts. As it happens none of these texts has survived. Consequently such major pre-Hippocratic figures as Thales, Pythagoras and Zeno are known either through cryptic and questionable fragments or through the words of a later and by no means disinterested commentator like Aristotle (384-322 B.C.). Although it is possible to see signs of scientific activity in other areas of the ancient world – Egypt and Babylon for example – in none of them has any earlier scientific texts been identified.[1]

Second, the *Corpus* has survived longer as an authoritative scientific text than any other work. By this is meant that the theory of disease expressed in the *Corpus* in the fourth century B.C. was broadly the same theory subscribed to by the physicians of western Europe over two thousand years later in the 18th century. Nor did the Hippocratic or humoral theory of disease completely disappear at this point but survived as a legitimate theory until well into the 19th century and can even be found in some quarters today. It has, so far, had a longer run than the Bible and its only rivals in terms of longevity are *The Almagest* of Ptolemy and *The Elements* of Euclid.

There is one important point of difference between the *Corpus* and the work of Ptolemy and Euclid. Although all three are inadequate as scientific theories, *The Almagest* and *The Elements* were approximations to the truth and have many useful features. The universe is not as Ptolemy described it but it is still possible using his cosmology to map the earth and accurately to predict eclipses of the sun and moon. So, too, with Euclid. Space may well not be Euclidean and his axiomatization of geometry, by modern standards, is far from rigorous, nontheless a multitude of basic geometrical facts and relationships can be adequately demonstrated in terms of *The Elements*. Hippocratic medicine radically differs, however, for here there is no approximation to the truth or preparation for later improved theories. The picture presented of man in the *Corpus* and the nature of the diseases that inflict him are not just misleading and inadequate but uncompromisingly and totally false. While it is true that the practice of Hippocratic medicine is unlikely to be lethal, it is equally true that no one can have been healed by its application. As a false theory it is hardly unique; as a theory commanding universal allegiance for two millenia, it raises fairly deep questions about the nature and confirmation of scientific theories.

Hippocrates

There is little hard data concerning the life and writings of Hippocrates. The *Oxford Classical Dictionary* merely notes that he was connected with the island of Cos, was a contemporary of Socrates (469-399 B.C.), of small stature, much traveled, and died at Larissa (Thessaly). The precise dates given for his life are speculative. Nor can he be linked directly with any specific item in the *Corpus*. Some scholars even claim that the *Corpus* is nothing more than the remnants of an early library and that the name Hippocrates is more likely to refer to the collector or owner of the library rather than the author.[2]

The Corpus

The bulk of the *Corpus* was composed during the late fifth and early fourth centuries, although some of the treatises were produced much later. The earliest references to the collection go back no farther than the third century B.C. when, it is assumed, the numerous treatises were assembled by scholars of Alexandria.[3] A surviving work of the first century A.D., a glossary to Hippocrates by Erotian, refers to as many as forty-nine Hippocratic works. The earliest surviving manuscripts date to the tenth and eleventh centuries and list about sixty items. For reference, the standard modern edition of Littré (1839-61) includes seventy distinct treatises.

The coverage is impressive. It is difficult to think of any substantive issue approachable with the techniques of the period that is not discussed at some length somewhere in the *Corpus*. Among the seventy items include:[4]

Case histories	*Epidemics* (in seven books)
Nature of medicine and disease	*Ancient Medicine; Humors; The Nature of Man*
Medical anthropology	*Airs, Waters, Places*
Monographs	*The Sacred Disease* (epilepsy)
Obstetrics and gynecology	*The Nature of Women; Sterile Women; Excision of the Foetus*
Medical ethics	*The Oath; The Physician; Decorum*
Therapy	*Regimen; Regimen in Acute Diseases; Regimen in Health*
Anatomy	*Places in Man; The Heart*
Surgery	*Fractures; Joints; Head Wounds*
Embryology	*The Seed; The Nature of the Child; Diseases IV*

Also worthy of mention is the unclassifiable *Aphorisms*, the most popular of all Hippocratic texts.[5]

The overriding impression conveyed by the most casual examination of the *Corpus* is the sheer scale of achievement. Virtually the whole of medicine is covered, at some length and often in some depth. It is not surprising that it should have so impressed physicians of many lands and numerous generations as to inhibit any radical innovation in medical theory and practice. The aim of many a physician over the centuries must have been to attain the level of Hippocratic medicine rather than to advance beyond it.

One further feature of Hippocratic medicine, quite startling in light of the times, calls for comment. Unlike the medicine of Egypt and a rival tradition in Europe that coexisted for centuries, the *Corpus* is a purely naturalistic work. It does not speak of spells, phylacteries, visions, enchantments, demons or gods. Such views were by no means alien to Greek thought. The *Odyssey*, for example mentions quite casually how Ulysses received a wound from a boar that was staunched by a spell (Odyssey XIX: 455) while *Oedipus Rex* opens in a Thebes afflicted by a divine pestilence.[6]

The *Corpus* proposes on the other hand an uncompromising and dramatic approach to sickness when it addresses for the first time without equivocation or hesitation the subject of "the sacred disease." Speaking of epilepsy, it begins, "I do not believe that the 'sacred disease' is any more divine or sacred than any other disease but, on the contrary, has specific characteristics and a definite cause." Scorn is poured on "faith healers, quacks and charlatans" who treat complaints by "prescribing purifications and incantations" or by imposing arbitrary bans on the wearing of black, goat skins, or "putting one foot on the other." The *Corpus* charges that the divine presence in a disease was particularly helpful to "quacks" because when the treatment failed, "they could screen their own failure . . . [by] explaining the gods are to blame" (Chadwick and Mann: 237-38). Nor is this the only work in the *Corpus* to find so clear an expression of naturalistic views. In *Airs, Waters, Places,* there is a discussion of just why impotence was so prevalent among the rich Scythians. Against the Scythian idea of "a divine visitation," the Hippocratic writer patiently but emphatically insisted "each disease has a natural cause and nothing happens without a natural cause." Thus the Hippocratic explanation saw impotence as a result of the Scythian way of treating varicose veins – induced by horse-riding, an exclusive habit of the wealthy – which was done by cutting the veins behind the ear. This is no doubt far-fetched; it presupposes the view that sperm passes through the veins from the brain to the loins. Still, the illustration helps to underline the insistence that medicine is a naturalistic discipline, which was the important contribution of Hippocratic thought and which earns the *Corpus* a place among the classics of science. Classics should not be judged by standards applied to modern texts, but rather by their attitude to theories. Unfortunately, it was the *Corpus*'s content, not the spirit of the Hippocratic approach, that impressed later generations.

Hippocratic Theory – Origins

What does it mean to be healthy or diseased? The basic insight supplied by Hippocratic medicine is remarkably simple. Greek philosophy had begun in the sixth century with a group of Milesian thinkers – Thales, Anaxamander and Anaximenes – who explored the possibility that nature basically came from a single substance. For some this was water; for others, air seemed more plausible. By the time of Hippocrates the early positions adopted by the Milesians were no longer considered tenable, and it is against this background that the first assumption of the new medicine is stated. In *The Nature of Man,* a work attributed by Aristotle to Hippocrates's son-in-law Polybus, it is argued that "generation cannot arise from a single substance" (Chadwick and Mann: 261). Rather, man was a composite, a mixture of a definite number of elements, and on this assumption a very general theory of health and disease emerged. If the mixture is properly proportioned, the body will be healthy; if poorly mixed, the body will be diseased. It resembled baking bread: get the ingredients in the right proportion and all goes well; too much or too little of an essential item produces a soggy mass or an indigestible rock. From this basic thesis derived virtually all the concepts of Hippocratic medicine. In a healthy man, all the elements were in their proper proportions, none dominated; in Greek, the term used was *isonomia,* literally equal rule, or health. Where one element was present in excessive amounts, sickness would result. The term used in this case was *monarchia,* literally single rule, the word from which the modern monarch was derived. It is a theory of great power, general enough to deal with any disease, yet flexible enough to provide a specific description of each. It was also simple enough to be readily comprehended by physician and patient alike.

One crucial problem remained to be solved before humoral medicine, as it was later called, could gain wide and prolonged support. If man is composite, out of what is he composed? The Hippocratic answer, given in *The Nature of Man*, was that "the human body contains blood, phlegm, yellow pile and black bile. These are the things that make up its constitution and cause its pains and health" (Chadwick and Mann: 262).[7] Yet why, one might ask, should this list of "ingredients" be given any more credence than the list in the nursery rhyme – "slugs and snails and puppy dog's tails" or "sugar and spice and all things nice"? Why is it that the Hippocratic list commanded such allegiance, stretching in time from the ancient Greeks to Europeans in the 18th century, including adherents among Christians, Muslims, Jews, and pagans?

There are two roots of the Hippocratic list. One may be traced to Alcmaeon of Croton, a Pythagorean of the early fifth century B.C. As a later commentator described his views, "Alcmaeon maintains that the bond of health is equal balance (*isonomia*) of the powers moist and dry, cold and hot, bitter and sweet, and the rest, while the supremacy (*monarchia*) of one of them is the cause of disease" (Kirk and Raven: 234). But it can be asked – and was asked – why just six substances and what constitute "the rest"? Why not more? Why not a different list? Praxagoras of Cos, of the late fourth century B.C., for example, spoke of eleven humors and included the sweet, the bitter, the acid, the glassy, the nitrous, the salty, the corrosive, the clotting, the yolk-colored, the leek-green, and the uniformly mixed

(Phillips: 136). In the *Corpus* alternative traditions can be found. *Ancient Medicine* declares, "there exists in man saltness, bitterness, sweetness, sharpness, astringency, flabbiness and countless other qualities" (Chadwick and Mann: 78). This reminds one of the seven vices or the four cardinal virtues, useful expressions of popular folklore but not to be confused with serious thought.

The second root of the Hippocratic list may be traced to Empedocles (c. 493-c. 433 B.C.). Born at a time of great intellectual crisis among the monists of early Greek philosophy whose two leading representatives, Zeno and Parmenides,[8] had developed a number of arguments designed to show that generation, change, motion and decay were all impossible, Empedocles proposed that all matter came from four eternal and irreducible elements: earth, air, fire and water. All objects and materials are compounds of these four elements and differ from each other only by the varying proportions of each element.[9]

If all ideas of a modern element are laid aside, then the four-element theory of Empedocles does have a certain plausibility. His elements appear to be basic in that they can be found anywhere, and that in most substances it is possible, with a bit of imagination, to identify an earthy, watery, fiery, and airy part. At the same time it must be noted that the same objections raised against Alcmaeon's system are just as apposite here. Why not six or eight or more elements? Why not fewer? Why not a different set? In China, for example, Tsou Yen (350-270 B.C.) insisted that there were five elements (earth, water, fire, wood, and metal) while chemists in 16th- and 17th-century Europe picked almost any number to support some pet theory. Adherents of two, three, four, five and more elements can be found without too much difficulty; some lists overlapped; in others, no one element was common. But if the history of science is to consist of more than the presentation of apparently arbitrary choices of elements, then it must be possible to show how out of such a multiplicity of alternatives one authoritative scheme could emerge.

There is a very simple way to show that a set of elements is not arbitrary: by deriving it from an already established set. In this way, to take a trivial example, the set A = [M, T, W, T, F, S, S] might at first sight seem to be an arbitrary sequence of letters. When told that it is generated from the days of the week by taking the first letter of their name it is seen in a different light. We can now see why it must have seven and only seven members, why if it begins with M it must end with S and so on. Something similar has often happened in the history of science, not always with happy consequences, and probably also happened with the structures of Alcmaeon and Empedocles.

Probably the first step taken was to tidy up Alcmaeon's list. The bitter/sweet contrast does not quite fit with the hot/cold and the dry/moist dichotomies; it clearly applies only to things tastable while the other two can be applied to almost anything physical. By dropping the bitter/sweet pair, one arrives at the same number as the elements of Empedocles (four).

Is it possible to pair them off in a rigorous way? A straight one-to-one link soon seems somewhat inept in that while fire is certainly hot and water undoubtedly moist it is less obvious that the earth is cold and the air dry. Air in the sense of breath or as the winds of the Aegean are normally moist. Again, it seems inadequate simply to say that fire is hot, it is also surely dry; water, too, is cold as well as wet. If then no sensible link can be made between single elements and humors, would it not be possible to link each element with a pair of qualities. Six such pairs are possible. Using initial letters in an obvious way they can be represented as: H-C; H-D; H-M; C-D; C-M; D-M. A moment's examination shows two combinations, H-C and D-M, to be impossible leaving just the four: H-D; H-M; C-D; C-M.

Bearing in mind the climate of the Greek isles, certain identifications between the two sets readily follow. Fire is clearly hot and dry and water must be cold and wet. Greek winds or air are hot and wet while the earth, below the surface layer, is indeed dry and cold. At this point it is well to appreciate that something quite remarkable has happened. Starting with two independent systems, each open to the charge of being arbitrary, the fact that they can somehow be mapped on to each other endows each set with a plausibility possessed by neither taken separately. It cannot be an accident, it could be argued, that the four elements and four humors could be expressed so exactly in terms of each other. The point can be brought out by taking the four identities: water = cold and moist, air = moist and hot, fire = hot and dry, earth = cold and dry and representing them by the simple matrix:

	dry	moist
hot	fire	air
cold	earth	water

Having seen this there is little incentive to seek a fifth element or for a further humor. The answer is clear: lists cease to be arbitrary when they become diagrams.[10] Many diagrams are possible, some better than others. The best of all is the matrix type used above; lists once fitted within the confines of such a diagram acquire a formidable plausibility. This is perhaps best seen in the most common of all matrices, the crossword. Anyone who has ever completed a crossword with all clues answered and all blanks filled has no need to look at the solution; no check on the matrix is called for. Completing the crossword is as good a criteria of correctness as the solution. In actual fact it is better. For if the solution did propose a wildly different set of answers to that in the completed crossword the natural response would be to suppose that it was full of misprints or even that the wrong solution had been printed. It is no accident that at the heart of much modern science there lies a number of simple diagrams with the same kind of elegance, symmetry and completeness that proved so attractive to the scientists of antiquity.[11]

In actual fact the diagram familiar to antiquity was not the matrix above but a geometrical figure in which the four elements occupy the corners of a square while the humors are represented by lines connecting them:

Diag. 2:1

Just when this and closely related diagrams first appeared is not known. No manuscripts from antiquity have survived incorporating the diagram; it is, however, found frequently in medieval manuscripts and has continued to appear in discussions of Hippocratic medicine ever since.

The points made about the matrix apply equally to the diagram above: the filling of the matrix cells corresponds to the exact match between the four corners and sides of the square with the elements and humors. The link between the elements and humors was emphasized by an alternative form of the diagram in which two independent squares with the minimum of manipulation come together and create an equivalent version:

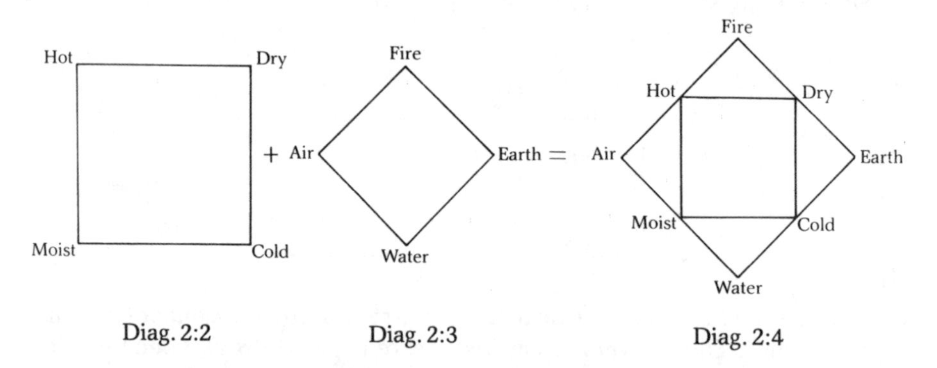

Diag. 2:2 Diag. 2:3 Diag. 2:4

The power of such diagrams should not be minimized. At their most trivial they serve as helpful mnemonics; at a deeper level their symmetry, exactness and the ease with which two independent structures can be fused together into an integrated and significant whole was too impressive to permit easy rejection.

Hippocratic Theory – Application

It was still necessary to link the elements and humors to disease, health and medicine. Initially this was done in terms of the seasonal variation in disease noted throughout the *Corpus*. The third section of *Aphorisms* notes, for example, that "autumn is worst for consumptives," "spring is the healthiest and least fatal time of the year," and while "every disease occurs at all seasons of the year some of them occur more frequently and are of greater severity at certain times." It was to these certain times that the Hippocratic theorist turned in *The Nature of Man*. Here it was noted that in winter the body fills with phlegm and "people spit and blow from their noses the most phlegmatic mucus." Spring, the hot and wet season, is the part of the year "most in keeping with blood . . . and people are particularly liable to dysentery and epistaxis." In short: "Just as the year is governed at one time by winter, then by spring . . . , so at one time in the body phlegm preponderates, at another time blood, at another time yellow bile and this is followed by the preponderance of black bile" (Chadwick and Mann: 265-65). If proof of this is wanted, the writer continues, give someone the same emetic at four different times of the year and note the result: "His vomit will be most phlegmatic in winter, most wet in spring, most bilious in summer and darkest in autumn" (Chadwick and Mann: 265-66).

The identifications being made are clear and in some cases plausible. Indeed, if one begins with winter the correspondences noted are trivial. Winter is cold and wet; wintery diseases are accompanied by cold and wet (phlegm) discharges. It is not unreasonable to suppose in this context that the cold and wet season is somehow producing in the body an excess of coldness and wetness. But an excess of coldness and wetness in the body is nothing more or less than a phlegmatic/watery sickness. In the same way, as spring is hot and wet it would tend to produce an excess of hotness and wetness in the body. As the discharge typical of spring is a bloody one then blood inevitably becomes identified with the hot and wet qualities of spring. Having started where the correspondences are unquestionable they are extended to less obvious ones. The full scheme can be represented as:

Winter	cold and wet (water)	phlegm/water
Spring	hot and wet (air)	blood
Summer	hot and dry (fire)	bile/cholic
Autumn	cold and dry (earth)	black bile/melancholic[12]

It is in this context that the opening of *Airs, Waters, Places* can be understood: "Whoever would study medicine aright must learn of the following subjects. First he must consider the effect of each of the seasons of the year." The last subject for the attention of the doctor is "the heavy eaters and drinkers" for it is not only the climate in the Hippocratic view which is capable of causing an imbalance in the humors (Chadwick and Mann: 148). Foods and drinks are, just like all else, composed of the four elements. Consequently the consumption of an inappropriate food or an excessive amount of any food should be able to disrupt the delicate humoral balance of a healthy body.[13] Thus food and drink with the climate serve as the two main causes of disease in the Hippocratic tradition. *Regimen,* for example, is just one of the many works that describes the powers of a large number of foods. Barley is cool and dry, wines hot and dry, while various meats, fruits and vegetables are listed with their appropriate properties.

One advantage of this approach to disease is to provide an evident therapy. If, in a simple case, sickness consists in an excess of cold, wet humors then it is the task of the physician to restore the balance. Thus, the advice in *Regimen for Health* for the winter is to eat foods which are hot and dry and thus neutralize the effects of the cold, wet climate. Consequently the regimen should consist of, among other items, bread, roast meat and fish. But with the coming of summer it becomes necessary to eat barley cake, boiled meat and fish in order "to make the body cool and soft, for the season, being hot and dry, renders the body burnt up and parched, and such a condition may be avoided by a suitable diet" (Chadwick and Mann: 272).

Such is the essence of the great allopathic principle of medicine extracted from the Hippocratic writings. It is scarcely an exaggeration to say that no new approach would be introduced into medicine for two thousand years and that the bulk of the medical texts, whether produced in the West or by Islamic scholars, consisted of little more than variations and reformulations of the basic Hippocratic theme. One further aspect of the Hippocratic tradition needs to be examined before the long history of humoral medicine can be briefly sketched.

A reading of a number of Hippocratic texts reveals that the cause and treatment of illness is notable only for its rarity.[14] In these texts the emphasis is on observation and classification. The best examples of this attitude can be found in *Epidemics,* sometimes dismissed as being mere case notes. (One might indeed ask just why the Hippocratic physician went to such lengths to record so many case histories.) Such cases are too long to quote in full, but several abbreviated samples illustrate what we mean (Chadwick and Mann: 113-16):

> Pythion. . . suffered from a twitching which began in the hands. First day: high fever, delirium. Second day: all symptoms more pronounced. Third day: condition unchanged. Fourth day: passed small, undigested bilious stools. . . . Sixth day: sputum. . . tinged red. . . . Tenth day: sweated, sputum rather ripe, the crisis reached. . . .

Pythion appears to have survived; no such luck for Philistes of Thasos who first complained of a persistent headache:

> First day: vomited small quantities of yellow bilious matter. . . . Afterwards rust coloured. Bowels were opened. An uneasy night. Second day: deafness, high fever; right hypochondrium was contracted and indrawn, urine thin. . . . He became mad about midday. Third day: uneasy. Fourth day: convulsions, a fit. Fifth day: died in the morning.

The various *Epidemics* contain a total of 567 case histories, and although there is a noted absence of either causal or therapeutic concern, comments like the following are frequently found: "I do not know of any in the circumstances who died if they had a good epistaxis" (Chadwick and Mann: 95); or, "All who suffered from disease of the belly were specially likely to die" (Chadwick and Mann: 124).

Epidemics 1 contains a long discussion of fevers. It is to this particular text that the terminology of fevers, still in use, can be traced. The author notes that fevers differ not only in their nature but also in their spacing: "Some . . . are continuous, others come at day and remit at night; others for the night, remitting by day. There are sub-tertian, tertian and quartan fevers, five day, seven day and nine day fevers" (Chadwick and Mann: 100).[15] Having successfully identified the different fevers the next task was to evaluate them; this was done in terms of severity and duration. Thus it was pointed out that "nocturnal fever is not especially fatal but it is long drawn out. . . . An exact tertian fever soon produces a crisis and is not fatal" while "the most fatal of all" is the sub-tertian (Chadwick and Mann: 100).

The Hippocratic physician must have been well aware that the very serious fevers, be they associated with malaria, dysentery, or tuberculosis, were beyond their therapeutic control. Although contemporaries peddled nostrums and offered cures of uncountable variety and number, the Hippocratics somehow managed to resist this particular corruption. But if they could offer no cure for malaria and other fevers it did not mean that they offered no service to their patients. They could offer prognosis. Given personal experience and access to works such as *Epidemics*, reasonable judgments could be made about the outcome of many serious complaints. Once the recurrence of the fever was established, and the characteristic features identified (for instance, did epistaxis occur or did gastric complications develop?), then the physician was in a position to provide a prognosis. Such knowledge should not be lightly dismissed. It is of considerable value to know that while an illness is severe recovery is likely, or that death is almost certain. Plans can be made or canceled, decisions taken and attitudes adopted. There were after all in this culture diviners and oracles of various kinds to which many Greeks approached to find out what the future held. In this context the physician was just one more specialist who could foretell events, in this instance not on the basis of a bird's flight or a goat's liver but rather by careful study of medical histories.

Medical traditions which are naturalistic, rational and honest leave themselves open to competition from less inhibited rivals. It was openly admitted in the *Corpus* "that not everything is possible in medicine" and further advised that "cases in which disease has already won the mastery" should not be pronounced curable (Chadwick and Mann: 140). It took little time for groups to appear who were much less aware of their own limitations.[16] The Hippocratic period was one in which a new irrationality swept through the Greek world. Before the Peloponnesian War had ended, "there had appeared in Athens the worship of the Phrygian 'Mountain Mother,' Cybele, and that of her Thracian counterpart, Bendis; the mysteries of the Thraco-Phrygian Sabazius, a sort of un-Hellenized Dionysius; and the rites of the Asiatic 'dying gods,' Attis and Adonis" (Dodds 1968: 193-94). Such gods did not lightly turn away business, not even from the incurable.[17] All a patient would be likely to receive from such gods, Hippocrates realized, was a rebuke from the priests for some presumed impurity inhibiting the therapeutic process.

In 420 B.C. the cult of Asclepius was introduced into Athens and the construction of a shrine at Epidaurus followed shortly thereafter. Much is known of the treatment practiced there through inscriptions of supposedly miraculous cures left by patients on temple walls. Many clients evidently suffered from blindness, paralysis, baldness, infertility, and lost property (incidentally, cases for which the Hippocratic doctor could offer no help).[18] Cure was obtained by various ritual observances and by sleeping in the temple. Asclepius would appear to the patient in a dream and cure him. A typical example is the case of Euphanes, a young boy suffering from a stone in the kidneys. While asleep, "it seemed to him that the god stood over him and asked, 'What will you give me if I cure you?' 'Ten dice,' he answered. The god laughed and said to him that he would cure him. When day came he walked out sound" (Edelstein: 231).

The Hippocratic Tradition

Despite such opposition the orthodox Hippocratic approach to medicine persisted without radical change and gained adherents. The manuscripts were collected and studied by the scholars of Alexandria. With Galen (A.D. 130-200), the greatest physician of the Roman world, the basic Hippocratic model was seized upon and deployed in new areas. Not only did Galen bestow his great authority on Hippocratic medicine and thus guarantee its survival into the medieval period but he also provided an anatomy and physiology to explain its operation. The physiology of Galen began with food. The stomach transformed food into "chyle" by a process known uninformatively as the "first coction." It matters little what "chyle" really was for it was immediately moved to the liver where by a second coction it was transformed into blood. Some of the food, the useless part, was absorbed by the spleen and there converted to black bile. In the same way, by processes not always clearly specified, yellow bile is produced in the gall bladder and phlegm in the pituitary (*pituita* is Latin for phlegm). The beauty of this vague scheme is that it suggests a workable mechanism of the etiology of disease. By taking in too much cold and dry food, the spleen would produce too much black bile and a bilious complaint of some kind would result; by eliminating cold and dry foods, the source of bile was reduced, thereby effecting a cure.

Galen also used Hippocratic humors to develop a theory of human personality, a theory known to most people through Elizabethan literature and the plays of Shakespeare. Why, Galen asked, "does an excess of yellow bile in the brain induce delirium, or of black bile melancholia, or why phlegm and chilling agents. . . cause lethargy?" (Brock: 233). Thus the jump was made from seeing physical states of the body as due to its humoral constitution to also explaining mental states in the same way. Some people have a natural excess of phlegm in their physical make up, they not unnaturally reveal a phlegmatic temperament. Similarly, others might show themselves cholic, melancholic or sanguine depending upon which particular humors predominated in their body. The idea of a true *isonomia,* a pure equilibrium of the four humors soon came to be recognized as a fiction; in each body, after an initial struggle, a harmony is obtained in which the humors are represented to various degrees. This state was known technically as the complexion[19] (from Latin *plectere,* to braid) and thus when Donne in *Anniversaries* talks of "She whose complexion was so even made," he was not referring to cosmetic skills, but to a good balance of humors. According to Shakespeare, the humors in Brutus were similarly balanced: "So mix'd in him that Nature might stand up / And say to all the world, 'This was a man.' " The sentiment was clearly a common one and was also found in *The Barrons' Wars* by Drayton: "He was a man. . . / In whom so mix'd the elements all lay, / That none to one could sovereignty impute." Nor were such expressions merely poetic. Falstaff, for example, could simply state, "I love not the humour of bread or cheese. . . ," but what a specific humor meant can be seen in Herbert's injunction in *The Church Porch*: "O England, full of sin, but most of sloth; / Spit out thy phlegm, and fill thy breast with glory." The phlegmatic character was still, as it had been in Galen's day, somewhat lazy. No longer an essential part of poetic imagery, humoral analysis has been revived by modern psychologists.[20]

So strong was the influence of Hippocratic writings that by the medieval period it is difficult to find signs of any other tradition. Thus, in *Etymologies* by Isidore of Seville (570-636), one of the main sources of ancient learning for medieval scholars, the Hippocratic beliefs are clearly stated. "All diseases arise from the four humours: blood, yellow bile, black bile and phlegm. . . . When any of the humours increase beyond the limits set by nature, they cause illness" (Grant: 701). Innumerable manuscripts, both Christian and Islamic, display similar passages with virtually the same wording.

With the collapse of learning in the West in the sixth century, it is to the Islamic world that one must look to trace the Hippocratic tradition.[21] In the ninth century the "sheikh of translators," Hunayn ibn Ishaq (809-73), translated a number of Hippocratic texts. Known in the West as Joannitius, Hunayn, a Nestorian Christian, translated for the Abbasid courts of Baghdad and Samarra much of Greek learning into Arabic or Syriac. He is a figure to be met with again when the transmission of Ptolemy and Euclid is discussed. He was so valued, according to one tradition, that Caliph al-Maᶜmun paid him in gold the weight of the books he translated. Much of our knowledge of this period derives from an invaluable tenth-century work, the *Fihrist* of al-Nadim, "an Index of the books of all nations, Arabs and foreigners alike, which are extant in the Arabic language and script, on every branch of knowledge" (Nicholson: 362). Al-Nadim reported editions of ten Hippocratic texts and attributed to Hunayn translations of *Oath; Aphorisms; Prognosis; Fractures; Airs, Waters, Places;* and *The Nature of Man.* Such writings were taken seriously and were fully absorbed by the writers of the great compendia of later centuries.

The greatest medieval text of the Islamic world was undoubtedly the *Qanun* of Ibn Sina (980-1037), an enormous work of a million words, and it is interesting to note that it, too, contains the Hippocratic humoral theory as set forth in earlier works by Galen and Isidore of Seville: "the humours are of four kinds: blood, phlegm, red bile and black bile" (Grant: 705).[22] It is also worth mentioning that an Arabic word for health, a word reported by some to be common in Turkey and Iran as well, is *mizaj,* from the root *mazaja,* meaning to blend.

Beginning in the 12th century, a great revival in western learning occurred, and for the first time many Greek scientific works were translated into Latin. This was mainly achieved thanks to highly motivated scholars who made their way to such centers of Islamic learning as Toledo, learnt Arabic, and began the difficult task of turning Arabic translations of Syriac versions of Greek texts into Latin. One of the most famous and industrious translators was Gerard of Cremona (1114-87) who was attracted to Toledo in search of *The Almagest* of Ptolemy. Once there, he remained to translate some seventy Arabic texts into Latin. It was through such scholars that Ptolemy, Euclid, Aristotle, Galen and Hippocrates were first heard of in the West outside inadequate compilations such as Isidore's. The demand for some Hippocratic texts was considerable; *Aphorisms* proved one of the most popular.

It is a curious fact that while the work of Ptolemy and Galen, the two great scientists of antiquity, was overthrown in the year 1543 by the publication of classic studies written by European scholars working outside the great cultural centers, the reputation of Hippocrates seemed to increase. Whereas after the mid-16th century scholars no longer were interested in reading Galen's anatomy or Ptolemy's cosmology, they continued to place high value on Hippocrates. In the 16th and 17th centuries, his collected works as well as individual texts poured off European presses in remarkable numbers. We must remember that we are referring to a demand for books over two thousand years old by physicians who were contemporary with Newton and Harvey.

The new mechanical philosophy that emerged in the 17th century presented little threat to the basic presupposition of Hippocratic medicine; the idiom may require modification but the spirit would remain the same. To the 17th-century scientist, matter was essentially corpuscular and the properties of individual corpuscles or atoms that mattered were size, shape, and motion. It was no longer acceptable to ask whether a thing were made from materials which were hot or cold, dry or wet; if it were asked, then these properties had to be interwoven with the new chemical, corpuscular terminology. An example of this approach can be seen in *Fundamenta medicinae* (1695) by Friedrich Hoffmann (1660-1742), a leading physician at the Brandenburg court and at the University of Halle. "All nature is mechanical, medicine included," he declared, going on to show how phenomenal properties could be derived from matter and motion. Thus if particles moved rectilinearly they would create a sensation of cold, but if they moved rapidly and curvilinearly heat would be produced. In this way a complex series of effects, regardless of whether a body were rigid, slippery, volatile, acidic, sulfurous, oily or dense, could be traced to a basic feature of its constitutent corpuscles.

With this background Hoffmann confidently dismissed traditional Hippocratic theory. "To declare all causes from phlegm, bile and pancreatic juice is quite inadequate," he wrote (Hoffmann: 1-41). Only however to restate the theory in a form acceptable to the 17th century? Health is defined as when the body fluids are "constituted in a suitable configuration, position and order" (Hoffmann: 3-2). In familiar language, disease is held to arise from disproportion: the "mobile, volatile, saline and sulfurous particles" can "predominate" or the "mucilaginous, acid and fixed particles"– in either case, specific diseases are supposed to result. The cause of such disproportion is inevitably diet–"viscid, salted . . . or acid food," leading to scurvy, severe inflammation and obstruction, and "spiritous, sharp food," causing gangrene and hemorrhage (Hoffmann: 2-28/29). Climate, too–not surprisingly–is singled out as a factor causing disease (Hoffmann: 2-6).

The plays of Molière (1622-73) are a rich source of information on the survival of Hippocratic beliefs in 17th-century Europe, and show clearly that the principles of humoral medicine were well enough known to be satirized on the popular stage.[23] In his *L'amour medecin,* Dr. Thomes, told of the death of a patient, responds, "That can't be. He can't be dead I tell you. . . .'Tis impossible. Hippocrates says that these sort of distempers don't terminate till the fourteenth or twenty-first, and he fell sick but six days ago." Although such remarks show that Hippocrates's influence continued to be publicly acknowledged throughout the 17th and 18th centuries, it was no longer the humoral theory that proved the most attractive aspects of his work, but rather his general approach to medicine and work on medical climatology.

Thomas Sydenham in *Medical Observations* (1666) spoke of the conditions needed for the advance of medicine. First, he demanded "a history of the disease . . . , a description that shall be at once graphic and natural." No easy matter for it must be full, written without philosophical prejudice, include reference to the seasons of the year favoring the complaint, and also aim at enumerating "the peculiar and constant phenomena apart from the accidental and adventitious." The inspiration behind the call for the careful observation and description of the history of diseases was Hippocrates: for, Sydenham declared, it was by this ladder that he "did . . . ascend his lofty sphere" (King 1971: 117-24).

The influence on man of climate in particular and geography in general—on morals, health, customs and politics—were topics much discussed in the 18th century. Inevitably, scholars tended to examine Hippocrates as a first authority. His *Airs, Waters, Places* was, for example, much read. The response was varied. Some made polite acknowledgements before proceeding with their own ideas; others sought to express the environmental ideas of Hippocrates in a language more acceptable to the 18th century. John Arbuthnot in *Concerning the Effects of Air on Human Bodies* (1731) wrote of the need "to explain the philosophy of this sagacious old man, by mechanical Causes arising from the Properties and Qualities of the Air." He attributed the differences between the Scythians of the north and the Egyptians of the south, as described by Hippocrates, to changes in atmospheric pressure, a phenomenon relatively new to this scientist (Glacken: 562-65).

But while there occurred a revival of interest in Hippocrates as medical anthropologist, approaches to medicine were emerging that would lead eventually to the final rejection of Hippocrates as physician. A crucial figure in these developments was G. B. Morgagni (1682-1771), an anatomist working in Padua, then the leading European university, who made a careful study of 640 autopsies. After many years' observation, Morgagni noted a link between the clinical picture presented by the occurrence of a disease and a characteristic lesion invariably found in a particular organ at post mortem. The results were presented in his *De sedibus (The Seats and Causes of Diseases Investigated by Anatomy)* (1761). The idea that diseases had a "seat" or were localized in a specific organ was in many respects the first new insight into the nature of disease since *The Sacred Disease.* Morgagni's book did not disprove humoral theory—as it was still possible to suppose that a lesion occurred as a result of some more basic imbalance—but it nevertheless made humoral theory less interesting since it showed the fruitfulness of research on the nature and pathology of the lesion and its organ. Still, humoral theory staged one last offensive.

Pathologists did seem able to link death with gross organic lesions normally detectable at autopsy. There remained a residual problem, for on some occasions either no lesion could be found or else one too trivial to be thought capable of causing death. One solution was to suppose the existence of some generalized disturbance affecting the whole body; as only one substance could be found present throughout the body—blood—there arose the new subject of hemopathology.[24] Thus K. Rokitansky (1804-78), in the first edition of *Handbuch der pathologischen anatomie* (1842-46), began to see humors in the blood which on this occasion were identified with the plasma proteins, albumin and fibrin, and which were thought capable of getting out of balance. Rokitanksy's ideas were soon overtaken by the emergence of the modern orthodoxy, the germ theory of disease.

When proposed by R. Koch (1843-1910) and L. Pasteur (1822-95), germ theory was far from new and could be traced back to Fracastorius (1478-1553) in *De contagione* (1546) in which he spoke of "seminaria contagium" or "seeds of contagion." It was difficult to see what use could be made of this insight at the time; consequently it exercised little influence. Throughout the intervening centuries comparable proposals often surfaced, but, again, with no impact on medical thought. It was not until the early 19th century that contagion theories reemerged with enough content to arouse substantial and lasting interest. The early evidence was epidemological and was marshalled by such physicians as John Snow (1813-58)[25] and William Budd (1811-80)[26] who studied the transmission of cholera and typhoid respectively. Later in the century the union of such epidemological work with the bacteriological studies of the Koch and Pasteur schools, together with the pathological tradition stemming from Morgagni, saw the consolidation and triumph of germ theory over humoral theory. Yet even in the present century it was still possible to find survivals of the Hippocratic tradition. One such was described by the Islamic scholar Snoek Hurgronje, writing of the days in Mecca at the turn of the century. There, physicians would recommend in their prescriptions "the avoidance of food of a 'hot,' a 'cold,' a 'damp,' or a 'dry' nature" (Hurgronje: 94).[27]

The Persistence of Hippocratic Medicine

The question posed at the beginning of this chapter awaits an answer. How could a theory of man, matter and disease that was so totally wrong command such widespread and prolonged support? By way of attempting an answer, we might point out that theories, like animals, have a tendency to survive. They are adaptable and capable of fitting in with changes made in other areas of science. As is clear from the discussion of the Hippocratic tradition, humoral medicine was certainly adaptable. The basic principles remained unchanged, yet in response to changes in other areas of science, new idioms and concepts were absorbed as they emerged. Thus, when science became mechanical and emphasized the presence and shape of corpuscles, so, too, did humoral theory; and when instead science came to be expressed in the language of biochemistry, humoral theory readily substituted a discussion of an excess or deficiency of plasma proteins for what used to be considered an imbalance of saline and sulfurous particles.

Second, Hippocratic theory appealed to a basic body of experience, familiar to all and constantly providing confirmation of its initial assumptions. There are indisputably close connections between climate, disease and diet. By emphasizing these connections the Hippocratic physician could find support for his theory by showing that epidemics of dysentery followed outbreaks of flu, and a wave of malaria closely followed dysentery epidemics. The theory was sufficiently imprecise to gain credit from seasonal changes without ever having to face more severe tests.

It should also be appreciated how for much of its history Hippocratic–or any other medicine–was seldom met with by the bulk of the population. Conversely, doctors of the period were unlikely to encounter anything like the full range of complaints and infections brought to doctor's offices today. Physicians were few and expensive,[28] offering their services to a limited segment of society and presented with a narrow range of cases. In earlier times, people sought aid from "an empiric, herbalist, wise woman, or other member of that 'great multitude of ignorant persons' so frequently denounced by the physicians" (Thomas: 14). Even such learned men as Francis Bacon and Thomas Hobbes, men with ample funds, admitted they preferred the advice of "an experienced old woman that had been at many sick people's bedsides, than from the learnedst but unexperienced physician" (Thomas: 17).[29] The fact was that physicians lacked exposure to a sufficient variety of disease ever to put Hippocratic theory into practice let alone to the test. Only with the growth of towns in the 18th century and the emergence of a new class of physicians who exposed themselves to all kinds of urban diseases did Hippocratic medicine receive its first substantial test. It is no accident that Snow and Budd, the two who saw the inadequacy of the traditional theory long before the emergence of germ theory, were urban epidemologists who had worked directly with the spread of disease in thickly populated areas.

Why, then, it could be asked, did not 17th-century physicians, with their knowledge of the spread of plague through great European urban centers, appreciate the inadequacy of the Hippocratic tradition? To the learned medical men of that age, plague and other contagious diseases mysteriously and spontaneously appeared in a community even though no one in it was diseased. When an infection arrived, it might infect only half a family; people who shared a bed might find that one partner was infected while the other remained free of the disease. Puzzlingly, it was observed that a person who virtually isolated himself could become infected while a physician, who came into regular contact with the sick, remained free of disease. Moreover, malaria, cholera, plague, flu, measles, or other diseased disappeared as abruptly and mysteriously as they appeared. To physicians, these facts did not suggest that disease was spread by contagion or infection. Without benefit of an immunologist, parasitologist or virologist and not equipped with germ theory, they could not understand how disease was spread.

Thus the picture presented in plague tracts of the time depicts a group of physicians moving away from the traditional Hippocratic account of disease to some kind of contagion theory only to be driven back into orthodoxy by the intractable nature of their material. Richard Mead (1673-1754), physician to Isaac Newton and George II, published *A Discourse on the Plague* in 1720 in which this approach is precisely demonstrated. While accepting that plague could be spread by "contagious atoms" found on people as well as on commercial goods and in clothes, he still found himself forced to refer to the Hippocratic notion of a "corrupted state of air." He wrote, for example, "for otherwise it were not easy to conceive how the Plague, when once it had seized any Place, should ever cease, but with the destructions of all the Inhabitants" (Winslow: 187). It would take a further century or more of thought on and experience of the manner in which cholera, typhoid fever and measles, among other diseases, spread to see the way through such recurrent difficulties.[30] Contagion theories had been well argued by such epidemologists as Snow and Budd long before the bacteriological work of Koch and Pasteur. With this later work, however, the way was at last clear to resolve the many difficulties which had so long delayed its acceptance.[31]

In summary, it was as much the difficulty of constructing a plausible alternative to humoral medicine as any of its own internal merits that preserved the medicine of Hippocrates for so long. The roots of germ theory are many and of a kind that required prolonged and gentle nurture. Properly cultivated, they would yield a rich harvest, but one that lay unsuspected and in the distant future. In the meantime, there was Hippocrates's humoral theory: adaptable, based on the undeniable correlation between climate and health, intelligible to all, and offering genuine comfort and service to its customers. Given these factors, one can understand why the theory lasted for so many centuries.

Publishing History

Ancient Commentators

Commentaries were known to have been written by the third-century Alexandrian anatomist Herophilus and his pupils Bacchios of Tanagra and Philinos of Cos. None of these works survive; Bacchios appears to have known twenty-three Hippocratic texts. Other commentators are known from the first century B.C., and there is an extant glossary to Hippocrates compiled in the first century A.D. by Erotian who refers to forty-nine works of the *Corpus*. The most significant commentator of antiquity was Galen who was so familiar with the literature that he could write on a subject that has fascinated scholars ever since: *The Genuine Writings of Hippocrates*; unfortunately, it is lost. He also wrote commentaries on seventeen Hippocratic works.

Medieval Editions

The earliest Greek manuscript dates only to the tenth century and contains no more than a dozen items; manuscripts of the 11th and 12th centuries increase this number to about sixty. Translations into Arabic of individual works are known from the ninth century when Hunayn ibn Ishaq translated *Aphorisms* and other works. The first translations into Latin from Arabic probably date to the 11th century and were done by Constantine the African. A North African Benedictine monk from Carthage, Constantine settled in later life at the great monastery at Monte Cassino where he produced translations of *Aphorisms, Prognosis* and *Regimen.* The great age of Latin translation started in the 12th century when, beginning with Gerard of Cremona, Adelard of Bath and a host of others, manuscripts of important Greek works poured out to meet the demands of the new universities. *Aphorisms,* for example, was required reading at the medical schools of Bologna and Montpellier. Vernacular translations are also known, and a French version of the ubiquitous *Aphorisms* was prepared by Martin de Saint Gilles of Avignon in the 14th century. Some idea of the demand for *Aphorisms* may be conveyed when it is realized that 483 manuscript copies exist in what Sarton termed "the learned languages": 140 in Greek, 232 in Latin, 70 in Arabic, 40 in Hebrew, and one in Syriac (Sarton 1961: 183). By contrast, the two Platonic dialogues, *Phaedo* and *Meno,* became available in the Latin West only in the 12th century in translations by Aristippus of Catania and are known in "perhaps a dozen surviving manuscripts" (Haskins 1957: 344).

Incunabula

Demand for Hippocratic texts did not diminish with the coming of printing. Studies of the scientific incunabula show Hippocrates to be, behind Albertus Magnus (c.1220-80) with 151 editions and Aristotle with 98 editions, the third most frequently printed author (Klebs: 13-342). None of the twenty-three titles in fifty-two editions is a text of the full *Corpus Hippocraticum.* Rather, individual editions of popular works were favored with *Aphorisms* in various Latin translations appearing at least eight times before 1500. In addition collections of the more popular Hippocratic texts were made and, like the *Articella* (Padua, 1476), were reprinted several times.

Collected Editions

It was not until the 16th century that complete editions of the *Corpus* began to appear. By 1800 over thirty editions had been printed. Of these nineteen were in Latin, two in Greek, seven in both Greek and Latin, two in French and one in German. Many of the later Latin editions are little more than pirated copies of earlier works and are distinguishable from their originals only by their title page. Consequently it will be necessary to list only the various *princeps* editions and also to survey the modern editions all of which derive ultimately from the Littré text of 1839-61:

Rome, 1525: The Latin *princeps;* translated by Fabius Calvus, published by Franciscus Minutius.

Venice, 1526: The Greek *princeps;* edited by Franciscus Asulanus, printed by Aldus. Contains fifty-nine Hippocratic texts.

Basel, 1538: A second Greek edition, edited by Janus Cornarius, printed by Frobenius. Described by Sarton as "more complete and accurate" (1927: I, 99).

Venice, 1588: The first Greek/Latin edition, edited by Geronimo Mercuriali, published by Juntae.

Frankfurt, 1595: A second and better known Greek/Latin version by Anuce Foes and reprinted in 1621, 1624, 1645 and 1647. Foes had earlier produced a 700-page folio volume, *Oeconomica Hippocratis* (Frankfurt, 1588), printed in small type in two columns: a Hippocratic dictionary and encyclopaedia described by Sarton as a "monument of medical learning" (1953: 354).

Modern Hippocratic scholarship begins with the great French lexicographer Emile Littré (1801-81):

Paris, 1839-61: *Les oeuvres complètes d'Hippocrate* (10 vols.), Greek text with French translation; so far the only complete modern edition in any language. It was republished in Paris, 1932-34 in four volumes and in a revised and corrected edition in 1955 in five volumes; it was also reissued in Amsterdam in 1961.

Leipzig, 1859-64: A three-volume edition of the Greek text by F. Z. Emerins.

Leipzig, 1894-1902: The Teubner edition of the Greek text, incomplete in two volumes by H. Kuhlewein.

London/New York, 1923-31: Loeb edition; four-volume Greek edition with parallel English translation. Vol. I, II, and IV were edited and translated by W. H. S. Jones and Vol. III (surgical) by E. T. Withington.

There are presently at least three long term projects for the publication of the entire *Corpus*:

The *Corpus medicorum Graecorum* (Leipzig and Berlin), aiming to print all extant Greek medical texts. So far three volumes have appeared: 1927, edited by Heiberg; 1968, edited by H. Diller; and 1970, edited by H. Grensemann.

Three volumes edited by R. Joly have also appeared in Paris produced by the publishing unit of the Universities of France (1967, 1970, 1972).

The first volume of *Ars medica* (Berlin) edited by H. Grensemann appeared in 1968.

Translations

The first English translation, containing only 17 works, appeared as surprisingly late as 1849 and was done by Francis Adams (*The Genuine Works of Hippocrates,* 2 vols.). It has been widely printed in the United States (New York: 1886, 1891, 1929, 1950; Baltimore: 1938 and 1939). It has also been reprinted in London (1939, 1964). A possible reason for its late appearance in English may be related to an attempt by physicians to draw attention to their status as readers of Latin and Greek. By the mid-19th century, there was no longer any practical need for Hippocratic works, so it was safe to make it available in the vernacular.

Paris, 1667: The much earlier French translation, *Les oeuvres du grand Hippocrates* (2 vols.), was done by C. Tardy. Other translations in French appeared in 1697 (by A. Dacier) and 1801 (by J. B. Gardeil), in addition to the already-mentioned effort by Littré.

Germany, 1781-92: The first German translation appeared during these years; later translations were published in 1814 and 1847. There is also a modern *Die Werke des Hippokrates* (1934) by R. Kappfere.

Lund, 1909-10: A two-volume Swedish translation.

Cremona, 1860: An Italian translation.

Contemporary Selections

The major texts of Hippocrates are available in most European languages in numerous editions. Among them, *The Medical Works of Hippocrates,* edited by J. Chadwick and W. Mann (Oxford, 1950), which has been conveniently reissued by Penguin (1978) with additional material and edited with a long introduction by G. E. R. Lloyd under the title *Hippocratic Writings.* Other notable selections include *Hippokrater-Schriften,* edited by H. Diller (Hamburg, 1962); *Hippocrate-Medicinegrecque,* edited by R. Joly (Paris, 1964); and *Opere di Ippocrati,* edited by M. Vegeti (Turin, 1963).

Further Reading

The standard history of medicine has been for many years F. H. Garrison's *An Introduction to the History of Medicine* (1960, fourth edition.). It remains useful as a reference work but sheds little light on problems of history, medical or otherwise. Much better is the more recent *A Short History of Medicine* (1961) by Charles Singer and E. A. Underwood. The background of Greek medicine can be studied in Brock (1929) and Phillips (1973), while for the general intellectual background against which the ideas of Hippocrates emerged two works of G. E. R. Lloyd should be consulted: *Polarity and Analogy* (1966) and *Magic, Reason and Experience* (1979). For most readers the extracts from Hippocrates provided in Chadwick and Mann (1978) in its new version with an introduction by Lloyd will suffice. To pursue the long history of Hippocratic medicine a certain persistence is called for. A monograph by Oswei Temkin, *Galenism* (1973), sketches in one aspect of the story and Grant (1974) contains some useful readings from the medieval period. The threat from Paracelsus is discussed in two works by Allen Debus: *The English Paracelsians* (1965) and *Man and Nature in the Renaissance* (1978). For the early 17th century C. K. Webster's *The Great Instauration* (1975) is essential while for the latter half two works of L. S. King should be consulted: *The Road to Medical Enlightenment* (1970) and his paper "The Transformation of Galenism" in A. Debus (ed.), *Medicine in Seventeenth Century England* (1974). The great story of how, beginning in the 18th-century, Hippocratic medicine became discredited and was finally replaced in the 19th century by the germ theory of disease has yet to be told adequately but can be seen in outline in Winslow (1980). For other aspects of the Hippocratic tradition Glacken (1976) should be consulted.

Notes

1. This claim may surprise some and probably needs clarification. There are earlier Egyptian medical texts – the *Ebers* papyrus, dating from 1500 B.C., for example – but such works are invariably lists of spells or a pharmacopoeia or a mixture of both. The chanting of spells and preparation of drugs may be compared to simple processes, such as smelting or pottery-making, which require little scientific knowledge. Similar comments can be directed against Egyptian mathematical and astronomical texts. The view of Neugebauer is definitive: "They are crude observational schemes, partly religious, partly practical in purpose. Ancient science was the product of a very few men; and these few happened not to be Egyptian" (Neugebauer 1969: 91). A better case can be made for Babylonian astronomical texts. Discounting the *Enuma anu enlil,* dating to 1000 B.C., a collection of about 7,000 celestial omens, we can mention the *Mul appin* of about 700 B.C., which exhibits a fairly advanced astronomical understanding. Such works consist mainly of tables – early ephemerides – and while scientific, they are impersonal, being the products of groups rather than individuals. The texts are also statements of results established elsewhere and their application to practical problems, like the prediction of eclipses, shows that they are not examples of scientific thought, wherein a position is taken and maintained against known and imagined objections.

2. "The bulk of the writings in the *Corpus* were associated with the name of Hippocrates, not because they were thought to be written by him, but because they originally belonged to him or to his school" (Jones 1945: 115). In this chapter the expression "Hippocrates said. . ." should be taken to mean no more than the views expressed somewhere in the *Corpus.*

3. There are earlier references to Hippocrates and his views. Plato, for example, in *Phaedrus* (270).

4. Figures given of the number of works in the *Corpus* vary considerably. This is mainly because some scholars take a work such as *Epidemics* as consisting of seven distinct books; others count it as a single work.

5. The attraction of this work to so many generations of physicians is difficult to appreciate today. Consisting of 412 aphorisms divided into seven books with little apparent organization, it contains such famous sayings as "life is short, the art long" and "desperate cases need desperate remedies." The bulk seems to be prognostic tips of a kind (sandy urinary sediment suggest a stone forming; a rigor on the sixth day of a fever is dangerous; epistaxis is a bad sign in quartan fever). There are also a number of more general propositions whose provenance probably lies in some long-forgotten folklore (eunuchs never get gout; people who lisp suffer from prolonged diarrhoea; women are never ambidextrous).

6. Only an examination of the magical literature will reveal the full achievement of the Hippocratic tradition. Here is an example from ninth-century England. A circle is divided into two with the numbers 1, 2, 3, 7, 9, 10, 13, 14, 16, 17, 19, 20, 22, 23, 26, and 11 written in the top half, and the numbers 5, 6, 8, 12, 15, 18, 21, 24, 25, 28, 29, 30 written on the lower half. The numbers 4 and 27 are not included. Around the circle a simple cipher alphabet is written assigning numerical values to the letters, in this instance A = 3, E = 12, X = 6. The text, which comes from a certain Willibrord, read, "The device is the sphere of. . . Pythagoras which Apollonius described for the discovery of anything concerning the sick. Thou shouldst determine the day of the week and of the moon (on which he fell sick) and (the numerical value) of his name according to the letters. . . . Add them together and divide by 30 and consider the remainder. Examine what is written below, and if it fall in the upper part, he will live and do well; if below, he will die" (Singer 1958: 144). Given this level of thought, it becomes clear why scholars became so enthusiastic about the *Corpus*. It really does stand out in a remarkable way against centuries of nonsense, credulousness and mindless superstition.

7. Of these bodily fluids, all linked with disease, phlegm and blood are obvious. Yellow bile is probably what we still loosely call bile and is known to most of us from its presence in vomit. Black bile is thought to be blood from an internal hemorrhage often found darkly-colored in feces, vomit and urine.
8. Both came from Elea in southern Italy in the fifth century B.C. Parmenides had argued in a rather oracular manner in a long poem that reality was indivisible, motionless, finite and spherical. Such views provoked not only opposition but, according to Plato, ridicule. Zeno, his pupil, expressed the Parmenidean position in a form less easily dismissed. He began by assuming that the views of Parmenides were false, accepting in their place a world of plurality, divisibility, and motion. From such an assumption he proceeded to derive conclusions more absurd than any espoused by Parmenides. Posed with great dialectical skill, they have continued if not to baffle then to bewilder generations of mathematicians ever since. One of his arguments, how the tortoise once given a start can never be caught by pursuing Achilles, has become part of folklore, and it is still worth repeating. The tortoise runs one hundred times slower than Achilles, and has a hundred-meter start. While Achilles runs those hundred meters, the tortoise advances a further meter; while Achilles runs that meter, the tortoise advances a further centimeter; while Achilles runs that centimeter, the tortoise advances a tenth of a millimeter, and so on. Achilles can thus appear to get ever closer to the tortoise but can never quite catch him.
9. The influence of Parmenides on Empedocles was specifically acknowledged. He accepted that nothing could come or cease to be, but went on to propose a scheme within these confines that would allow for change, development, and decay. Four eternal irreducible elements – earth, air, fire, and water – through their various combinations, separations, and recombinations, could thus account for the facts of nature apparently dismissed by Parmenides.
10. No one seems to realize this better than anthropologists in general and Levi-Strauss in particular. Scarcely a page of his *Mythologiques* is free from some form of diagram or equation. Levi-Strauss, in this formidable work, has taken as his raw material several hundred myths from the Americas and aimed to show that such "apparently arbitrary data," "spontaneous flow of inspiration," and "seemingly uncontrolled inventiveness" are governed by "laws operating at a deeper level." One of his key techniques is to extract from the myths a number of matrices with their cells all neatly occupied like a complete crossword. The impression conveyed is that if the contents of the myth can be so arranged then it certainly cannot be a collection of "apparently arbitrary data." The link with Hippocratic humoral theory is evident. Unfortunately the material of Levi-Strauss is too interconnected and bulky to allow easy summary; an examination of the first volume of his four-volume *The Science of Mythology*, namely, *The Raw and the Cooked* (1970), will indicate what is meant.
11. In molecular biology, for example, a checkerboard grid links the four nucleotides to the twenty or so amino acids. In elementary particle physics, the various multiplets are convincingly displayed in symmetrical diagrams very similar to that used in antiquity. For details, see R. Clowes, *The Structure of Life* (1967) and J. S. Trefil, *From Atoms to Quarks* (1980).
12. The correspondences represented here turned out to be only the begininng. The zodiac, for example, was divided into four triplicities. This involved assigning in an arbitrary manner Aries, Leo and Sagittarius to *Fire*; Taurus, Virgo and Capricorn to *Earth;* Gemini, Libra and Aquarius to *Air* and Cancer, Scorpio and Pisces to *Water*. By the early Renaissance, a complex system of correspondences had been built up in which anything of consequence – metals, minerals, parts of the body or whatever – was linked with one or other of the elements.
13. This delicate balance, according to Galen, could be upset in two ways: one element could gain ascendency or all the humors could be augmented in roughly the same proportion. The first case was known as *cacochymia* and the second as *plethora* or *plenitude.* It was important to distinguish between them as the required treatment could differ.

14. There are references, in *Epidemics* for example, to bleeding. Thus in one case a patient with "continuous aching in the right side and a dry cough" was recorded as having been bled at the elbow on the eighth day (Chadwick and Mann: 132). It is said to be most efficacious in the spring and recommended for dysuria. Equally, there are references to bathing and diet throughout the *Corpus,* conveying the impression that a visit to a Hippocratic physician would more likely involve a therapy of dietary restrictions and hot fomentations than it would cauterization and venesection.
15. Counting here is inclusive so that quartan fever falls on the first, fourth, and seventh day–every third day in fact. Tertian fever, therefore, comes on alternate days; quotidian, daily; and sub-tertian, irregularly from a few hours to every other day.
16. According to one authority, they were aided by the spread of malaria and its sequelae throughout the Greek world after 400 B.C. Unable to offer any satisfactory treatment, the Hippocratic physician found that many of his patients deserted him and his rational methods of cure for "dream oracles, charms and other superstitions" that began to flood into the Greek world about this period (see W. H. S. Jones, *Malaria and Greek History* [1909]).
17. The competition persists. In much of the Third World the two traditions are still found. In Ghana, for example, the sick seek aid either from those who have sworn the Hippocratic oath or from priests who are attached to shrines like those dedicated to Asclepius as Epidaurus. See M. J. Field, *Search for Security* (1960). Nor is such competition to orthodox medicine restricted to the Third World; it can be found in present-day Los Angeles or London as easily as it can in Accra or Lagos. However, there is a tendency to dismiss Third World practitioners as witchdoctors and fetish priests, while their Western counterparts are said to practice "alternative medicine," a euphemism for quackery which seems to have crept into the language in the last decade or so.
18. The location of lost property might at first sight seem a strange power claimed by a god of healing; the Epidauran inscriptions nonetheless contain a number of examples. In the absence of newspapers or any other means to advertise the loss, it was a form of business too lucrative to be ignored. It was in fact a skill claimed by oracles, cunning men, magicians and astrologers over the centuries. Thus Thomas (1973: 364-66) has documented how in Tudor and Stuart England customers would consult astrologers "in search of mislaid bits of crockery and stolen washing. . . missing silver or other valuables." Together with miracle cures it was a business shunned by the Hippocratics.
19. Other words containing traces of the old humoral theory are *temperament,* from the Latin *temperare,* to mix; and *idiosyncracy,* from the Greek *idio*, peculiar or private and *syn-*, with, together; and *krasis-*, a mingling.
20. In the late 19th century Wilhelm Wundt (1832-1920), founder at Leipzig of the first modern psychological laboratory, proposed to analyze personality along changeable/unchangeable and strongly emotional/weakly emotional continua. The four possible combinations just happened to fit the four classical temperaments displayed neatly in the following diagram:

	emotional	*non-emotional*
changeable	choleric	sanguine
unchangeable	melancholic	phlegmatic

Under the influence of the well known psychologist Hans Eysenck this reformulation of Galen's theory was once more revised:

	neurotic	*non-neurotic*
extrovert	choleric	sanguine
introvert	melancholic	phlegmatic

and in this form can be seen frequently displayed as the most modern, scientific analysis of personality modern psychology can provide. Once more the seductive powers of a diagram with its completed grid is very much in evidence. Full details can be found in Eysenck, *The Scientific Study of Personality* (London, 1953).
21. In the Muslim world, Hippocrates is known as Bukarat.
22. Medieval scholars were obsessed with the need to number. In the *Isagoge* of Hunayn ibn Ishaq, for example, nothing can be mentioned without it also being indicated how many kinds exist. Thus there are three kinds of drinks, two kinds of food (good and bad), and two kinds of tissue in surgery. Everything else exists in fours: four humors, elements, temperaments, gospels, apostles, cardinal points, seasons, and winds. There are also four hair colors (black, red, grey, white), four geographical locations (high, low, mountainous, coastal), and four kinds of swellings.
23. In the 16th century an equally valuable source is the work of Rabelais (1494-1553). In a typical passage, he talks of the people of the Island of Ruach: "They do not shit, piss or spit on this island . . . [but] fart and belch most copiously. They suffer from all sorts and varieties of diseases. For every malady originates and develops from flatulence as Hippocrates proves. . . . But the worst epidemic they know is the windy cholic. . . . They all fart as they die, the men loudly, the women soundlessly, and in this way their souls depart by the back passage" (Penguin edition, 1955: 541).
24. Hence the appearance at this time of books like G. Andsal, *Essai d'hematologie pathologique* (1843).
25. Snow had a practice in Soho (London). In *On the Mode of Communication of Cholera* (1849), he proposed that cholera was water-borne. Powerful support for his views came in 1854 when he was able to trace the outbreak of a new cholera epidemic to a single pump on Broad Street. Removal of the pump handle stopped the spread of the epidemic immediately.
26. In a series of papers, collected in his *Typhoid Fever* (1873), Budd showed how the spread of typhoid in the West country of England was related to the movement of people. Bad sanitation by itself could not produce typhoid; nor could it in conjunction with any kind of atmospheric condition. Bad sanitation could, however, spread an already present typhoid.
27. Nor was it just experimental psychologists who continued to accept humoral psychology well into this century. Ibn Saud, founder of modern Saudi Arabia, was reported to divide human beings into four classes: "Damawi, or plethoric; Saudawi, or melancholic; Belghami, or phlegmatic; and Safrawi, or bilious" (see H. St. John Philby, *Arabian Jubilee* [London, 1952]). The source of Ibn Saud's classification is obvious. Less clear is the origin of the humoral pathology apparently found among Indians of Mexico and Central America. It has been reported that among some cultures, "most foods, beverages, herbs and medicines are classified as 'hot' or 'cold'. This classification is usually independent of . . . physical temperature. . . . Illness is often attributed to imbalance between heat and cold in the body, and curing is likewise accomplished by the restoration of the proper balance." See R. Currier, "The hot-cold syndrome and symbolic balance in Mexican and Spanish-American folk medicine, *Ethnology* 5:3 (1966): 251. Did such a belief travel, as Currier proposes, from Greece to the Arab world and thence via Spain to the New World? The proposal has found little support.
28. In England in more modern times the Royal College of Physicians deliberately kept its numbers small. There were only thirty-eight members in 1589 and no more than forty in 1663, despite the growth in the population. One estimate gives one licentiate of the college to each 5,000 Londoners for the 17th-century period. As for their charges, the usual figure was £1 per day, an enormous sum for the time.
29. References in contemporary literature to physicians and their skills tend to be dismissive if not ruder. Burton complained of their inadequacy: "Many diseases they cannot cure at all. . . . A common ague sometimes stumbles them." Even physicians appeared to think none too highly of themselves: Sydenham commented that the

poor often owed their lives to their inability to afford physicians, while Dr. Ridgely is said to have observed that "if the world knew the villainy. . . of the physicians. . ., people would throw stones at 'em as they walked in the streets" (Thomas: 16-17). Nor was their reputation better in France if Molière is taken as representative of popular opinion. In a line from *L'Amour medicine,* he writes, "What will you do, sir, with four physicians? Is not one enough to kill any body?" Or one may mention the dying Charles II in 1685 who, according to Macaulay, was tortured like an Indian at the stake. Dosed with fifty-eight powerful drugs over a period of five days, he was given emetics, bled, cauterized, blistered, clystered and scarified. Few could envy the rich their access to such expensive medical advice.

30. A third major epidemological study was made during this period by the Dane, P. L. Panum (1820-85). The Faroe Islands had been free from measles since 1781, but in 1846 a visit from an infected carpenter reintroduced the disease with appalling consequences. In an area of isolated islands and communities, where visits between homesteads were rare enough to be remembered reliably, Panum was able to carefully document the spread of the disease and produce conclusive evidence that it occurred via personal contact.

31. Some of those difficulties have been revived recently by the cosmologist Fred Hoyle and his collaborator C. Wickramansinghe in their books *Diseases from Space* (1979), *Space Travellers* (1981), and *Evolution from Space* (1981). Like the physicians of the 17th century, they, too, are struck by the apparent ease with which certain epidemics seem to appear from nowhere and to disappear as suddenly and mysteriously as they came. How could influenza, they ask, infect isolated shepherds in the wild of Sardinia – to take just one of their examples – unless the disease were not transported there by extraordinary means? They hypothesize that the earth is being intermittently bombarded with bacteria and viruses from outer space.

CHAPTER 3

The Elements
(circa 300 B.C.)

by Euclid

Aristippus, a companion of Socrates, shipwrecked and cast up on the shore of Rhodes, noted geometrical figures drawn on the sand. "Let us be a good hope, for indeed I see the traces of man."

–Vitruvius, *De architectura* (first century A.D.)

He was forty years old before he looked on Geometry; which happened accidentally. Being in a Gentleman's Library, Euclid's *Elements* lay open, and twas the 47 E1. libri 1 (Pythagoras theorem). He read the Proposition. By G–, says he, . . . this is impossible! So he reads the Demonstration of it, which referred him back to such a Proposition; which Proposition he read. That referred him back to another, which he also read. *Et sic deinceps* that at last he was demonstratively convinced of the truth. This made him in love with Geometry.

–Aubrey, *Brief Lives,* written about Thomas Hobbes

He studied and nearly mastered the six books of Euclid since he was a member of Congress. He began a course of rigid mental discipline with the intent to improve his faculties, especially his powers of logic and language. Hence his fondness for Euclid, which he carried with him on the circuit till he could demonstrate with ease all the [theorems] in the six books.

–Abraham Lincoln (writing of himself), *Short Autobiography*

At the age of eleven, I began Euclid. . . . This was one of the great events of my life, as dazzling as first love. I had not imagined there was anything so delicious in the world.

– Bertrand Russell, *Autobiography*

A book to be read in bed or on holiday, a book as difficult as any detective story to lay down when once begun.

–T. L. Heath, *Euclid*

It is difficult to think of anything more characteristic of civilized man than his ability to think deductively. As *Homo axiomaticus,* he has been able to start with a few self-evident truths and, by applying only established principles of reasoning, to derive conclusions of great originality, depth and authority. The roots of this intellectual craft, like most others, go back to ancient Greece and the time of Euclid.

No non-imaginative work has had a longer history of a more profound impact on the mind and civilization of man than *The Elements* of Euclid. While most scientific texts of antiquity after the Renaissance were of interest only to scholars, editions of Euclid in the mid-19th century were churned out in their dozens as working texts for schools, universities and factories. Shortly after 1900 interest in Euclid rapidly waned and the number of new editions dropped steeply. Nonetheless, from 300 B.C. to A.D. 1900, *The Elements* had a longer life than has any other non-fictional work.

Nor is it just the enormous time-span that makes *The Elements* unique among books. There is also the wide geographical spread it achieved. One may cite as an example the experience of Lieutenant Hugh Clapperton, R.N., who in 1824 traveled through what is today northern Nigeria in search of the source of the Niger River. When he met the ruler of the Hausa kingdom, Muhammad Bello, he presented gifts, among them an Arabic edition of *The Elements.* Visiting the ruler later, Clapperton recalled, "Saw the Sultan this morning, who was sitting in the inner apartment of his house, with the Arabic copy of Euclid before him, which I had given him as a present. He said that his family had a copy of Euclid brought by one of their relations who had procured it in Mecca; that it was destroyed when part of his house was burnt down last year; and he observed, that he could not but feel very much obliged to the king of England for sending him so valuable a present" (Hodgkin: 215).

Virtually nothing is known about the author of *The Elements.* He reportedly flourished during the reign of Ptolemy I (306-283 B.C.) and is said to belong to the generation between the pupils of Plato and those of Archimedes. Even his place of birth is not known. It is known, however, that he founded a school and taught in Alexandria. Two well-known stories are associated with him. He was asked by Ptolemy I if there was an easier way to learn geometry than by studying *The Elements*; the monarch received the proverbial response that there is no royal road in this discipline. There is another story of a student asking, as students often do, what value the study of geometry had. Euclid gave the lad a small coin so that he need no longer feel he had wasted his time.

It is clear *The Elements* was not created entirely from Euclid's own resourcefulness. It is not quite a collection of the works of others, but rather a systemization of earlier and perhaps cruder works. Unfortunately, the very success of Euclid meant that these earlier treatises were abandoned, and they have failed to survive. Thus, Proclus, an important commentator of the fifth century A.D.,[1] described Euclid as "arranging in order many of Eudoxus's theorems, perfecting many of Theaetetus's, and also bringing to irrefutable demonstration the things which had been only loosely proved by his predecessors" (Midonick: 410). Other names have been mentioned by other authorities and in some cases theorems and even whole sections of *The Elements* have been attributed to earlier mathematicians. The theory of proportion in Book V is known from other sources to have come from Eudoxus of Cnidus (fourth century B.C.) while the arithmetical books, VII-IX, undoubtedly originated with the Pythagoreans.

"The Elements" in Antiquity

Proclus says that both the text and contents of *The Elements* were much discussed in his day, and on more than one occasion he complains of interminable discussions among the commentators over what he considered trivial points. Earlier commentaries are known to have been produced by Heron of Alexandria (fl. A.D. 62), Porphyry (A.D. 232-305), and Simplicius (sixth century A.D.), none of which has survived. Consequently much of our understanding of the attitude of antiquity to Euclid derives from Proclus whose commentary on Book I of *The Elements* has survived complete.

An important consequence of the numerous ancient commentaries made on *The Elements* is that Euclid's text has been thoroughly edited. Ancient editors and scholars are known to have made extensive additions and deletions to manuscripts they thought in need of improvement (a habit, incidentally, followed by medieval editors). Examples can be found in the work of the Alexandrian editors of Homer who objected to and considerably improved upon certain lines, sometimes removing those they considered unseemly, and in the work of the Byzantine scholar Maximus Planudes (c. 1255-1305) who edited an astronomical poem of the classical author Aratus (third century B.C.); so conscious was he of his own superior astronomical knowledge that Planudes could not resist the temptation to replace some thirty lines of Aratus with his own verse.

There are many signs that Euclid failed to escape completely this process of textual revision. All pre-19th-century editions of *The Elements* ultimately derive from a text prepared by Theon of Alexandria (fl. A.D. 364). The evidence for this is based on a casual admission contained in his commentary of Ptolemy, in which he refers to a theorem "proved by me in my edition of *The Elements* at the end of the sixth book" (Heath: 46).[2] A number of complete manuscripts of the Theonine recension have survived dating from the 9th-12th centuries. The earliest, known as *B*, is found in the Bodleian Library and was written by a clerk named Stephen in the year 6397, or 888 of our epoch. It was acquired by Arethas of Caesarea (c. 860-c. 935), who was archbishop and a bibliophile, for fourteen gold pieces (a vast sum by the standards of the day; but for his copy of Plato he was prepared to pay half again as much in 895!).

During the Italian campaign, Napoleon appropriated large numbers of Vatican manuscripts and on examining them in Paris, the scholar F. Peyrard discovered a manuscript of *The Elements* that lacked Theon's familiar interpolations. The added theorem of Book VI was specifically missing. Dating to the tenth century and known as *P*, it has been identified as an earlier recension than Theon's.[3] The relationship between the various manuscripts, scholia and other fragments is fearfully complex, requiring the skills and knowledge of a paleographer,[4] but the textual problems nonetheless bring up a couple of points of non-scholarly general interest.

The bulk of changes introduced by Theon and others was minor and for the most part designed to make the text clearer. Intermediate steps were inserted into proofs, alternative versions were sometimes offered, arguments were repeated, and words of clarification added. It should also be pointed out that this practice did not begin with Theon, for there is a report of a third-century manuscript containing many such additions. As a second-century manuscript is known to have been free of interpolations, it has been assumed by Heiberg, an authority on Euclid, that most changes to the text date to the third century.

It is nonetheless possible that significant changes in the axioms of *The Elements* were made by commentators. They certainly sought to add extra ones, some of which were discussed by Proclus. Others, like Heron, sought rather to reduce the number of axioms. Are the axioms contained in the modern edition of *The Elements* those used by Euclid? We will return to this point later.

In truth, the ancients had a very different attitude toward Euclid than did latter-day mathematicians. Contrast what Proclus wrote about the fifth postulate—"This ought to be struck from the postulates altogether"—with what the mathematician Augustus de Morgan (1806-71) wrote in 1849: "There never has been, and till we see it we never shall believe there can be, a system of geometry worthy of the name, which has any material departures . . . from the plan laid down by Euclid" (Heath: v). Rather than revering *The Elements,* as the Victorians did, the ancients viewed geometry as a living tradition, something to be modified, changed, added to, reduced, and, where necessary, improved.

The Medieval Period—The Islamic World

Most works, important or otherwise, failed to survive the medieval period; others are known from a single manuscript. Latin works fared better than Greek, but in all cases the toll was heavy. Sometimes if a manuscript was not physically destroyed, the text was erased in order to free the parchment for more pressing use. It is from one such palimpsest, as they are known to paleographers, that Cicero's *De republica* managed to survive. Copied initially in the fourth or fifth century, it was covered with and survived as a seventh-century edition of a commentary on the Psalms by St. Augustine (353-430). But the works of Euclid proved no exception, and important books of his, such as *Conics,* failed to survive.

Survival in this context meant being copied. The chance of an original text dating from the classical period surviving into the medieval period was minimal; it depended upon whether there was any call for a scribe to undertake the prolonged and expensive task of copying. *The Elements* is a lengthy work. Fortunately, there was a limited demand from rich collectors, such as Arethas of Caesarea. But as the ability to read Greek was far from common in the Latin West in the medieval period, the bulk of the copying and preservation was carried out by scholars of Byzantium and the Islamic world.

Much of our knowledge of early Arabic translations comes from the indispensable *Fihrist* of al-Nadim who flourished in the tenth century. He probably came from Baghdad and may have been a bookseller, a copyist or a librarian. His *Fihrist,* composed in ten volumes, is a priceless index of all books written in Arabic or in non-Arabic languages, together with basic biographical data. The seventh volume concerns ancient science.

Spurning masterpieces of Greek literature, the Arabs eagerly turned to works on medicine, astronomy and geometry. Many Greek manuscripts were obtained during this period through diplomatic exchanges with Constantinople, and three independent translations of Euclid into Arabic were made. Sometime during the early ninth century, al-Nadim reports, Euclid was twice translated by al-Hajjaj ibn Yusuf ibn Matar, the second version of which he claimed to be more reliable. Only a part of this latter version survived. Soon, however, it faced competition from a new translation done by Ishaq ibn Hunayn around 900. Ishaq's translation was soon modified and improved upon by the well-known astronomer Thabit ibn Qurrah (836-901). In addition, a large number of commentaries and partial translations also appeared. A dozen or so are mentioned by al-Nadim, while an even longer list can be found in Heath.

A new translation was made by al-Tusi (1201-74), director of the important Maragha Observatory near Tabriz (Iran), sometime in the 13th century. It has been described by Heath as "not a translation. . . , but a rewritten Euclid based on the older Arabic translations" (Heath 1956:I, 78). In this respect it differed a little from some of the Latin translations.

Moreover, an extraordinary number of commentaries on *The Elements* were produced by Muslim mathematicians. Looking at the al-Nadim and Heath lists, it seems as if every major figure in Islamic science felt the need to write a commentary. Among them were Ibn Sina, al-Haitham, al-Kindi, al-Farabi, and al-Razi. Heath's by no means complete list contains thirty-one entries of which about twenty date from the ninth and tenth centuries. Some were works of little importance; others were probably mere copies. Regardless, the lists testify to the amazing interest shown in Euclid's geometry by Muslim scholars.

Although Euclid and geometry proved popular subjects for the Arabs – as opposed, for example, to the anatomy of Galen and the Alexandrians, which did not – *The Elements* was not an object of fascination wherever it traveled. The Chinese, according to Needham were largely indifferent to the charms of deductive geometry while pursuing other areas of mathematics with persistence and skill (Needham 1959: 91-112). Nor was the response of the Japanese more than polite when the first translation of *the Elements* arrived in 1733: "They failed utterly to see that it was an exact logical system, thinking it much too elementary and inelegant to define with so many words things intuitively evident or to prove self-evident propositions with such laboured demonstration" (Sugimoto and Swain: 370).

The Medieval Period–The Latin West

An alternative tradition began to emerge in the West in the 12th century, one of the great periods in the intellectual history of the world. After centuries of being "lost," translations of Euclid, Ptolemy, Hippocrates, Plato, Aristotle, and other major thinkers of antiquity appeared within a few years of each other, often in more than one edition. They were certainly needed. Before 1100, scholars unfamiliar with Greek or Arabic could only acquire the science of antiquity from encyclopedic works like those produced by Cassiodorus in the sixth and Isidore in the seventh century.[5] There are references to a supposed Latin translation by Boethius (475-524), but, judging from the surviving manuscripts, this was only a partial translation covering no more than five of the thirteen books of *The Elements* and, in any case, it dates from many centuries later than Boethius.

When it is possible to glimpse the level of geometrical knowledge possessed by the learned of the period, the result is bizarre. It is not backwardness which is revealed but total incomprehension of basic geometrical terms. Thus, medievalists have drawn attention to a remarkable correspondence between Ralph of Liège and Reginbald of Cologne in the early 11th century. They had apparently come across a reference in a commentary on Aristotle to the proposition that the interior angles of a triangle are equal to two right angles (Euclid I: 32). What on earth, they ask each other, are interior angles? Were they, Reginbald asked, the angles formed by a line dropped from one angle of a triangle to the opposite side? (He cited as his authority Fulbert, bishop of Chartres, one of the most learned men of the day) (Southern: 192-93). It would appear that even the learned had only a small idea of the content of Greek geometry and lacked completely any appreciation of its subtleties. It must have been appallingly frustrating for those who sensed the riches of thought that lay beyond the crude summaries then available in various compendia and encyclopedias. Eastern scholarship lay out of reach of the West for a variety of religious and political reasons. The reconquest of Spain, begun in the 11th century, finally made available to the Christian West the major scientific texts of the Greeks so fortunately preserved by Islamic scholars.

The seizure of such centers of learning as Toledo, Seville, and Cordoba permitted the crucial intellectual breakthrough for the period. From all over Europe frustrated scholars crossed the Pyrenees in search of the ancient texts; John of Seville, Plato of Tivoli, Robert of Chester, Gerard of Cremona, Adelard of Bath, Rudolf of Bruges, Hermann of Carinthia,

Hugh of Santalla, translators whose names read like a gazeteer of Europe. First, it was necessary to learn Arabic if only well enough to check the quality of text produced by hired helpers. Two versions of *The Elements,* both translated from Arabic texts, appeared in the 12th century. The first was made by Adelard of Bath about 1120; Gerard of Cremona (1114-87) did the second later in the century. Gerard's version is described by Heath as "much clearer" and "neither abbreviated" nor "edited" as Adelard's had been. A third translation, by Johannes Campanus, appeared in the 13th century. It was not a new translation but rather a reworking of Adelard's earlier attempt, and the definitions, postulates, axioms, and enunciations are word for word identical in each.[6] The Campanus translation is, according to Heath, clearer and more complete because Adelard had a curious habit of placing the proofs before the enunciations, a style neither adopted by Campanus nor by later editors and translators.[7]

A growing familiarity with the basic Euclidean concepts was evidenced in the 13th century as geometry monographs and textbooks began to appear for the first time. Campanus wrote a treatise on the sphere and another on the problem of squaring the circle. Earlier, Leonardo of Pisa, also known as Fibonacci (1170-1250), the scholar famed for introducing Arabic numerals to the Christian West, had written a *Practica geometriae* (1220). Geometry began to be taught in the new universities, and sections of *The Elements* became familiarly known among students. Roger Bacon (1214-92) picked up the term "Elefuga" which in 1250 was used by students to describe Euclid I:5 (the theorem that the angles at the base of an isoceles triangle are equal).[8] According to Bacon, the term meant "flight of the miserable" (from the Greek *eleos,* miserable, and *phuge,* flight.) The theorem of Pythagoras (Euclid I:47) was obscurely known as "Dulcarnon," a term derived by Heath from the Persian *du,* two, and *karn,* horn. This is a supposed description of the familiar diagram used to illustrate the theorem's proof with its two small squares sticking up like horns from the larger square. It is more likely, although precisely how and why is not clear, that the term derives from the familiar Arabic description of Alexander the Great as *dhu'l-qarnayn* from the Arabic *dhu,* owner of, and *qarnayn,* two horns. The term appears in the Koran (Sura 18) and probably entered the West sometime after the 12th century through the popular romance literature in which Alexander was a frequent subject.

Whatever its origin the term was sufficiently available for Chaucer (1345-1400) to use, with Elefuga, pointedly and somewhat inaccurately, in an otherwise unintelligible passage (*Troilus and Criseyde*):

> I am, til God me better minde sende,
> At dulcarnon, right at my wittes end.
> Qoud Pandarus, 'ye, nece, wol ye here?
> Dulcarnon called is "fleminge of wrecches".

The reference here is presumably to the widespread belief that Pythagoras's theorem (I:47) or Dulcarnon was sufficiently difficult to drive many a student to their wit's end. The identification in the final line of Dulcarnon with "fleminge of wrecches," or the fleeing of the miserable, in other words Elefuga (I:5), is thus likely to be a mistake.

There is ample evidence that by the 14th century geometrical knowledge was sufficiently widespread to be incorporated in the slang of students and included in the popular poetry of the day. It should not, of course, be expected that the medieval period was a time in which advances of any substance would be made; to understand the complexities of Euclid and prepare a sound text was challenge enough. Improvements would come but not before the advent of printing in the 15th century.

The Modern Period and the Fall of Euclid

As *The Element*'s publishing history testifies (see further on), Euclid enjoyed a rare renown in the Old World as well as new ones. But judging by the frequency and extravagance with which he was defended, Euclid's role in mathematical education – at least in the English-speaking world – came under attack in Britain as early as the 1830s. Newer mathematical methods in analysis and algebra developed on the continent during the 18th century threatened the sovereignty of Euclid. Such methods were slowly introduced into British university syllabuses and a growing number of mathematicians began to form pressure groups for the necessary innovations.[9] The resistance was formidable.

Members of educational establishments seldom have any difficulty in justifying, at least to their own satisfaction, why the authorities and techniques of the past should be preserved. It was argued against them that the newer, analytical methods were more efficient in that they could solve harder problems in a simpler manner. The point was conceded by Augustus de Morgan (1806-71), first professor of mathematics at University College, London. But, he wondered, was the new mathematics as effective as Euclid in "teaching the mind and in exercising thought?" About the same time, another powerful figure in the educational establishment, William Whewell (1794-1866), Master of Trinity College, Cambridge, also noted that if all anyone wanted from his mathematics was the correct answer, then, indeed, modern algebra and calculus should be adopted. But could they, he asked, "provide an educational discipline," could they ever be part of the "gymnastics of education"?

Two particular objections were made to the new mathematics. With Euclid the student had understanding. In Whewell's words, "He knows that all depends upon his first principles, and flows inevitably from them; that however far he may have travelled, he can at will go over any portion of his path, and satisfy himself that it is legitimate." By contrast, the use of calculus and algebra was an exercise in which the student was required to seek the correct solution by using arbitrary rules that meant nothing to him (Garland: 34).

It was secondly objected that the new material was too concerned with symbols. According to de Morgan, the modern obsession with symbols dismissed "from our minds altogether the conceptions of things which the symbols represent, whether lines, angles, velocities, forces or whatever." A similar point was made by G. B. Airy (1801-92), a leading astronomer, when he declared that he had not the "smallest confidence in any result which is essentially obtained by the use of imaginary symbols"; he preferred what he termed "strictly logical methods" (Garland: 36).

If such arguments failed to carry the point, a nationalistic element was then introduced. The Cambridge don J. R. Crowfoot argued in 1849 that "the unsoundness in theology of our German neighbours arose mainly from their not being familiar with the rules of strict demonstration" (Garland: 50).

But why was Euclid so important to "educational discipline?" it was asked. Why wouldn't any competent geometry textbook do? Isaac Todhunter, the Cambridge mathematician, raised this objection in the preface to the 1862 edition of his Euclid. Conceding that "nearly every official programme of instruction or examination explicitly includes some portion of this work," he pointed out that all attempts to find an "appropriate substitute" had been in vain. Yet once Euclid was abandoned, he believed it unlikely that an alternative text could ever be found. Furthermore, while there were no doubt many faults in Euclid, he thought it was idle to suppose that any other work would be free from defects and difficulties.

Shortly after, signs of an organized opposition to Euclid as a universal textbook began to appear in the professional literature. In 1870, *Nature* noted the information of an Association for the Improvement of Geometrical Teaching (AIGT) by a group of mathematical masters. The leading figures behind the association's formation were R. Levitt and E. F. MacCarthy of King Edward's School, Birmingham; J. M. Wilson of Rugby; and R. Tucker of University College School.[10] Their first conference was held in London on 17 January 1871, during which they expressed their aim to "induce all conductors of examinations to frame their questions independently of any particular textbook."

A particular target of AIGT was the widespread practice of requiring pupils not only to be familiar with Euclid but also to reproduce from memory the actual words and numbers used in specific texts of Euclid. Thus C. Smith and S. Bryant in their school edition of Euclid (1901) complained that "many teachers think it meritorious to insist upon their pupils rendering the very words of Simson"; they also condemned the imposition on students of the "heavy and useless burden of learning how to give accurately numbered references to the different propositions of Euclid." One way to break such habits, AIGT argued, was to have a variety of texts. "I:47" would not then automatically refer to the theorem of Pythagoras, as every student must have been only too aware, but to a variety of theorems depending upon which book was being used.[11]

A more radical answer was proposed. Wilson of Rugby pointed out that botany students dissected flowers before they learned how to classify them. So, too, with geometry a practical emphasis should be followed. Geometry should be taught by "interesting and varied applications of geometrical methods to measure and copy actually existing things" (*Nature* 4 [1871]: 366). This was too much for one reader who, signing himself simply as "a father," objected that Wilson's proposals "seemed rather suited to teachers of geometrical drawing rather than of math." What he wanted for his son was a book, like *The Elements,* "suitable for being committed to memory," for only in this way could a child learn (*Nature* 4: 404).

Undeterred, AIGT continued holding conferences and setting up subcommittees. Some early help came from the British Association in 1874 when it, too, established a committee to study methods of improving the teaching of mathematics.[12] Not all it had to report was progress, however. To its "distress," the second annual conference in 1872 noted the decision by the Italian government in 1867 that "in order to give to the instruction in geometry its maximum mathematical efficiency . . . [it proposed] to follow . . . the example of England and return to the *Elements* of Euclid" (*Nature* 5 [1872]: 400). Opposition in England to the Association also began to emerge. Lewis Carroll, for example, produced a defense of *The Elements* in *Euclid and His Modern Rivals* (1879). The modern rivals are not Riemann and Bolyai, two of the first to see the possibility of a non-Euclidean geometry, but those like Wilson of the AIGT who sought to develop modern and practical alternatives to Euclid.[13]

A major problem facing AIGT was deciding what text could be used as a substitute for Euclid. If a common text were removed, geometry teachers would face an avalanche of competing works, each proposing different techniques and approaches, each adopting a different terminology. The fact that this danger was avoided and that AIGT did not sink into trivial squabbles over the choice of competing textbooks was largely due to the efforts of John Perry (1850-1920), professor of engineering mathematics from 1881 to 1896 at City and Guilds, London.

Perry generalized the dispute, bringing a depth and urgency that transcended origins. In a series of addresses to the British Association and in articles in *Nature,* Perry repeated the charge that schools were producing generations of pupils who were unable to use simple mathematical tools and who had developed so powerful a loathing for mathematics that they were incapable of learning any new skills. The reason for this situation, he averred, was the traditional approach to the teaching of mathematics. Long ago, schools had dropped the Greek alphabetical notations for numbers, yet, according to Perry, they retained aspects of Greek mathematics equally archaic. In particular, he objected to the teaching of algebra through Books II and V of Euclid, books which could be replaced by "a page of simple algebra." The emphasis on proof was also declared to be misguided. More important was the ability to deploy the techniques learned in a wide variety of situations.

Perry sketched what he hoped for in an address to the British Association in 1902. He looked forward to a time when a fifteen-year-old student was trained in "experimental mathematics," able to "use logarithms. . ., have a working power with algebra and sines and cosines;. . . be able to tackle at once any. . . new problem with squared paper . . . have no fear of the calculus." Perry was successful in his goal of removing Books II and V from the elementary syllabus, although it is doubtful whether his second goal–a school syllabus in which fifteen-year-olds are trained in "experimental mathematics"–has yet been realized.

The first signs of success were visible as early as 1903. It was then that the powerful Cambridge Local Exam Board announced it would accept any proof of a proposition that was "part of a logical order of treatment." In the same year the Oxford Board conceded that it, too, would accept "any solution which shows an accurate method of geometrical reasoning" and, further, that "geometrical proofs of the theorems of Book II will not be insisted upon."

The impact of these decisions was almost immediately reflected in the publishing world. Whereas publishers' lists had been full of "Euclids," new titles now began to appear with a distinct premium placed on the term "practical." Among such works were Harrison's *Practical Plane and Solid Geometry* (1904) and *Lessons in Experimental and Practical Geometry* (1903) by Hall and Stevens. The new texts began with chapters on instruments and scales where earlier they would have started with Euclid's definitions and axioms.

Hall and Stevens proved particularly adaptable. Not only did they almost immediately produce a range of works on practical geometry, but with a minimum of effort converted their earlier *Euclid* into a more acceptable text titled *A School Geometry* (1903, twenty-four editions by 1918). This work remained a basically axiomatic geometry done in the Euclidean manner to which additions were made. Indeed, for all their enterprise, Hall and Stevens remained in the Euclidean tradition. When in the 1920s a new generation of geometry textbooks emerged, as, for example, those by the prolific author C. V. Durrell, the final break with the tradition of Euclid was made. His *Elementary Geometry* (1925 and many subsequent editions) contains no axioms. The early theorems are presented without proof and were then used not so much to prove later theorems but to show how reasonable they are. It is a geometry of demonstration and persuasion rather than one of rigor and proof. However sensible and useful it might be, it is not the geometry of Euclid.

Evidence for the fall of Euclid can also be seen in the British Museum catalogue. The Todhunters that appeared every year were by the 1920s a distant memory. The final Todhunter was published in 1903; thereafter, it is only found in a series like Everyman's Library which included it in 1933. The Everyman's Todhunter and the Heath edition, which had been available since 1956 in a Dover Books paperback, are now the only two editions of Euclid still available in the English language. Today, the Todhunters, the Hall and Stevens, the Baker and Bournes, names once familiar to all students, can be found in profusion only as neglected and dusty volumes in the backrooms of secondhand bookshops.

"The Elements"

Oddly enough, Euclid never used the term "geometry" even though it was in currency in his time. Then, it referred to the work of the surveyor rather than to the mathematician. (Euclid spoke of *stoicheia* or "elements.") The term only came to be applied in the 16th century with such titles as *The Elements of Geometry* used by Billingsley in 1570. It extended to thirteen books although some early editions, including the Greek *princeps,* mistakenly added two further books.[14]

The Euclidean text begins with twenty-three definitions, five populates, and five common notions. The definitions for the most part present no difficulties and consist of explanations of what is meant by such terms as rhomboid, obtuse angle, isoceles triangle, and the like. The distinction between postulates and common notions is that the former are more distinctly geometrical while the latter express more general rules of inference. They are brief enough to state in full:

> Postulate 1: To draw a straight line from any point to any point.
> P2: To produce a finite straight line continuously in a straight line.
> P3: To describe a circle with any center and distance.
> P4: That all right angles are equal to one another.
> P5: That if a straight line falling on two straight lines makes the interior angles on the same side less than two right angles, the two straight lines, if produced indefinitely, meet on that side on which the angles are less than the two right angles.

As for the common notions:

> Common Notion 1: Things which are equal to the same thing are also equal to one another.
> CN2: If equals be added to equals, the wholes are equal.
> CN3: If equals be subtracted from equals, the remainders are equal.
> CN4: Things which coincide with one another are equal to one another.
> CN5: The whole is greater than the part.

Were these foundations Euclidean or had they been added at a later date? The point is not insignificant because, as will be seen shortly, whoever introduced the parallel postulate (P5) had considerable insight into the structure and demands of an axiomatic system. Those who have denied that the common notions were Euclidean have based their arguments mainly on stylistic and literary grounds, and in the absence of any alternative tradition in antiquity, they are no longer considered sound.[15] It is believed that some early commentators attempted to reduce the number of common notions. Heron, for example, tried to manage with only the first three, but later critics pointed out that he smuggled into his proofs, without acknowledgement, the content of the two remaining common notions. It is also known, from Proclus, that several mathematicians tried to add to the Euclidean foundation. He mentions, for example, the proposal that two straight lines do not enclose a space. This postulate was actually included in some manuscripts of *The Elements*; later, it was dropped as unnecessary. Of the five acknowledged postulates, Heath firmly states that the "formulation. . . was Euclid's own work" (Heath 1956: I, 202).

In later works little resemblance to the Euclidean foundations can be seen. Texts in the Simson-Todhunter tradition contain as many as thirty-five definitions, but only three postulates – the first three of Euclid – and twelve axioms. These include many identity postulates, such as "If $a = b$, then $a + c = b + c$" and "If $a = b$, then $a - c = b - c$." But in this way the so-called fifth postulate of Euclid inevitably becomes the twelth axiom of Todhunter.

Before the contents of *The Elements* are considered, it is perhaps appropriate to give some indication of the power and special virtues of the axiomatic method. Its source, presumably, lay in the distant realization that intuitions and accepted knowledge however plausible they may seem and however well supported in experience they may be can turn out to be hopelessly wrong. Such an insight was gained by the Pythagoreans who, long before Euclid, discovered the incommensurability of the diagonal of the unit square. Begin with the square *ABCD* below, of unit size with diagonal *AC*. How are the diagonal (b) and a side (a) related? What could be more obvious than the reply b is some multiple of a; not an integral multiple perhaps, but a multiple nonetheless. The precise numbers cannot be known without actually measuring the lines when, it would confidently be claimed, b would be shown to be 1½ perhaps or 1⅓ times the length of a.

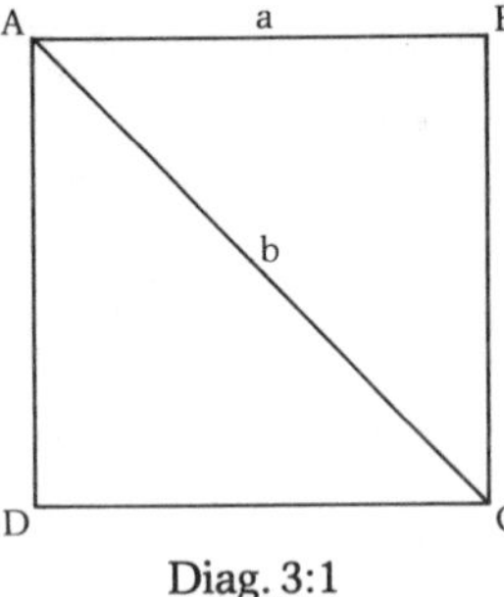

Diag. 3:1

While few would have been prepared to state the exact ratio between a and b without precise measurement, all, before the time of Pythagoras, would have insisted with considerable force that despite their ignorance the ratio did exist. It was as real as the number of uncounted grains of sand in a bucket. Using some fairly straightforward reasoning, the Pythagoreans showed this assumption to be naive and false and consequently shattered forever the permissibility of developing mathematics on such foundations.

From the theorem of Pythagoras the length of b can be determined accurately and simply:

(1) $b^2 = 2a^2$, but by definition $a = 1$, therefore $b^2 = 2$; or

(2) $b = \sqrt{2}$. Thus to show that b can be expressed as an exact ratio of a, it will only be necessary to express $\sqrt{2}$ as an exact ratio of a. But if it can be shown that $\sqrt{2}$ cannot be expressed as the exact ratio of any *number,* it would follow that $\sqrt{2}$ and therefore b cannot be expressed in terms of a. Begin by

assuming that $\sqrt{2}$ has been expressed as a ratio of two number p,q already reduced to their lowest terms so that they have no factor common other than 1;

(3) $\sqrt{2} = p/q$ or, by simple algebra, $p = q\sqrt{2}$. Squaring both sides yields

(4) $p^2 = 2q^2$, in which case *p must be even* (note, twice any number produces an even number, therefore $2q^2$ must be even. Also what it equals, p^2, and therefore p). But if p is even, *q must be odd* as it has no factors common with p. But if p is even it can be expressed in the form $2r$; substitute, therefore, $2r$ for p in (4);

(5) $(2r)^2 = 2q^2$ which becomes $4r^2 = 2q^2$, and which reduces to

(6) $2r^2 = q^2$. But by reasoning identical with that used in (4) it must follow that *q is even* whereas it was shown above (4) that *q must be odd.* This is a clear contradiction for no number can be both odd and even and therefore the initial assumption that $\sqrt{2}$ or the diagonal of the unit square can be expressed as a ratio of any other number has been shown to be false. The number 2 has no rational square root, there is no number b such that $b^2 = 2$; nor does the line b in the square above have any rational length. These are genuine conclusions and not the result of any sophistry.

They are also genuinely puzzling and, to the Pythagoreans who sought to construct the universe out of number, alarming as well. A new class of number had been discovered, one it is traditionally reported that the Pythagoreans tried to keep secret by threatening with death any who released news of it. There were, however, other than cosmological problems to face. How could such errors be avoided in the future? Were there perhaps other contradictions awaiting the unwary mathematician? It is at this point that the axiomatic approach reveals its strength. Do not begin with assumptions that seem plausible and obvious for intuition has been seen to be unreliable. Begin rather with those propositions which are more than obvious, self-evident in fact; propositions which are so clear as to be trivial and which it is impossible to conceive of as being false. Thus may be explained the attraction for such postulates as the whole is greater than the part and all right angles are equal to each other. Once these initial axioms are accepted and once only well-accepted principles of reasoning are used, then whatever is derived from them can never be in conflict with the foundations of the discipline.

So far the axiomatic approach has been seen as a prudent method preventing the distant appearance of absurdities and contradictions. But it soon showed itself genuinely creative by allowing the deduction of consequences which would otherwise have remained unsuspected. To use an example, it is unlikely that an architect or engineer, doodling with pencil and paper, would ever discover how a regular polygon of 65,537 sides could be constructed. Euclid had shown how to construct regular polygons of three, four, five and 15 sides and, of course, any polygons derivable from the above by straightforward bisection of the sides. For the next two thousand years mathematicians remained unsure whether other polygons could be constructed in the traditional Greek way. But in 1796, the Göttingen mathematician, C. Gauss (1777-1855), showed how a regular 17-sided polygon could be constructed. More importantly, Gauss proved that a regular polygon with n sides could be constructed with only a ruler and compass (where n is prime if, $n = 2^{2^m} + 1$.) (When $m = 0, 1, 2, 3$ and 4 respectively, then $n = 3, 5, 17, 257$ and 65,537.)[16] Thus, against all intuition a regular polygon of 14 sides cannot be constructed with a ruler and compass while one with 17, 257 or 65,537 sides can be. It is a remarkable example of to what unexpected conclusions axiomatic reasoning can lead – conclusions that are totally outside the range of practical experience, yet demonstratively true. What began as a rather restrictive technique, a means of preventing unwanted conclusions unexpectedly appearing became the most creative of all modes of reasoning. Besides the consequences derivable in some of the more important axiomatic systems, imaginative flights of poets and novelists appear somewhat pedestrian.

* * *

To a generation no longer familiar with the works of Euclid, the full range of work is surprising. The contents can be briefly summarized:

Book I: This is by far the most familiar to the modern reader. It comprises definitions, postulates and common notions and enunciates forty-eight propositions. 1-26 are concerned mainly with triangles, 27-32 with the theory of parallels, and the remainder with the areas of squares, triangles and parallelograms. I:5 is the so-called *pons asinorum* and I:47 the theorem of Pythagoras.

Book II: This is likely to be the oddest to the modern mind simply because it deals with a familiar topic in ways which can only seem bizarre to today's mathematician. At first sight, it is difficult to see the point of II:4 for example: "If a straight line be cut at random, the square on the whole is equal to the squares on the segments and twice the rectangle contained by the segments." The theorem is illustrated by the figure below where the line *AB* has been cut at random at *C*. The theorem states that the square on *AB* (*ABDE*) is equal to the squares on *AC* (*HGDF*), *CB* (*CBGK*) and twice the rectangle contained by *AC*, *CB* (*ACHG* and *GKFE*). Thus, what is claimed and proved in II:4 is: $AB^2 = AC^2 + CB^2 + 2(AC \times CB)$. However, let the line *AC* in the figure be represented by a and the line *CB* by b and a more familiar equation emerges: $(a + b)^2 = a^2 + b^2 + 2ab$.

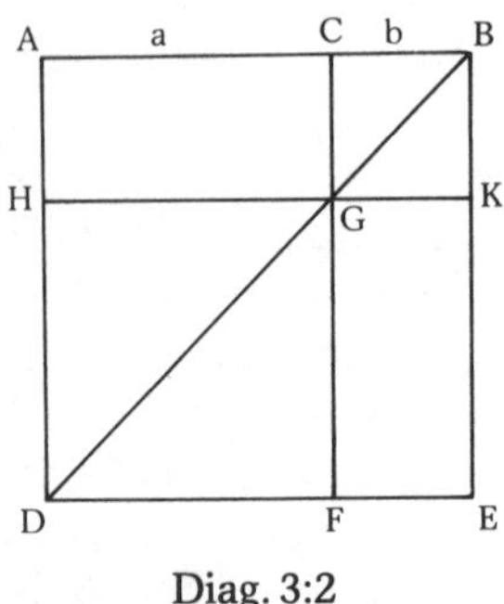

Diag. 3:2

Lacking all algebraic insight and symbolism, Euclid formulated and *proved* in geometrical terms truths automatically seen today through the variables of algebra rather than the lines of geometry. It is a salutory thought that until about 1900 schoolchildren were encouraged to see algebra in the same way! Numbers can be represented just as generally and as arbitrarily by lines as they can by the variables *x, y, z* etc. Consequently Euclid could prove many truths and state many problems which have since been normally approached through algebra.[17].

Books III and IV: Here, Euclid returns to a more orthodox aspect of his subject, the geometry of the circle. Topics dealt with include the properties of chords and tangents, intersecting and touching circles, the angle in a semi-circle. The problem of inscribing and circumscribing regular figures is dealt with in Book IV. He gives details on how to treat in these ways regular triangles, squares, pentagons, hexagons and fifteen-sided polygons.

Books V and VI: A scholiast claimed that the subject of Book V, the theory of proportion, was derived from Eudoxus. The theory is applied to plane geometry in Book VI.

Books VII-IX: These are the Pythagorean books which deal with number theory. One hundred and two propositions are introduced, some of which are fundamental. It is here, for example, that it is proved that there are an infinite number of primes (IX:20), and that every number can be expressed as the product of primes in a unique manner (VII:32). The treatment is geometrical throughout. Thus, consider the proof of IX:2: "Given two numbers not prime to one another, to find their greatest common measure." Euclid begins by taking two lines, *AB, CD,* and proceeds to find their greatest common measure. Other proofs are constructed in a similar fashion around lines of arbitrary length.

Book X: With 115 theorems this is more than twice as long as any other book; it is also the most difficult and ambitious. In it, the incommensurability of the diagonal of the unit square is given (X: 117), which is considered by Heiberg and Heath (1956: vol. III) to be an interpolation. Much of the material is supposedly derived from the earlier work of Theaetetus and Eudoxus. The subject matter is the irrationals approached through a study of incommensurable line segments. The treatment is extremely elaborate and involves a forgotten classification of the irrationals into the medial, binomial and apotome. The basic technique used in Book X is based on an analysis of the properties of the line by constructing a square on it and analyzing the geometrical properties of the square.

Books XI-XIII: The seventy-five theorems here concern solid geometry. Topics include the five Platonic regular solids, and techniques for working out areas and volumes of spheres and cylinders.

* * *

The Elements had an enormous effect on virtually all aspects of mathematics, but it registered its sharpest impact on the western intellectual tradition as a model for thought. It offered a paradigm, clearly successful in its own field, which over the centuries many have sought to employ in other areas. Much of *The Elements* has, inevitably, been discarded as new techniques became available and new approaches pursued. Examine any modern textbook of geometry, such as the volumes of H. Coxeter (1961) or the more recent Dan Pedoe (1970), and the casual reader will be forgiven for failing to recognize in them any connection with the work of Euclid. The symbolism, the language, the problems tackled – all seem different. In one aspect, however, even the casual reader can recognize a strong affinity between the various texts: they are all axiomatic works, deductively and rigorously seeking to develop the subject. All else might have changed in geometry, but this aspect remains the enduring legacy of Euclid to geometry, mathematics, science, and to whatever intellectual pursuit that attempts to develop rigorous, abstract thought.

The impact of this model on western thought has not always been beneficial. Whenever a new discipline emerges and its devotees justify its existence, the chances are great that someone, somewhere will claim Euclid as the authority for the new enterprise. For example, Descartes spoke explicitly in *Discourse on Method* (1637) of the model he proposed to adopt: "The long chains of simple and easy reasonings by means of which geometers are accustomed to reach the conclusions of their most difficult demonstrations" had shown him "that there is nothing so far removed from us as to be beyond our reach. . . , provided only we abstain as accepting the false for true, and always preserve in our thoughts the order necessary for the deduction of one truth from another." Another who sought to impose a geometrical model on his subject was Baruch Spinoza (1632-77) who in *Ethica* (1677) tried to develop a morality as a set of theorems deduced from Euclidean-style axioms and concluded with a genuine Q.E.D. Elsewhere, one may consider those late 17th-century scholars who, in attempting to establish the new discipline of economics, faced charges of bad faith. How could their advice be trusted when their own financial position could well be affected by any policy they might recommend? Again, the same answer was heard: Euclid, Roger North pointed out in 1691, had never been accused of "maintaining any of his theorems because they suited his self interest." Consequently he suggested as a solution that "economics" be established as "a taut deductive system that infers its conclusions from a set of single principles" (Letwin: 97). In this way the burden was shifted to the critic who was now called upon to challenge the deduction of the conclusion rather than assume the egoism of the financial adviser.

In fact, *The Elements* is far from rigorous by modern standards. Rather than explicitly formulating all his axioms, Euclid frequently relied upon implicit assumptions whenever he needed them. Some of these were so simple that their incorporation into the system could be effected without difficulty. Postulate 1, for example, states that it is possible to draw a straight line from any point to any other point; no mention is made of the fact that only *one* such line can be drawn. Nonetheless, proofs in *The Elements* on more than one occasion implicitly assume the uniqueness of such lines. Later editors simply modified Postulate 1 to recognize this omission. Hall and Stevens (1903), for example, express Postulate 1 as "there can be only one straight line joining two given points." Over the centuries mathematicians were able to do a good deal of such tidying up. As they became more familiar with the nature of an axiom system and what truly constituted a mathematical proof, it became increasingly clear that, even ignoring Postulate 5 and the problem of non-Euclidean geometries, the system of *The Elements* itself was too deeply flawed to be made rigorous by mere tinkering. Consequently, towards the end of the 19th century, mathematicians began to search for clearer and more rigorous foundations for the geometry of Euclid.

The most authoritative treatment was provided by David Hilbert (1862-1943) of Göttingen in *Grundlagen der Geometrie* (1899). The key insight of the modern axiomatic approach has been to reduce to a minimum not so much the axioms – redundant axioms are harmless – but the primitive notions. The aim is always to use as few primitive notions – terms used to define all other terms but which are themselves indefinable – as possible. If difficulties, paradoxes or inconsistencies do arise, their source will be clear. This general move was noticeable throughout the mathematics of the period. Thus in the field of logic Bertrand Russell (1872-1970) and A. N. Whitehead (1861-1947) in *Principia mathematica* took just the notions of negation and disjunction as primitive, defining all else in terms of them; in number theory, G. Peano (1858-1932) in 1889 introduced only three specifically numerical terms: "is a number," "0," and "is a successor of;" set theory was built by G. Cantor (1845-1918) in terms of the primitive notion of "is a member of."

In the same spirit Hilbert began to examine *The Elements* with its profusion of terms, definitions, postulates, and common notions. In their place, he proposed three primitive terms: "point," "line," and "plane;" and six primitive relations that could hold between them: "being on," "being in," "being between," "being congruent," "being parallel," and "being continuous." Twenty-one assumptions or axioms were also laid down which were subdivided into right axioms of incidence, four of order, five of confluence, three of continuity, and the parallel postulate.

One significant feature of Hilbert's system was its introduction of whole classes of axioms the need for which was totally unsuspected by Euclid. Hilbert proposed four axioms of "betweenness," a concept not found in *The Elements*. Axioms state symmetric properties so that one may naturally conclude that if *b* lies between *a* and *c*, it must also lie between *c* and *a*. Such statements may seem trivial and obvious, but they are still necessary if the axiomatic program is ever to be taken seriously and completed.

The problems of axiomatic rigor were ones that only emerged late in the history of mathematics; in contrast, the difficulties over Postulate 5 (P5) were appreciated right from the beginning. Axioms must be self-evident and command unquestioned allegiance if they are to retain support when conclusions about the comparative possibilities of polygons of 14 and 65,537 sides are being derived. Yet, a glance at the postulates is sufficient for doubts to arise over one of them. P5 is manifestly longer and more complex than the others; it is also less intelligible. Admittedly, no great mental skills are required to appreciate its plausibility, but it lacks for most the instant intelligibility and necessity possessed by the other postulates. In considering P5, a moment's thought and even a diagram is called for when first met. In diagram 3:3, the line *XY* falls on the two lines *AB* and *CD* in such a way that the interior angles marked *I* are less than two right angles. A brief glance makes it clear that the lines *AB* and *CD* will meet on the side of the line *XY* on which the angles marked *I* are less than 180°. But is this undoubted truth also a self-evident one? To many the answer was that it could not be as self-evident as the other postulates.

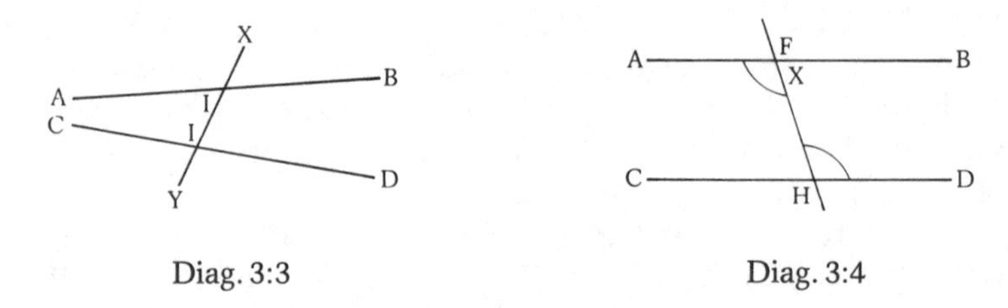

Diag. 3:3 Diag. 3:4

It is not, however, a redundant luxury; without it *The Elements* would not be able to advance further than I:29 when it is first used. Here Euclid was seeking to show that when a straight line falls on two parallel lines, *AB* and *CD* at *H* and *F*, the two alternate angles, *AFH* and DHF, will be equal.

The proof begins by assuming that the angles are not equal, that one, *AFH* for example, is larger than the other. Add to both the angle *BFH* and consequently *AFH + BFH* should be greater than *DHF + BFH*. But, as *AFH + BFH* equals two right angles, *BFH + DHF* must be smaller than this sum.

At this point it is tempting to continue the argument: As *BFH + DHF* equal less than two right angles, the lines *AB* and *CD* will meet; it follows that *AB* and *CD* are not parallel; this, however, is incompatible with the initial assumption. It thus follows that we cannot assume, without falling into contradiction, that either of the alternate angles is greater than the other. They must, in other words, be equal: Q.E.D.

The above proof, at the first step, clearly needed to assume P5. It says much for the acumen of the early Greek geometers that they could spot the need; it also says a good deal of their grasp of the axiomatic method that, however reluctantly, they accepted it as a postulate. That it was accepted reluctantly is known from Proclus who, commenting in his *Commentary* on the need to prove P5, noted that it was "alien to the special character of the postulates." He further indicated that this was not an insight of his, but had rather been the preoccupation of mathematicians for centuries. He mentions an attempt by the astronomer Ptolemy to prove the parallel postulate in the second century A.D.

In general, when presented with an offending axiom of this kind, mathematicians can adopt one of three strategies. They can, firstly, accept the inadequacy of the axiom and reject it. Such purist attitudes are not unknown in mathematics, although the price paid is too high for most. In this case, it would mean the rejection as unproved of I:29 and all further theorems dependent on it. While a geometry could be developed without P5, it would nonetheless be a very restricted geometry. It would not be possible to prove in it such obvious truths as I:30, which states that straight lines parallel to the same straight line are also parallel to one another; or I:32, which holds that interior angles of a triangle equal to 180°; or I: 47, the Pythagoras theorem.

A less restrictive response would retain the deductive power of P5 while removing its offensive features. This could be achieved in one of two ways. Although offensive as an axiom, there is nothing wrong with P5 as a theorem. Many theorems in *The Elements* are a good deal more complex and less intuitively obvious; nor do they raise doubts. If P5 could be derived as a theorem from the remaining axioms of the system then the problem would be solved. Attempts made by Ptolemy and Proclus have survived from antiquity, and there were no doubt many others. The same effect could be achieved not by proving P5 but by substituting for it an equivalent axiom which was as simple and self-evident as the other postulates.

This latter approach proved most attractive, and yet it was not more successful than any other approach. No serious attention was paid to the possibility of eliminating P5 altogether, while Ptolemy's and Proclus's attempts to prove it were quickly seen to be invalid. So, for centuries mathematicians sought an acceptable substitute for the parallel postulate. It is a remarkable example of how a scientific problem persists over the ages. It preoccupied Proclus, al-Tusi in the 13th century, and Gauss in the 19th.[18]

While mathematicians had little difficulty in finding alternatives to the parallel postulate – Heath lists nine proposals without exhausting the field – none proved to be any more simple or self-evident than Euclid's original postulate. Some were made by mathematicians of great power: Legendre, Laplace and the great Gauss himself all made determined efforts in the 19th century. Here is a short-list of alternatives:

(1) If a straight line intersects one of two parallels, it will intersect the other (Proclus).

(2) There exists a triangle in which the sum of the three angles equals two right angles (Legendre). (This is part of Euclid I:32, but its proof depends on P5.)

(3) Given any figure, there exists a figure similar to it of any size (Wallis).

To pursue and analyze all the proposals would require a treatise of which there are already several. Gauss's response cannot be ignored for not only did it show the inadequacy of the above proposals, but it also indicated where the solution to the problem would ultimately lie. In a letter to Farkas Bolyai (1775-1856) in 1799 Gauss wrote, "If we could show that a rectilinear triangle is possible, whose area would be greater than any given area, then I would be ready to prove the whole of Euclidean geometry absolutely rigorously. Most people would certainly let this stand as an axiom; but I, no! It would indeed be possible that the area might always remain below a certain limit, however far apart the third angular points of the triangle were taken."

The reluctance of Gauss to accept his proposed axiom requires an explanation. If we restrict ourselves simply to triangles drawn on paper, then the axiom should present no difficulty. It is also possible, by stretching the imagination a little, to imagine triangles drawn on the surface of the earth of continental magnitude. Again it is not too difficult to see how a triangle with an area greater than that of Texas or even of the whole United States could be constructed. But let us now remove all restrictions and imagine triangles marked off by laser beams of perfect straightness and billions of light years long. Are our intuitions so clear in such a case? Could it not be the case that "cosmic" triangles do not behave as terrestial triangles? There could in fact be a limit to the area a triangle could cover; the limit is so vast that it would only be approached by triangles constructed on the galactic scale. With them, however, increasing the length of a triangle's side would have no effect on its area; it could well be the case that with such constructs, as Gauss suggested, "the area might always remain below a certain limit." If the axioms of geometry do not appear self-evident and necessary on both the cosmic and terrestial level then, for Gauss, they could not be accepted as axioms.

Unless, as Gauss speculated in 1817, we are prepared to accept that the "necessity of our geometry cannot be proved," that rather than seeing geometry as ranking on a par with arithmetic we rank it rather with mechanics.[19] Gauss was on the verge of reaching a momentous conclusion. For centuries, mathematicians were presented with the two premises (a) all geometrical truths are either self-evident axioms or derived theorems; (b) Euclid's P5 is not a self-evident axiom; and had always drawn the conclusion (c) P5 is a derived theorem. As P5 clearly could not be derived from the remainder of the postulates, the problem had to be redescribed as what must be added to the original axioms to permit P5 to be deduced as a derivable theorem. Hence: the search over the centuries for an acceptable alternative to P5. The search for the elusive proof of P5 could perhaps have been avoided if it had been realized that the premises (a) and (b) also support the conclusion (d) P5 is not a geometrical truth.

There are hints that such an idea had occurred to earlier scholars, but none had done more than mention the possibility until the work of the Italian Jesuit Gerolamo Saccheri (1667-1733) appeared in 1733. Its title, *Euclides ab omni naevo vindicatus* or *Euclid Freed of Every Stain,* accurately indicates his motivation. Saccheri began by assuming the first twenty-eight propositions of Book I, those independent of P5. He also took the figure below

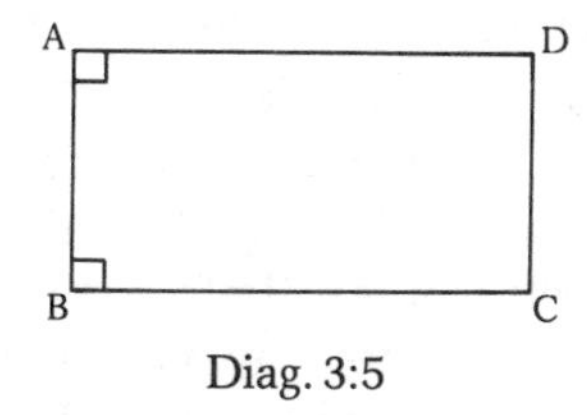

Diag. 3:5

in which it was assumed that *A* and *B* are right angles and *AD* and *BC* are parallel. Using the first twenty-eight propositions it is simple to show that *D* and *C* equal each other. Three possibilities can be distinguished: (1) *D* and *C* are both right angles. This is equivalent to assuming P5; (2) *D* and *C* are both acute angles conveniently known as the acute angle hypothesis; (3) *D* and *C* are both obtuse angles conveniently known as the obtuse angle hypothesis. The three possibilities are exclusive and exhaustive. One of them must be true but only one of them can be. Saccheri's ploy was to assume in turn that assumptions (2) and (3) were true. If, as he confidently expected, both assumptions led to a contradiction, then the remaining assumption must be true. This would in fact have amounted to an indirect proof of the parallel postulate. Known also as the method of *reductio ad absurdam*, it was a form of argument used and recognized by Euclid himself. His proof of the infinity of primes, for example, had been argued in this manner.

To his own satisfaction, Saccheri's ploy was successful. The contradictions he derived, however, would not be so recognized by a modern mathematician. He was the first, nonetheless, to consider the possibility of denying P5. Despite the originality of his approach, his work remained virtually unknown until it was rediscovered in 1889 by Eugenio Beltrami (1835-1900). Saccheri was far from being the only 18th-century mathematician to work extensively on the fifth postulate. G. S. Klugel, for example, published in 1763 a treatise in which he examined thirty "demonstrations" of P5. Three years later a further substantial treatment came from H. J. Lambert (1728-77) in *Theorie der Parallelinien* in which he developed a position comparable to that adopted by Saccheri.[20] A more powerful mathematician, A. M. Legendre (1752-1833), also tackled the problem in successive editions of *Elements de geométrie* (1794; twelfth edition 1823). Legendre also collected his lifetime's writing on the subject in a substantial monograph published shortly before his death in 1833. There is little doubt that the status of the parallel postulate was something mathematicians continued to feel strongly about; it remained a major issue.

What Saccheri had actually seen, without realizing it, was a first glimpse of what would soon come to be called non-Euclidean geometry. It was first publicly revealed by Janos Bolyai (1802-60) and Nikolai Lobachevksy (1793-1856) (and later to be popularized in a Tom Lehrer song). Janos Bolyai was a Hungarian and the son of Farkos Bolyai already mentioned. His father had warned him against paying too much attention to P5: "You should detest it just as much as lewd intercourse, it can deprive you of all your leisure, your health, your rest, and the whole happiness of your life" (Pedoe: 162). He knew, for he, too, had once pursued the path which had "extinguished all light and joy." "I entreat you," he concluded, "leave the science of parallels alone." Bolyai inevitably took no notice of his father.

He worked with a version of P5 known as Playfair's axiom. This simply states that through a given point one and only one line parallel to a given straight line can be drawn; that is, in the diagram, through P it is possible to draw only one line parallel to AB. In the spirit of Saccheri and others before him Bolyai denied Playfair's axiom by asserting that through P more than one line parallel to AB can be drawn.

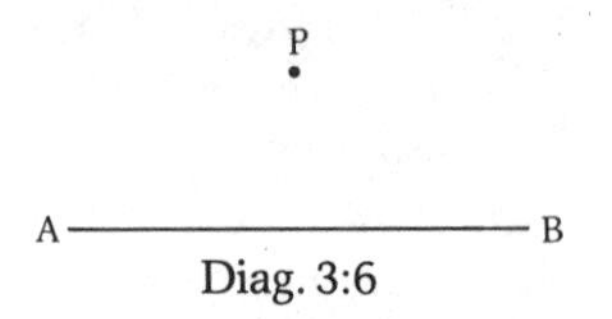

Diag. 3:6

The consequences which followed from P5 with Bolyai's axiom were perhaps unusual, but they were not contradictory. Thus, Bolyai showed that there could be triangles with angles less than 180° but, in the absence of P5, the Euclidean proposition that the interior angles of a triangle equal 180° (I:32) does not follow. Instead of an *inconsistent* geometry being produced by the denial of P5, it began to look as if an *alternative* system was beginning to emerge. Bolyai saw the significance of his work and in late 1823, while still only twenty-one he excitedly wrote to his father, "I have made such wonderful discoveries that I am myself lost in astonishment." When worked out in more detail they were published as an appendix to a work of his father in 1831.

He sent a copy to Gauss. To Bolyai's horror Gauss replied that the results of his son "coincide almost exactly with my own meditations which have occupied my mind for 30-35 years."[21] He even expressed relief that someone else had made the same discovery; it would relieve him from the burden of publishing it himself and facing inevitable controversy. Bolyai had also been anticipated, this time in print, by Lobachevski who had published similar results in 1829 in Russia. It was only much later, in 1840, when republished in German, that his work came to the attention of leading mathematicians like Gauss.

As the century developed further alternatives emerged. G. B. Riemann (1826-66), one of the most creative and influential of all 19th-century mathematicians, proposed to deny P5 in its Playfair form not by allowing a multitude of parallels through P but by denying that there were in fact any.[22] Again, the new system, later to be known as elliptic geometry, permitted theorems inconsistent with those of Euclid.

Before these new systems could be taken completely seriously, one further point needed to be cleared up. Were they consistent? The axioms appeared to be compatible and no contradiction had actually been derived but, in the absence of any proof or guarantee of their consistency, this was hardly sufficient. Failure to detect contradictions was no proof of their absence; mathematicians had perhaps not looked hard enough, been unlucky, or the contradictions could be present but obscure and subtle ones only likely to appear after the concerted effort of decades rather than months. The way out of this impasse was shown by Beltrami in 1868. Basically, he managed to so link the hyperbolic geometry of Gauss and Bolyai, proposition by proposition, to those of the plane geometry of Euclid—known technically as a mapping—that they all hung together.

Beltrami had, in fact, produced a relative consistency proof showing that if Euclidean geometry was consistent then so were the others. This, however, offered no guarantees for Euclidean geometry.

The final step was taken with a further reduction of Euclidean geometry to plane analytic geometry in which propositions of *The Elements* become propositions about real numbers. Thus, in another relative consistency proof, it had been shown that if the arithmetic of real numbers was consistent, so, too, was the geometry of Euclid and, consequently, the non-Euclidean geometries.[23] But this is as far as it is possible to go along this road, for not only is there no consistency proof of the real number system but also there are good reasons for believing that no such proof could ever be found.

For most, however, worries about the validity of the real numbers are as sensible as anxieties caused by the ultimate heat death of the universe. That there can be non-Euclidean geometries as valid as those of Euclid was something which may not have alarmed many people, it certainly puzzled them.[24] If anything had ever seemed certain before it was propositions of the kind that "parallel" lines do not meet and the angles of a triangle equal 180°. Yet it now began to seem that mathematicians could pick and choose their geometry and could elect almost whimsically what properties a triangle could have. Could the same thing happen with arithmetic? Could mathematicians now tell us that they had discovered a system in which 2 + 2 did not equal 4 and 7 was not a prime? Perhaps even historians would be able to join in the game and talk of alternative histories in which Caesar failed to cross the Rubicon and Truman was defeated by Dewey.

Scientists, mathematicians, philosophers were thus forced to consider in a very specific form the nature of geometry, thought, experience and the relations between them. Some sought refuge in crude conventionalism. One such was the great mathematician Henri Poincaré (1854-1912) who declared the query, "Is Euclidean geometry true?", to be meaningless. "We might as well ask if the metric system is true and if the old weights and measures are false. . . . One geometry cannot be more true than another: it can only be more convenient" (Poincaré: 50). Thus, the engineer laying roads and cutting canals will presumably adopt the geometry of *The Elements* as the most convenient; those like Einstein dealing with bodies moving over cosmic distances at the speed of light might find the geometry of Riemann more suited to their needs.

Something of this attitude can be noted in other fields. It is about this time that the great undigested outpourings of the cultural anthropologists began to appear in multi-volume works from the presses of the world. Ethical and cultural relativity was much discussed, very heavily documented in such works as *The Origin and Development of the Moral Ideas* (1906-08) by the Finnish social philosopher Edward Westermarck (1862-1939), and began to appeal to intellectuals of all kinds. There was nothing new in the idea of cultural relativity and nothing Westermarck had to say could not be found in such earlier works as Montesquieu's *Per-*

sian Letters (1721). The ideas began, however, to look more attractive. For now it could be argued that the claims of a universal morality, religion or social custom had no more force than those for a universal geometry. If mathematical physicists could select the geometry which most suited them, then how much more reasonable would it be for the Eskimo, the desert nomad and the forest dweller to make comparable choices over their moral rules? If there were no absolutes in geometry, they could hardly survive in morality. The link between the hyperbolic geometry of Bolyai and the permissive society may well be a tenuous one, but it is still a real one.

Publishing History

Printed Editions

To describe and list all printed editions of *The Elements* would involve the labor of a lifetime and require a substantial volume. Few books have remained in print for five hundred years and even fewer can ever have been made available in so many countries and in so many editions. Once such obvious candidates as the Bible and other sacred works are discounted, together with the works of such national poets as Homer, Dante, and Shakespeare, there are few rivals. Although Plato's *Republic*, Herodotus's *Histories*, and Aristotle's *Ethics* can indeed rival Euclid's *Elements* in terms of durability, they nevertheless fail markedly when compared on the basis of number of editions produced.[25] Only an outline history of the publishing history of *The Elements* since 1482 can be sketched here.

Venice, 1482

The first printed edition, in Latin, was published by Erhard Ratdolt and was based on the Campanus text. It was reprinted in Ulm (1486) and Vicenza (1491). Early printers faced a difficulty in representing geometrical diagrams. Ratdolt had clearly solved the problem as his text contains several hundred geometrical figures clearly printed in his two-and-a-half-inch margins. He claimed, without revealing any details, to have found a method by which he could print diagrams as easy as he could print letters.[26] Whether they were produced exclusively from woodcuts or metal rules bent into the appropriate shapes is still a matter of dispute between the experts. Whatever his method, the result was a most handsome and rare work.

Venice, 1505

The first printed Latin translation from Greek was made by Bartolomeo Zamberti who objected to the corruptions introduced by Campanus. The work took seven years to produce, spawning several related editions. Luca Paciuolo, for one, responded by issuing a revised Campanus (Venice, 1509) whom he praised as Euclid's most "faithful interpreter." A few years later, another edition appeared (Paris, 1516) which allowed scholars to judge for themselves the relative merits of the two translations: both were issued together in a single volume. Reissued in Basel in 1537 and 1546, this edition is noteworthy for its curious assumption that only the enunciations came from Euclid, the Latin proofs supposedly coming from Campanus and the Greek from Theon!

Basel, 1533
The *editio princeps* was printed by Simon Grynaeus, who used two manuscripts which Heiberg later identified as among the worst available. Although deficient, Grynaeus's text formed the basis for all other Greek editions until the 19th century.

Pisa, 1572
This Latin translation by Commandinus of Urbino served as the basis for most other Latin versions before the 19th century.

Oxford, 1703
An edition of the complete works of Euclid in the original Greek with a Latin translation by David Gregory (1661–1708)[27] was based mainly on Commandinus (1572) and Grynaeus (1533) although Gregory claimed to have examined other manuscripts. Gregory's work served as the only complete edition until the late 19th century. The frontispiece contains an etching of the shipwrecked Artistippus on the Rhodian shore examining geometrical diagrams in the sand.

London, 1756
The first edition of Robert Simson appeared in both English and Latin,[28] and it proved the most popular and pervasive of the English texts, whether in the English or Latin version. Between 1756 and 1846 the English Simson went through thirty editions. Revised by Todhunter, it began a new life in the heyday of the British empire (see further on).

Paris, 1814–1818
An edition in Greek, Latin, and French published by F. Peyrard was based on the newly discovered Vatican manuscript looted by Napoleon in 1808. Peyrard actually used the new manuscript to correct the Basel edition of 1533.

Leipzig, 1883–1916
The *opera omnia* in eight volumes by J. Heiberg and H. Menge, this is the definitive modern text and is based on the pre-Theonine Vatican manuscript *P.*

Notable Modern Translations
Many translations are available in many languages; of particular interest are the first editions in specific modern languages.

Italian: The first was published in 1542 by Nicolo Tartaglia (1506–57), a leading mathematician of his day and the first to solve cubic equations.

Euclid's translators were not hacks. It was based on Campanus and reissued in 1565 and 1585.

German: The arithmetical books (VII-IX) were translated by Johann Scheybl in 1558 while the first six books appeared in 1562 in a translation made by Wilhelm Holtzman (1532–76), also known as Guilielmus Xylander, professor of Greek at Heidelberg. The first complete translation was made by J. F. Lorenz and appeared in 1781.

French: Books I-IX were translated by P. Forcadel in 1564 while the first complete edition was made by D. Henrion (1590–1640), compiler in 1626 of the first French log tables.

English: The earliest English Euclid was made "into the Englishe toung by H. Billingsley, Citizen of London. . . With a very fruitful Praeface made by M. I. Dee, London, 1570." This is certainly one of the handsomest of Euclids. In 928 folio pages it is also one of the largest and contains, in addition to the text, notes by commentators from the Greek period onward. Billingsley was a wealthy merchant who in 1596 served as Sheriff and Lord Mayor of London. Dee (1527–1608), a mathematician, merchant and book collector, was best known as a magus who actually tried to summon demons and who served as the model for Shakespeare's Prospero. His preface is a most important document giving expression to Dee's unusual insight into the importance the understanding of geometry would have for the trade and industry of Britain. ". . . How many a Common Artificer," he wrote, "is there in the realms of England and Ireland, that dealeth with Numbers, Rule and Cumpasse: Who, with their owne Skill and experience, already had, will be hable (by these helpes and informations), to find out and devise, newe works, straunge Engines, and Instruments?"

Spanish: The first six books were translated by Rodrigo Camorano in 1576.

Arabic: The al-Tusi translation was printed by the Tipografia Medicea, Rome, in 1594. It should be borne in mind that the printing press did not arrive in the Muslim world until much later. The first press to arrive, for example, in Egypt was brought by Napoleon in 1798. There are consequently no early printed Euclids to be found in the Muslim world.

Dutch: The first six books were issued in a translation by J. P. Dou in 1606; the first complete version was made by Frans van Schooten and published in 1617.

Chinese: Needham (1959:105) mentions the possible existence of a 13th-century translation, using an Arabic text, but concluded, "the evidence does not absolutely indicate that the books in question were translations of Euclid into Chinese, though it seems probable." Others have thought the title alone may have been translated while the text remained in Arabic. The first undoubted translation was made by the remarkable Matteo Ricci (1552–1610), Jesuit scientist and missionary, and Hsu Kuang Chhi in 1607. Only the first six books were translated at that time while the remaining books were not translated until 1857 by A. Wylie and Li Shan-Lan. The first "complete corrected" edition was made by Tseng Kuo-Fan in

1865. Of the relative importance of Euclid in China, Needham has commented, "In China there never developed a theoretical geometry independent of quantitative magnitude and relying for its proofs purely on axioms and postulates accepted as the basis of discussion" (1959:91).

Russian: A translation was made from the Latin in 1739 by Ivan Astaroff; the first translation from the Greek was made by P. Suvoroff and Y. Nitkin and published in 1789.

Swedish: The first six books were translated by M. Stromer in 1744.

Danish: The first six books wre translated by E. G. Ziegenbalg in 1745; the first complete Euclid in Danish was issued during 1897–1912.

More briefly, other languages in which Euclid has appeared with the date of its first appearance are Portuguese (1768), modern Greek (1803, in a translation by Benjamin of Lesbos), Polish (1817), Finnish (1847), Hungarian (1865), Sanskrit (1901) and Gaelic (1908). Not all are complete editions.

Deserving of special mention and of unequalled value for the English speaker is the three-volume translation by T. L. Heath. First published in 1908 with comprehensive and informative notes, it has been available since 1956 in a handsomely produced Dover Books paperback.

* * *

A list of printed editions and translations fails, however, to do justice to the sheer quantity and variety of editions that have been produced. The ninety-seven titles issued between 1482 and 1820 listed by Heath do not include reprints and give little indication of the explosive growth in publication during the remainder of the 19th century to meet the demands of the spread of popular education in Britain and its developing empire. A number of traditions can be distinguished. The most important, edited by Todhunter, is traceable to the Simson edition of 1756. Revised and modernized, it first appeared in 1862 and was reprinted in various forms at least fifteen times in the following forty years. The other great Euclidean entrepreneurs were H. S. Hall and F. S. Stevens whose numerous editions began to appear in 1888. For the most part publishers concentrated on Books I-VI plus XI and XII with a preference for the first six. The basic text was added to or split up in a variety of ways; each book could be issued separately, or exercises and answers might be included; or, if some algebra were added, a mathematics text emerged. Presented in a slightly different way to meet changes in fashion Euclid's text could be sold as an introduction to mensuration or practical geometry.

It was no simple matter for a Victorian publisher to issue an edition of Euclid. Hall and Stevens, for example, was published by Macmillan and available in its fullest form as Books I-VI plus XI. In addition each book could be bought separately as could the first two, three, four or five books. They were cheap, ranging from sixpence in 1918 for a single book to five shillings for all six, and clearly very much in demand. An edition of Books I and II published in 1903 went through twenty-four editions by 1918.

There was in fact little new in the approach of Hall and Stevens to the problem of marketing Euclid. For centuries publishers had been seeking to present what was basically a very difficult work as a highly desirable book. One ploy often adopted by publishers was to claim that their edition was specially designed for some group or other. Thus, in 1694 von Pirckenstein packaged a version for "generals and engineers." Even earlier, Xylander's German translation of 1562 was claimed to meet the special needs of artists, goldsmiths, builders and other artisans. Three hundred years later new markets had emerged; to meet them British publishers began to produce numerous editions "for the use of Indian schoolboys."

Other common marketing strategies stressed the particular accuracy, reliability or comprehensiveness of an edition. Earlier editions might be billed as the first to be translated from the Greek or, when this no longer impressed, strategies were modified and an edition might be advertised as the first prepared from some especially authoritative Greek manuscript, or, the first to be based on all known manuscripts. Publishers also offered commentaries in addition to the text. Editors of Euclid never seemed short of a compelling reason why their edition was the only truly faithful text available or, as a variant provided the reader with an edition in which the errors of a Theon, a Campanus, a Commandinus or whoever were "at last" excised.

Mostly, those who bought Euclid were neither generals, goldsmiths, Indian schoolchildren or even concerned with the interpolations of Theon. For them *The Elements* was a difficult and boring work which was read only because it was a compulsory textbook. To overcome this resistance editions of Euclid were advertised as providing, if not instant enlightenment, at least a shorter and smoother path to understanding than rival editions. To this end proofs were lengthened, shortened, rearranged or dropped altogether. Some editors introduced symbols, others shunned them completely. So, too, with axioms. While one editor might increase their number, another, such as R. Howard in 1835, advertised his as a "geometry without axioms."

Arising from this tradition is Oliver Byrne's 1847 edition of the first six books of Euclid, one of the strangest geometry texts ever produced. Byrne was "surveyor of Her Majesty's settlements in the Falkland Islands" and also the author of a large number of elementary works, mainly pocket books of various kinds designed for engineers, mechanics and surveyors in which the essential data and formulae of their trades were listed.

In place of the usual symbols used in *The Elements* Byrne employed "coloured diagrams. . . for greater ease of learning." Thus, the line *AB* or the triangle *ABC* was shown as appropriately shaped patches of color. A triangle, for example, was shown not as the lines *AB, BC* and *AC* but as red, blue and yellow lines and with equally distinctively colored angles. Throughout the proof, therefore, reference was made to a particular line or angle by simply presenting the appropriate colored line or blob. A proof of any length soon became brilliant patches of color separated bizarrely by "buts," "therefores," ">"s and "="s. As the theorems grew more complex and circles and parallelograms appeared the effect became ever more dazzling. No words can convey the shock of opening a volume of Euclid and noting on page after page little more than complex arrays of color.

Byrne sternly pointed out that he had not introduced color "for the purpose of entertainment, or to amuse." With a quote from Horace to the effect that stronger impressions are conveyed by the eye rather than the ear, Byrne claimed, "Such is the expedition of this enticing mode of communicating knowledge, that *The Elements*. . . can be acquired in less than one third the time usually employed, and the retention of the memory is more permanent" (Byrne:ix). The absence of a second edition or any imitators indicated clearly the unreality of his approach. There does, however, appear to have been a predecessor for, according to Aubrey, Seth Ward (1617–1689) "did draw his Geometricall Schemes with black, red, yellow, green and blew Inke to avoid the perplexity of A, B, C."

Further Reading

The general mathematical background can be found in two recent excellent histories: *An Introduction to the History of Mathematics* (1969, third edition) by Howard Eves and C.B.Boyer's *A History of Mathematics* (1968). Both works are more concerned with elucidating the mathematics of the past than relating it to its historical background. For Euclid himself students are doubly fortunate to have the two works of Heath (1921; 1956) which, between them, present a detailed picture of the Greek mathematics from which *The Elements* emerged, trace the history of the work, provide a commentary and contain also a wealth of bibliographical, textual and miscellaneous information. Also for Euclid, the entries in the *Dictionary of Scientific Biography* by Ivor Bulmer-Thomas and John Murdoch are particularly generous and informative. For the development of non-Euclidean geometries Greenberg (1980) provides a lucid account of the actual geometry involved while some of the more general assumptions behind them can be found in Jeremy Gray, *Ideas of Space* (1980). Much relevant data can be found in the chapters on J. Bolyai, N. Lobachevski, B. Riemann and C. F. Gauss in E. T. Bell's *Men of Mathematics* (1937) and in the relevant entries of the *Dictionary of Scientific Biography*. Some of the wider influences of Euclid through the ages can be seen in Dan Pedoe's valuable work, *Geometry and the Liberal Arts* (1976).

Notes

1. The *Commentary on Euclid, Book I* by Proclus (410–85) is one of the main sources for the history of Greek geometry. Like most Greeks, he assumed that many elements of his culture ultimately derived from Egypt and consequently began his brief history of geometry with the false claim, "Thales travelled to Egypt and brought geometry to Hellas." More attention should be paid to his account of

Pythagorean geometry: "Pythagoras. . .examined this discipline from first principles and he endeavoured to study the propositions without concrete representations by purely logical thinking" (van der Waerden 1963:90). If it had taken Pythagoras to do this, what then had Thales brought with him from Egypt? Presumably it was techniques of surveying or land measurement *(ge metros)*, a discipline very different in fact from that of deductive geometry.
2. The actual theorem of Theon was: Sectors in equal circles are to one another as the angles on which they stand (VI:33).
3. Actually, Peyrard did not issue an edition based on *P*; he merely corrected the *editio princeps* of 1538 on the basis of *P*. It was left to Heiberg to publish the first pre-Theonine text.
4. Some idea of the complexity of the task facing Heiberg can be gained by noting that he had at his disposal eight manuscripts of the text and as many as 1,500 scholia.
5. Isidore (c.570–636), bishop of Seville, author of *Etymologies* or *Origins*; Cassiodorus (fl. 500), author *De artibus*.
6. The now unfamiliar term *enunciation* is a translation of the Greek *protasis* and refers to what is given and what is to be proved, what would now be loosely referred to as the theorems. It is a reminder that *The Elements* also contains a fair amount of terminology, expressions like *porism* and *lemma* for example, which though once known by every student are no longer part of our common vocabulary.
7. Heath believed that on textual grounds "there must have been in existence before the 11th century a Latin translation" or a fragment of one. An old English verse may refer to this translation when, in talking of the "clerk Euclide," it concludes

> Mony erys afterwarde y understonde
> Yer that the craft com ynto thys londe
> Thys craft com into England, as y sow say,
> Yn tyme of good Kyng Adelstone's day.

This would put "the introduction of Euclid into England as far back as 924–940 A.D." (Heath:95). If so, it can have been of little consequence for no trace is discernible until centuries later.
8. The better known name for I:5, *pons asinorum* or bridge of asses, is in fact much more recent, dating no earlier than Murray's *English Dictionary* (1780): "If this rightly be called the bridge of asses / He's no fool that sticks but he that passes" (Heath:415).
9. Among these was the Analytical Society, formed in Cambridge in the early years of the century. Centered on such figures as Charles Babbage (1792–1871), father of the modern computer, it sought to convert British mathematicians to "pure D-ism, in opposition to the Dot-age of the University." The reference was to the progressive analytical mathematicians of the continent who had adopted the calculus notation of Leibniz (*dx* for the differentiation of x) in contrast to traditional mathematicians of Cambridge who persisted with the dotted letters of Newton ($\dot{x}$ for dx).
10. Wilson provides a brief account of the origin of the campaign in his *Autobiography, 1836–1931* (London, 1932). The Taunton or Endowed Schools Commission was set up in 1864 to investigate English public schools. It noted an inadequate knowledge of geometry among students compared with their peers on the continent. Wilson, a mathematics teacher at Rugby 1859–79 carried out his own investigation and reported (*Autobiography*, p. 37):

> I obtained a set of geometrical text-books used in France, Belgium, Germany, Italy and the U.S.; all totally unlike Euclid in arrangment. What determined the difference? It was something so obvious that I can only marvel that it was

> new to me. It was no axiom or principle. It was a mere convention: that 'no hypothetical construction may be assumed.' In plain English. . . , you must not assume that a line has a middle point, or an angle a bisector, until with a straight-edge ruler, unmarked, and a pair of compasses which close automatically when they are taken from the paper, you can find the bisecting point and the bisecting lines and prove that they are the bisecting point and line.

Shortly afterward Wilson produced his own *Elementary Geometry* (1868; 1869; 1873; 1878) and began the campaign that led to the formation of AIGT in 1870.

11. Another example of the restrictions current in the teaching of geometry can be seen in the diary of Lewis Carroll. He had published his *Euclid I, II* in 1882. On 1/11/82, he noted in his diary, "Began today preparing the 2nd edition. . . of Euclid I, III by erasing "+", "=", "> ", "< " signs, which seem likely to make the book less useful, as all algebraical signs are supposed to be forbidden in Cambridge examinations."

12. The British Association for the Advancement of Science was founded in 1831 as both a pressure group in which some of the younger and more radical scientists could express their views and as a public platform for the discussion of new theories, controversial or not, and the announcement of new results. It came to exercise a powerful influence on the content of science and its public organization. The American equivalent, the American Association for the Advancement of Science, was formed in 1848.

13. This is the most interesting of all the Carroll works on Euclid. There were altogether at least thirteen distinct works, from eight-page pamphlets to his 300-page *Euclid and His Modern Rivals*. Some have not survived. *Euclid* is, at least in form, a four-act play. The "rivals" appear before Minos and Rhadamanthys, judges of the underworld, to defend their presumption in attempting to usurp Euclid. A dozen "Euclid wreckers," including J.M. Wilson and Legendre, are duly found guilty and dispatched to Hades.

14. Thus the *editio princeps* of 1533 carried the title *Eukleidos stoixeion*. Ratdolt's first printed edition (1482) has no title page but begins with the heading "Preclarissimus liber elementorum. . . Euclides." It was thus as *Stoikeion* or *Liber elementorum* that Euclid was initially known. The tradition persisted and is found in Simson (1756) with his *Euclidis elementorum*. . . and in Heath's translation (1908). A Paris edition of 1516, carrying the title *Euclidis megarensis geometricorum elementorum libri xv* must have been one of the first to use the term geometry in the title.

15. The fact that Apollonius (fl. 210 B.C.) tried to prove the common notions suggests strongly that some at least were Euclidean. However, it has been argued against CN4 and CN5 that they are not actually used by Euclid even when, as in I:4, they were needed. Heath was inclined to conclude that they were later insertions.

16. There are reports in the literature of recreational mathematics of some obsessive or, perhaps, sceptical types who have devoted much of their lives to the construction of such a figure. The difficulty is in finding someone to check the work.

17. For addition and subtraction of quantities, lines can be extended or reduced; multiplication can be shown by constructing a rectangle of the lines to be multiplied; while the division of one line or quantity by another can be expressed as a ratio between the lines.

18. The persistence of recalcitrant problems over centuries is well-known in mathematics. One famous instance, due to Pierre Fermat (1601–65) and known as Fermat's last theorem, asserts for no n other than 2 does $x^n + y^n = z^n$. This problem still awaits a proof or disproof. Fermat claimed to have a proof, but writing shortly before his death, he complained that the margin of the book in which he was writing was too small to contain it. More commonly, a conjecture might be made in

one century (as for example, Fermat's claim that $2^{31} - 1$ is a prime number) and be confirmed in the following century (as, in this case, by the 18th century mathematician, Euler.)

19. It was in this spirit that Gauss in 1827 measured the triangle formed by three widely separated mountain peaks. He found they exceeded 180° by a mere 15", less than the experimental error.

20. Lambert came very close to the discovery of non-Euclidean geometry as the following quotation makes clear: "I am almost inclined to draw the conclusion that the third hypothesis (the acute angle) arises with an imaginary spherical surface" (Heath:213).

21. As Bolyai's father feared, the result of his son's work on P5 was disastrous. Despite success, he could not accept that his work must be shared with Gauss. He wanted the work to be recognized indisputably as his own and not as the unpublished leavings of another. He seems to have suffered a breakdown of some kind and publishing nothing else in his lifetime.

22. Riemann, a student of Gauss, expressed his views in "On the Hypotheses Which Lie at the Foundations of Geometry," a lecture delivered in 1854 to the Göttingen faculty. It was published posthumously in 1868 after his premature death from consumption at the age of forty.

23. Such a designation, it has been suggested, is the highest of all honors in science. There are certainly few of them: non-Euclidean geometry and non-Aristotelian logic are the only two clear ones; non-Newtonian physics may, perhaps, also be included. The term is first recorded in English in 1878. Earlier pioneers used a variety of terms, such as "anti-Euclidean geometry" (Gauss) and "astral geometry" (Schweikart, on the grounds that deviations between the alternative geometries would only become apparent over cosmic distances). Beltrami used the phrase "geometria non-euclidea" in the title of a work published in 1868, and it must have been one of the earliest times it appeared.

24. It is about this time that under the influence of the new geometries writers of fantasy sought to depict worlds governed by strange geometries. One of the earliest and well-known is A. Square (real name: Edward Abbott), author of *Flatland* (1884). He imagined a world populated by two-dimensional beings. Women, the lowest form, appear as lines; workers as isoceles triangles; and the middle classes as equilateral triangles. The upper classes begin as hexagons, and as they increase in status so do the number of their sides until, as priests, they are indistinguishable from circles. The same theme is explored in a differed way by C. H. Hinton in *An Episode of Flatland* (1907).

25. Times have changed. While today few homes or bookshops contain copies of *The Elements*, many have a *Republic*, a Dante or a Herodotus. In the 17th and 18th centuries, few scholar's libraries lacked a Euclid. Newton owned five copies, some undoubtedly presentation copies.

26. According to Thomas-Stanford (1926: 3), Ratdolt was not the first to incorporate geometrical diagrams into a printed text. This honor belongs to Archbishop Peckham, author of *Perspectiva communis,* published in Milan by Petrus of Corneno, updated, but probably in 1481. It contained seventy-seven "complicated and well executed diagrams."

27. In addition to his various mathematical writings, Euclid also wrote on astronomy, optics, and musical theory.

28. Robert Simpson (1711-61) was professor of mathematics at Glasgow. The thirty or so editions between 1756 and 1855 was just one strand of publication. At the same time John Playfair's *Elements* went through ten editions between 1795 and 1846, while between 1828 and 1855 Dionysius Lardner's rival text was published eleven times.

CHAPTER 4

The Almagest
(Second Century A.D.)[1]

by Ptolemy

Perhaps the most influential treatise on astronomy ever written. . . .
–Marie Boas Hall, *The Scientific Renaissance*

Copernican models require about twice as many circles as the Ptolemaic models and are far less elegant and adaptable.
–Neugebauer, *The Exact Sciences in Antiquity*

Ptolemy is not the greatest astronomer of antiquity, but he is still something more unusual: he is the most successful fraud in the history of science.
–Newton, *The Crime of Ptolemy*

I do not profess to be able thus to account for all the motions at the same time; but I shall show that each by itself is well explained by its proper hypothesis.
–Ptolemy, *The Almagest*

Life

As with Euclid and Hippocrates, and indeed with almost every major scientist of antiquity, virtually nothing is known of the life of Ptolemy. There was unfortunately no Plutarch or Diogenes Laertius to record the lives of the scientists. From *The Almagest* it is known that Ptolemy made a number of observations "in the parallel of Alexandria," the earliest of which was recorded in A.D. 127 and the last in A.D. 141. As with Euclid, the name no longer refers to the individual but to a book.

Astronomical Background

Ptolemy was heir to a rich tradition and if he was the greatest astronomer of antiquity, this was because he completed a thriving tradition rather than established a new one. It was a tradition distinct from the astronomy practiced by any other Western people. Its source remains a mystery. It is known that Greeks were worried about certain calendrical problems: How could they link a lunar month of 29.5306 days (synodic month) with a solar year of 365.2422 days (tropical year)?[2] If twelve months are allotted to the year, it will have 354.36 days; if thirteen months are allotted, then the year will contain 383.89 days. Either way, the seasons of the year and the months with their traditional religious festivals and agricultural commitments would soon get out of phase.

Some few communities have been prepared to accept such a consequence and permitted the civil calendar to travel over the centuries through the seasonal year. The best known example is the Islamic calendar. The Koran states, "To carry over a sacred month to another, is only a growth of infidelity."[3] The price of such simplicity is the inconvenience of having the months of the 354-day year rotate through the seasons approximately every thirty-three years. For this reason, the duties of fasting imposed during the month of Ramadan can be extremely rigorous in those parts of the cycle when it falls in the hotter times of the year. The simplest alternative is to adopt some system of intercalation by which a number of extra days are added at carefully worked out intervals and in precise amounts to the 354 days of the lunar year, in order to bring it into line with an integral number of months. Such a task presupposes a fairly detailed knowledge of the length of the year and month. Meton, the fifth-century Greek astronomer, had noted that 235 lunar months equalled nineteen solar years if 110 of the months were hollow ones of twenty-nine days and the remaining 125 all contained thirty days – 6,940 days in all. This actually gives a year of 365.263 days. His calculations also meant that some seven months needed to be intercalated sometime during the nineteen-year "Metonic" cycle.

This was by no means trivial work. It called for a familiarity both with an observation of the heavens and with computational techniques; nor was it in any way a Greek monopoly. The contrast between the calendrical skills of the fifth century and the advances made in astronomy in the fourth century is enormous and apparently unbridgeable. The utility of calendrical work is easy to grasp and one understands just why scholars would take it up. But the study of astronomy is more complex: it is astronomy and science at their purest; not even the obvious link between navigation and the heavens can be deployed in this particular context. That was a link forged in the Renaissance; it had no relevance for either Greek merchants or mariners. Nor, perhaps, should this be particularly surprising. Millions of views in Britain, Europe, America and most other parts of the world will regularly watch television programs about the structure of galaxies, quasars, and all variety of stars without it making one jot of difference to their working lives.

Where does the study of astronomy begin? Sometime in the century following Meton, Eudoxus (c. 400–350 B.C.), a native of Cnidus and pupil of Plato, took the major step of offering a theoretical explanation of planetary movements. Others before Eudoxus had tried to understand the movements of the heavens, on a naturalistic rather than a theological basis, but all invariably suffered from one of two main defects. Some astronomers remained completely speculative, freely adding or subtracting from the heavens whatever they pleased. Such is the Pythagorean tradition in which the center of the universe is occupied by fire. Why cannot we see the fire? Because, the Pythagoreans answered, there was another body, the counter-earth, in between the central fire and the primary earth, invisible but opaque enough to shield the central hearth from terrestial eyes. The conventions and constraints inherent in this cosmology are so little understood that it is difficult to know whether we are dealing with facile numerology or an incipient but scarcely perceptible science.[4]

An alternative approach, pursued by Assyrians and Babylonians, was to construct tables of astronomical events. Over several centuries astronomer-priests would observe, calculate and note for each month the time and position of the first lunar sighting and full moon. They prepared planetary tables giving, among their data, daily heliacal risings and settings.[5] Systematic reports date back to about 700 B.C. but by 300 B.C., according to Neugebauer, the mathematical astronomy of the Babylonians was fully developed. The construction of ephemerides and the development of techniques for the calculation and prediction of future planetary and lunar positions are impressive achievements; the fact remains, however, that for all their data and skills the Babylonians did not take what would have appeared to have been an obvious step and work out the orbits and systematic relationships of the heavenly bodies. Rather, they were interested in noting and predicting and not picturing and connecting. They resembled those generations of explorers who studied continents in detail and who assembled volumes of co-ordinates for all the main topographical features yet failed to construct a map. The Babylonians were adept gazetteers but lacked the ambition to be cartographers.

In contrast, gazetteers rather bored the Greeks and, with the barest encouragement, they sought to construct an atlas. That the earth was spherical was reported to have been a commonly accepted belief of the Pythagoreans in the sixth century B.C. It was also known by this time that the movements of the celestial bodies were, in broad outline, periodic. Thus the poet Hesiod in *Work and Days* (c. 700 B.C.) was aware that when the Pleiades could be seen rising on the eastern horizon just before sunrise, it was time to gather the harvest (11.383–87):

> When the Pleiads, Atlas' daughters, start to rise
> Begin your harvest; plough when they go down.
> For forty days and nights they hide themselves
> And, as the year rolls around, appear again.

There is nothing particularly special about associating the appearance of certain stars with certain times of the year: such insights were probably held, explicitly, by most peoples. Another feature of the elementary Greek cosmology was the notion that heavenly bodies moved in a circular fashion. Empedocles in the fifth century B.C. certainly spoke of the moon as circling the earth. He also described a universe which circles about an immobile central earth. The claim at this time was metaphorical and was based on the belief that the earth was held immobile by the heavens rotating around it just as "water remains in a goblet which is swung quickly round in a circle" (Dreyer:26). It remained for a later generation of Greeks to deploy these insights in a more literal and less metaphorical way. Comparisons with the chariot wheels, goblets of water, and upturned basins frequently encountered in early Greek cosmology are highly suggestive ideas, but real science begins when metaphors are discarded and models are introduced.

As already mentioned, the model was first assembled by Eudoxus. Ignoring the complexities and minutiae of his system, we find a simple and not unattractive model in which each heavenly body moves in a circle around the earth. As earth was the common center of the universe, the system can be represented by a set of concentric or homocentric spheres. The fixed stars from the point of view of the model are taken to rotate as a unity and consequently occupy a single sphere; the remainder from the stars inward are occupied by Saturn, Jupiter, Mars, Sun, Venus, Mercury, and the Moon, all moving around a central, immobile Earth. Four points were essential to the system in its purest form: it was concentric and had a common center; it was geocentric or had earth as that center; heavenly bodies moved in a circular orbit; and, finally, this latter motion was uniform. All the points seem perfectly natural and obvious, supported both by observation of an elementary kind and some basic assumptions about nature. No one, for example, had actually observed the planets moving around earth with uniform motion but rather had asumed that such a motion would take place. Why, it could be asked, should a planet move faster at one place and slower at another? To it, all sectors of the orbit are indistinguishable.

As it happened, a physics and metaphysics emerged in the works of Aristotle to make this assumption inevitable. (This was largely justification after the event.) Each of the four assumptions is questionable; the process of demonstrating their falsity and replacing them with more realistic hypotheses was one that would take almost two millenia to complete.[6]

Long before this, however, two particular sets of observations showed that the Eudoxian scheme was untenable in its initial form. The first concerned the movement of planets around the earth. Nightly, when visible, the planets move westward across the sky. There is also a less obvious motion, eastward, against the background of the fixed stars.[7] If, at exactly the same time each night, the position of the sun, moon, and planets is noted relative to a particular constellation in the ecliptic, it is observed that on the following and all subsequent nights they will have moved backward or eastward along the ecliptic by a variable amount. If observation is continued for any length of time, it will be found that the planets eventually move completely round the ecliptic and return to the first constellation they had been adjacent to. Each planet has its own orbital period: The moon will take a month, the sun a year. This is the astronomic meaning of the terms sidereal month and year. Observation will show, in broad terms, that Saturn takes thirty years to complete its orbit, Jupiter twelve, Mars two, and Venus about seven and a half months. So far such observations merely re-inforced the original model by showing that it contained unsuspected riches – not only do the planets orbit daily around the earth in one direction but also they move in an opposite and longer orbit around the ecliptic. Such an additional feature was compatible with and easily represented in the system of Eudoxus.

Further observation, however, revealed a more recalcitrant feature. At certain points along the ecliptic the motion of the planets appears to halt and for the period of several nights advance no further; then they appear to reverse their direction and move westward through the constellations before resuming their normal eastward path. This unpredictable, unexpected, almost perverse planetary behavior, known as retrograde motion, seemingly destroyed the harmony and simplicity of the original Eudoxian model. The contrast between the expected behavior of, for example, Jupiter, moving around the earth in a circular orbit (fig. 4:1) and the actual observed behavior in its twelve-year passage around the ecliptic (fig. 4:2) is, at first sight, a puzzling one. Known to the astronomers of antiquity as the "second anomaly," it was also recognized by them as being dependent

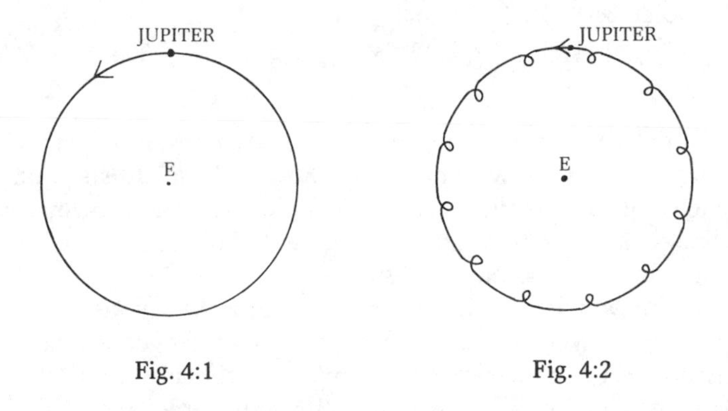

Fig. 4:1 Fig. 4:2

on the position of the sun. For Saturn, Jupiter and Mars, it occurred only when they were in opposition; in contrast, Mercury and Venus needed to be at inferior conjunction before they moved retrogressively.[8]

Equally puzzling, given the Eudoxian assumptions, was the "first anomaly." This referred to the well known fact that celestial bodies did not appear in their orbits to move with uniform speed. Some days they would move faster than at other times and consequently cover more of the ecliptic. The moon's daily motion, for example, varied between 10° and 14°. Unlike the second anomaly this problem was independent of the position of the sun.

Indeed, it was with the sun that such "anomalistic motion" was most evident for it was here that variations in daily motion made themselves felt in seasonal inequalities. Astronomically, the seasons are marked by the four natural "sunmarks" of the solar year: the summer and winter solstices, when the sun rises and sets at its most northerly and southerly point from the equator respectively; and the vernal and autumnal equinoxes, the two days of the year when the sun rises and sets exactly on the equator. If the sun moved uniformly about the earth in a circular path, then it follows that its journey between the four natural sunmarks must be performed in equal times. However, Euctemon, an astronomer who lived in the fifth century, reported seasons of different lengths. His figures given in days, may be compared with those provided by Callipus, a third-century astronomer. In fig. 4:3, the results are contrasted with modern estimates.

Fig. 4:3

	Euctemon	Callippos	Modern
Vernal equinox – summer solstice (21 March) (22 June)	93	94	94.1.
Summer solstice – autumn equinox (22 June) (23 September)	90	92	92.2.
Autumn equinox – winter solstice (23 September) (22 December)	90	89	88.9.
Winter solstice – vernal equinox (22 December) (21 March)	92	90	90.4.
Total	365	365	365.6.

With these two problems the concentric system lost much of its intitial plausibility, but they show, remarkably, how early scientists grasped the inadequacies of models that were too simple and neat. However attractive, the models had to be amended, augmented or restructured in a way to accommodate the anomalous results.

The true answer lay in the realization that the basic assumptions of Greek cosmology were wrong. The universe was *not* geocentric, nor did heavenly bodies orbit the earth with uniform circular motion. But rather than challenge and replace such assumptions, the Greeks instead chose to tamper with them and so preserve them. Thus, the notion of an epicycle entered into ancient astronomy.

To later generations epicycles and the like have often seemed like geometrical solutions to geometrical problems. Such an attitude is not without its plausibility nor without historical foundation. It is for example found explicitly expressed in Plato. Heavenly bodies in *The Republic* are dismissed as "visible" and "material"; no "exact truth," it is declared, will ever be gained from them. The recommended approach was to ". . treat astronomy like geometry, as setting us problems for solution" (Plato:298). In a related tradition Plato is held to have enjoined astronomers to seek uniform and orderly motions by which the apparent motions of the planets can be accounted for, or, in the phrase which has since echoed down the centuries, to "save the appearances." It is a long and in many ways an ambiguous tradition, still to be found with all its difficulties in the science of today. What geometrical constructions can be adopted which will both preserve the initial assumptions of uniform, circular, geocentric motion and also permit the appearance of retrograde planetary motion? Presented in this way the question is pure geometry; as will be seen, though, more is involved.

The solution of epicyclic motion was proposed by the mathematician, Apollonius of Perga, in the late third century. Sources are scarce for this period, but it can be confidently assumed that the theory would have been further developed by the second-century B.C. astronomer Hipparchus. The basic trick of the epicycle is to show that if planetary motion cannot be represented by a *single* circle then perhaps it can be represented by *two*. As before the daily motion of the planets around the earth can be represented by a simple concentric circle centered on the earth (Fig. 4:4); in addition, however, the planet as it moves around its orbit moves in a secondary circle, the epicycle, centered on its orbit around the earth, known as the deferent (fig. 4:5).

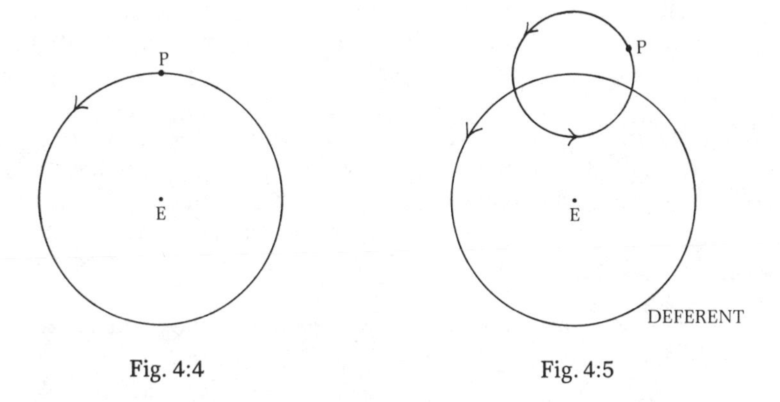

Fig. 4:4

Fig. 4:5

The effect of such motions, both perfectly uniform and circular, on an observer at the center, *E*, would be to make it look as if the planet *P*, as it revolved around *E*, occasionally stopped and regressed for some way before continuing on its normal path. Just how frequently such retrograde movements take place and for how long can be controlled by the radii and velocity assigned to the epicycle. In fact, given suitable parameters, orbits of any size, shape and perverseness can be provided. Toulmin and Goodfield (1963:154) actually go so far as to construct and illustrate a square orbit.

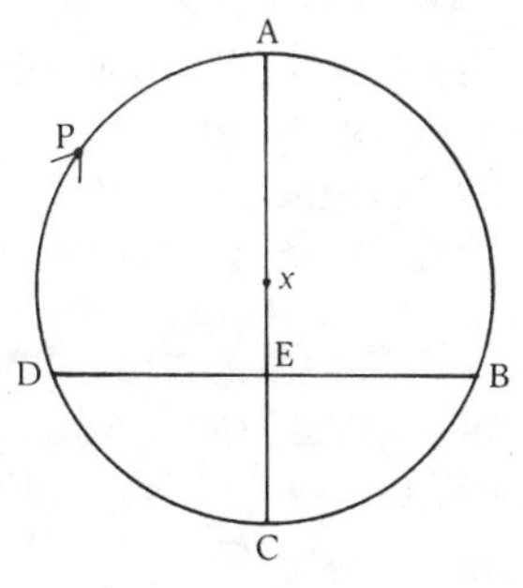

Fig. 4:6

A second construction introduced by Apollonius and Hipparchus was the eccentric. With this, the orbiting planet still moves with a uniform, circular motion only its orbital center is no longer the earth but a point *x* displaced some distance from it. One gain from such a scheme, as can be seen in fig. 4:6, is to show how variations in seasonal length can occur. To an observer on *E*, the planet *P* moving uniformly in its orbit will obviously seem to take longer to traverse the *A-B* quadrant than the *B-C* one. Again, different seasonal lengths can be obtained by adjusting the degree of eccentricity or the distance *E-x*. Here the center of the planet *P*'s orbit is not the earth *E* but an empty point in space *x*. The deferent represents the sun's annual path through the ecliptic with the *A* and *C* representing solstitial points and *B* and *D* the two equinoctial points. The arcs *AB, BC, CD,* and *DA* therefore represent the four seasons. Although the sun continues to move with uniform circular motion, it will now be clear why the seasons can appear unequal to an observer on earth.[9]

In addition to the constructions of the epicycle and eccentric, Ptolemy also inherited a good deal of other essential material from Hipparchus. First, we may mention the observations listed by Hipparchus in his star catalogue. It has not survived but from other sources it has been estimated that it contained about 850 stars, given in terms of their co-ordinates along the ecliptic, their latitude and longitude. The motives for this work are described by Pliny as arising from the appearance of a new star. To determine whether such an event was common, Hipparchus realized it was first necessary "to do something that would be rash even for a God, namely to number the stars for his successors and to check off the constellations by name."

A second advance made by Hipparchus and available to Ptolemy was the determination of certain features of the heavens for the first time and of others with an improved accuracy. The most notable advance in this field was the discovery of the precession of the equinoxes. Hipparchus had noted in his catalogue that Spica was some 6° from the autumnal equinox while earlier observers 150 years before had placed it 8°. Hipparchus concluded that the equinoctial points, where the planes of the ecliptic and equator intersect, are not fixed but move around the ecliptic. If it had taken 150 years to move 2°, its rate of precession was about 45"–50" a year, or, a degree in about 70 years.[10]

A third and important mathematical advance is also found in Hipparchus. So far the reader may have gained the impression that astronomy for the Greeks was indistinguishable from geometry; that it was, in fact, something more than this, and that is was possible to apply geometrical structures with precision and accuracy to the heavens and the actual orbits of planets was due mainly to the development of certain mathematical tools. Today, these would loosely be described as trigonometry, but they were already apparent in the lost work of Hipparchus.

The development of such concepts as sines and cosines lay centuries in the future. The Greeks worked rather with the lengths of chords as functions of the angle they subtend at the circle's center as expressed as a fraction of a diameter conventionally divided into 120 parts. To illustrate with a simple example. In this case $A = 90°$; therefore, by Pythagoras's theorem the chord *BC* can be calculated by solving the equation $AB^2 + AC^2 = BC^2$ where *AB* and *AC* as radii must equal 60 parts. This yields $BC = 7200$ which, expressed as a sexigesimal part of a diameter divided into 120

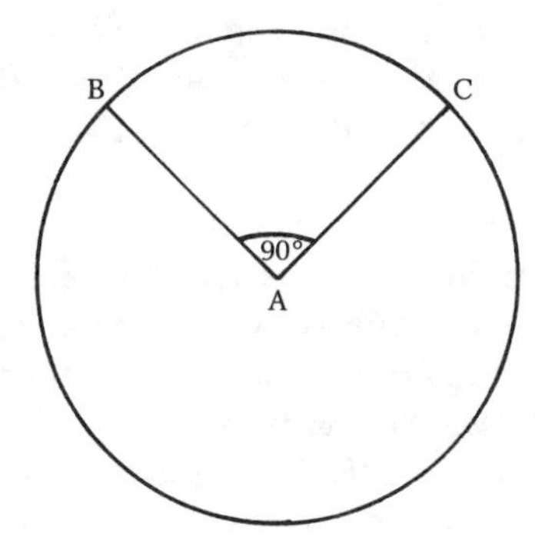

Fig. 4:7

parts, produces the answer 84^P 51′10″. This simple example gives some idea of the power and comprehensiveness of *The Almagest* in which Ptolemy gives lengths of chords for all central angles from ½° to 180° at ½° intervals.

Ptolemy's actual procedure was to begin with chords which subtend angles of 36°, 72°, 60°, 90°, and 120°. These are in fact the angles subtended at the center of a circle by inscribed regular decagons, pentagons, hexagons, squares and triangles respectively. The advantage of beginning with these is that their chords can be calculated by using straightforward Euclidean geometry. Having got these their supplementary angles of 108° and 144° can be quickly and simply derived. The rest was a little more difficult and involved techniques not found in *The Elements*.

Given the values of certain chords, Ptolemy showed how other values could be calculated. Thus, given the values for *chords 72°* and *60°*, Ptolemy defined a function over them which enabled *chord 72°–60°* or *chord 12°* to be calculated.[11] Applications of the function to the seven values initially obtained would lead to the calculation of all chords from 1° to 180°. The process was in fact simplified by the introduction of formulae which determined the value of a chord by the addition of known chords. In this way *chord 132°* could be obtained once *chord 60°* and *chord 72°* had been obtained. It only remained to show how ½° of any calculated value could be derived, and the rest was simply hard work.

The Almagest

Like *The Elements* of Euclid, *The Almagest* of Ptolemy was the culmination of a centuries-old tradition. The table of chords, the catalogue of stars, many of the constructions, to mention some of the contents, were largely taken from the work of such predecessors as the second-century B.C. Hipparchus. It is difficult to be too precise for, like *The Elements*, the success of

The Almagest virtually destroyed the labors of Ptolemy's rivals and predecessors. What survived, in Neugebauer's words, is mainly "elementary works. . .for teaching purposes. . . .The rest was obliterated" (Neugebauer 1969:55).

It is certainly an impressive work to examine. The first impression gained is of sheer comprehensiveness. In addition to table and catalogues, there are accounts of astronomical instruments, eclipses and their computation, calculations of planetary latitudes and longitudes, lunar parallaxes, solar apogees, stationary points of planets, their first and last points of visibility, and much other data. All this is presented in a precise, mathematical manner. Little allowance was made for the merely curious; Ptolemy was under no pressure to make *this* work acceptable to the general public.

The first part of Book I – sections 1.2–1.8 – contains the only discursive passages in the work and is concerned with justifying traditional Greek cosmology. In putting forth a generally accepted position, Ptolemy did not think it expected, or even desirable, to produce new and unfamiliar arguments. Rather, he little more than summarizes the main features and arguments of Aristotelian cosmology. Ptolemy lists them as follows: "(1) that the heaven is spherical in form and rotates as a sphere; (2) that the earth too. . .is spherical; (3) that it is situated in the middle of the whole heaven; (4) that. . .the earth bears to this sphere the relation of a point; (5) that the earth does not participate in any locomotion" (Hurd and Kipling; I, 65).

Purely astronomical arguments are advanced for the above assumptions. The spherical heavens are proved by noting how the sun, moon, and other heavenly bodies move from east to west "in circles always parallel to each other." The alternative possibility, that the stars do not actually circle the earth but rather are "kindled when they rise from earth and again are snuffed out when they return to the earth," is dismissed as "quite contrary to reason." Ptolemy maintained it was absurd to hold that stars could kindle or die out because what was observably true at one latitude would be untrue at another.

Of special interest to modern readers are the reasons advanced for the specifically Ptolemaic views: the central position of the earth in the universe and its immobility. Ptolemy was well aware that some of his predecessors had suggested that the Earth was neither central nor immobile. Heraclides of Pontus in the fourth century B.C. had, while believing in a central position for earth, assigned to it a daily rotation. In the following century, Aristarchus of Samos advanced this position and, according to Archimedes, maintained that the "fixed stars and the Sun remain unmoved, that the Earth revolves about the Sun in the circumference of a circle, the Sun lying in the middle of the orbit" (Heath 1932:106). Neither of these works survives and consequently the reasons given for such unorthodox views are unknown.[12]

That the earth rotates Ptolemy declared ridiculous. If it did, the effect would be unmistakeable. "Everything not actually standing on the

Earth. . . clouds, and any of the things that fly or can be thrown could never be seen travelling towards the east, because the Earth would always be anticipating them" (Hurd and Kipling: I,71). He believed equally absurd the idea that the earth could orbit some other body. In Aristotelian physics the speed of an object was proportional to its weight, consequently if the earth was in orbit ". . . it would clearly have got ahead of everything as it fell because of its vastly greater size; and the animals and all separate weights would have been left behind floating on the air. . . This sort of suggestion has only to be thought in order to be seen to be utterly ridiculous." The earth would, if allowed to move in this way, "have fallen completely out of the Universe itself" (Hurd and Kipling: I,70).

The contents of *The Almagest* can be briefly described:

Books I-II: In Chaps. 3-9 of Book I, Ptolemy describes what he took to be the basic shape and structure of the universe. Chaps. 3 and 4 argue for the spherical nature of earth and the heavens, and that all celestial motions are circular. Chaps. 5-7 establish that the universe is geocentric, vast, and stationary. Chap. 8 is on the principle motions of the heavens. In Chap. 10, Ptolemy constructs a chord table in the following manner:

Arcs	Chords	Sixtieths
½	0 31 25	0 1 2 50
1	1 2 50	0 1 2 50
1½	1 34 15	0 1 2 50
⋮	⋮	⋮
179½	119 59 56	0 0 0 9
180	120 0 0	0 0 0 0

Here the column of "sixtieths" is the interpolation if chords of arcs not listed are required. (Ptolemy was fond of tables, and *The Almagest* contains thirty-two major tables and hundreds of minor ones.)

The remaining chapters of Book I and all of Book II are concerned with problems of spherical astronomy and with an explanation of the mathematical tools that are employed in the remainder of the text.

Book III: Ptolemy here develops his solar theory which largely derives from Hipparchus. He points out that the inequality in the seasons could be represented "by that of an epicycle when the movement of the sun is in the direction of the movement of the heavens on its arc at the apogee. But it would be more reasonable to stick to the hypothesis of eccentricity which is simpler and completely effected by one and not two movements" (Chap. 4). Ptolemy consequently adopts an eccentric orbit for the sun and no other concept more clearly underlines the theoretical and geometrical nature of his astronomy. He never questions which of these two constructions, eccentrics or epicycles, of the solar orbit is actually adopted by the sun; instead, a choice is made on purely geometrical grounds. It had not yet occurred to astronomers like Ptolemy that their constructions were applicable to the heavens and not just convenient tools for the calculation and description of the movements of celestial bodies. Only in the work of Copernicus was there expressed the complaint that the constructions of astronomers failed to represent celestial reality.

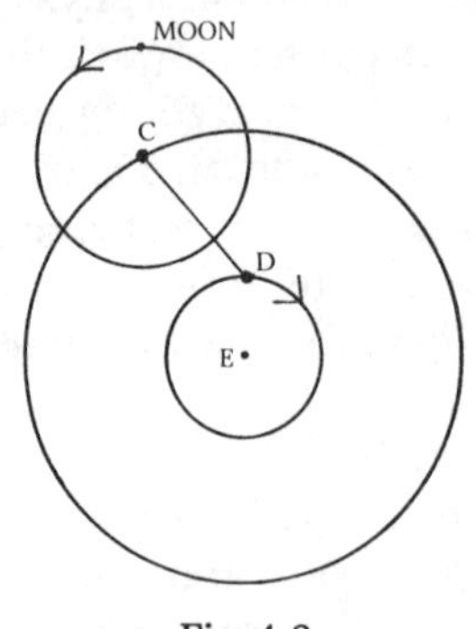

Fig. 4:8

Books IV-VI: Ptolemy presents his lunar and eclipse theory. It showed a distinct advance on the lunar theory of Hipparchus who had used a simple epicycle to represent the moon's motion. Further anomalies had emerged since the days of Hipparchus and to deal with them, Ptolemy introduced a further complication.[13] For Hipparchus, the moon moved on an epicycle as in fig. 4:5 above; Ptolemy, however, found it necessary to make the center of the moon's deferent not the point *E* in fig. 4:5 but the moveable eccentric of fig. 4:8 in which the moon's epicycle (center *C*) orbits around D which in turn describes a circle around the earth *E*.

Books VII-VIII: These deal with the fixed stars and contain Ptolemy's famous catalogue of 1,028 stars. As an example of how the catalogue is laid out, consider his description of some stars in the constellation Perseus:

(Constellation of Perseus)	Longitude	Latitude	Magnitude
The bright star in the Gorgon's head	Ram 29 2/3°	N 23°	2
The star east of this	Ram 29 1/6°	N 21°	4
The star west of the bright one	Ram 27 2/3°	N 21°	4
The star left further west than this	Ram 26 5/6°	N 22¼°	4

and so on for the rest of the constellation and all other constellations. This clumsy system was not improved upon until 1603 when Johann Bayer (1572–1625) introduced in *Uranometria* the modern system of naming stars with the letters of various alphabets.

Thus, Ptolemy's bright star in the Gorgon's head, as the brightest star of the constellation, *Algenib* in fact, was renamed *alpha* Persei; the next brightest star, *beta* Persei, and so on until the Greek alphabet was exhausted; thereafter, the Roman alphabet was used. Ptolemy's stellar magnitudes ran in descending order from 1 to 6. The accuracy of the catalogue was checked by Peters and Knobel in 1915. Assuming a date of A.D. 100, they concluded that the mean error was about 51′ in longitude and about 26′ in latitude. It is interesting to compare Ptolemy's assignment of magnitudes with a modern classification based on sensitive photometric instruments:

Magnitudes	1	2	3	4	5	6
Ptolemy	15	48	208	474	217	49
Modern	14	48	152	313	854	210

From this, we can gather that Ptolemy overestimated a number of brighter stars and considerably underestimated the fainter ones.

Books VII-XIII: The remaining five books deal with the orbits of the five planets. The actual constructions proposed by Ptolemy, with the exception of Mercury, were basically the same. It was necessary to incorporate three factors into each model: Firstly, that all planets moved around the ecliptic; secondly, that they retrogressed; and thirdly, that their motion around the ecliptic was variable. The motion around the ecliptic could be shown in an obvious way as the motion around a deferent; their retrograde motion could be shown by an epicycle. To show the variable motion of the planets Ptolemy adopted a new technique, the equant, one which over the centuries was to prove the most controversial of all. It can be seen in the figure below:

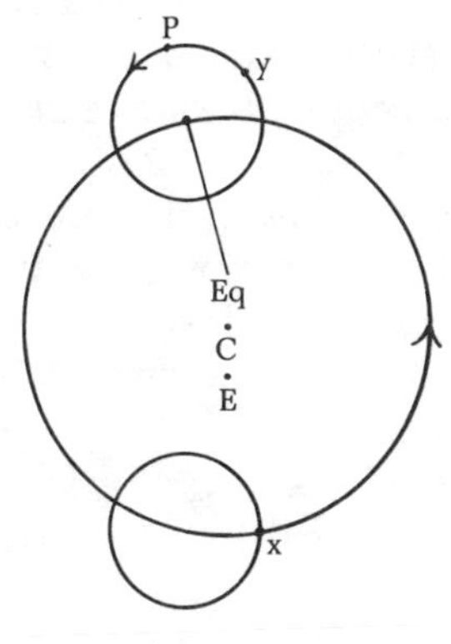

Fig. 4:9

Here the planet *P* moves around its deferent (the ecliptic) and is also carried around its epicycle in order to account for its occasional retrograde motion. The center of the epicycle does not move around the Earth *(E)*, but rather around the point *C*. Thus at certain points in the planetary orbit to an observer on the eccentric Earth the planet will appear larger and brighter *(x)* than at other points where it will appear smaller and dimmer *(y)*. But what of its variable velocity? It will not appear to move uniformly to an observer at *E* or one at *C*; it is possible, however, to select a point *Eq* such that *CE* = *CEq* and to an observer at *Eq* the planet *P* would move around the deferent with a uniform speed. The point was known to astronomers as the equant point, or, simply equant.[14]

So much for the geometry. The astronomer had then to work out what the appropriate radii, directions, and distances were for the four planets, Mars, Venus, Jupiter and Saturn. For Mercury he adopted the construction used for the lunar orbit: an epicycle and a moving eccentric.

Reception – Antiquity

Three commentaries are known to have been written on *The Almagest*: the Pappus and Theon commentaries date from the fourth century; Proclus's work dates from the fifth. Pappus confined his commentary to the first six books and only his comments on the fifth and sixth books have survived. Proclus's *Outlines of Astronomical Theories* is still available in a modern edition edited by C. Manitius (1909).

The commentary of Theon is almost complete; for the full text, however, it is necessary to go to the Greek *princeps* of Ptolemy, published in Basel (1538). G.J. Toomer dismissed Theon's work as "a trivial exposition..., explaining obvious points at excessive length...never critical, merely exegetic" *(Dictionary of Scientific Biography)*. Of more value was Theon's edition of Ptolemy's *Handy Tables*.

It seems reasonably clear that Ptolemy was anxious for his work to receive wider transmission than the technicalities of *The Almagest* would permit. Consequently, he culled material from the major work that could be issued in a handier form. One such work was his *Planetary Hypotheses*, a popular summary of the main features of Ptolemaic cosmology. In the same spirit, he extracted the numerous tables from *The Almagest* and issued them separately, with an introduction, under the title *Procheiroi canones* or *Handy Tables*. It is known only in the version by Theon who wrote two commentaries on it. In this form it passed to the Arabs and returned to West in the 12th century.

As to the geographical spread of *The Almagest* in antiquity, Neugebauer (1969:187) refers to a Greek copy that arrived at the Persian court of Shapur I in about A.D. 250; Sarton mentions a sixth-century Syriac version produced by Sergeios of Resaina. By this time, however, astronomy was largely transmitted through the encyclopedias of Pliny or Isidore of Seville: it was sketched in the barest outlines, not well understood by the compilers, and mixed indiscriminately with wonders and inaccuracies. Mathematical astronomy returned to Europe only with the growth of Islam.

Reception – The Medieval Period

After Proclus the only references to Ptolemy's astronomy is found in the pages of the encyclopedists. Their knowledge of science and the texts themselves was limited. For example, Isidore of Seville (c. 570–636), author of *Etymologies*, mistook Ptolemy for one of the ruling Ptolemies of Egypt, and for this reason Ptolemy was often shown in medieval manuscripts wearing a crown. It was not until the ninth century that references can be found once more to the serious study of astronomy and, consequently, to *The Almagest*.

As with Euclid and Hippocrates, it is in the Abbasid capital that the medieval study of Ptolemy can first be discerned. In the late eighth century, Indian astronomical texts appeared in Baghdad; they were translated into Arabic, as were certain Persian texts. The first reported translation of *The Almagest* is supposedly that done by al-Hajjaj ibn Yusuf ibn Matar (786–833) from a sixth-century Syriac version in 829/30. These early translations of major Greek texts tended to be quickly superseded by more reliable ones. In this way, a more professional translation was supplied by Hunayn ibn Ishaq in the ninth century; revised by Thabit ibn Qurra, Hunayn's translation was accepted as the authoritative one and was used as the text for the first Latin translation.

But, as was to happen later in the West, *The Almagest* proved too complicated for those being initiated into the subject of astronomy, and consequently a need arose for an introduction and summary. This was provided by al-Farghani, known by his latinized name as Alfraganus, who produced between 833 and 857 a work entitled *Jawami*[c] or *Elements*. It was a straightforward summary of Ptolemaic astronomy, descriptive and nontechnical. It proved popular and went through a number of editions over the following centuries. It was translated into Latin twice in the 12th century, once by John of Seville, and once by Gerard of Cremona. Known variously as *Rudimenta astronomica* and *Liber differentiorum*, it even survived the advent of printing and was issued in a printed version as late as 1546. Among many, Dante (1265–1321) acquired his knowledge of astronomy from al-Farghani. In contrast, Dante was unfamiliar with *The Almagest*, for on the three occasions that he quoted Ptolemy's opinions, he erred in each case. He had apparently copied earlier errors made by such scholars as Albertus Magnus or Averroes (Orr:159-60).

It should not be thought that Islamic scholars were uncritical agents for the transmission and vulgarization of Ptolemaic astronomy; some improved upon Ptolemy. Al-Battani (fl. 900), for example, in *Zij al-Sabi (Sabean Tables)* actually corrected and improved a number of Ptolemaic measurements. Thus while Ptolemy had argued for a precessional rate of 36" a year or 1° a century, al-Battani proposed a figure of 54.5" a year or 1° each sixty-six years. (A modern textbook gives the figure of 50.2564" a year or 1° every seventy-two years.) Al-Battani's other corrections tended to be equally justified.

Others sought to oppose rather than correct Ptolemy. It was his basic assumptions and not just his calculated parameters that Ibn Bajja, known as in West as Avempace, the 12th-century Spanish scholar, attempted to replace. A contemporary, Jabir ibn Aflah, actually wrote a book with the title *Islah al-Majisti (The Correction of The Almagest)*[15], containing objections to the elaborate geometrical constructions favored by Ptolemy. Ibn Bajja objected to epicycles but allowed eccentrics; his pupil, Ibn Tufayl, went further and refused to admit the latter construction. Their work was not specifically against Ptolemy's geocentric cosmology; rather, it was in favor of a purer astronomy, such as that proposed by Eudoxus who only used homocentric spheres.

A center of dissatisfaction was located at Maragha Observatory in Azerbaijan where in the 13th century, under the leadership of Nasir al-Din al-Tusi (1201–74), highly competent astronomers sought to "construct models in which only constant speed circular motions were involved" (see Kennedy 1966). Much of this literature was known in the West. The *Islah* of Ibn Aflah, known to Albertus Magnus as Flores, was translated into Latin by Gerard of Cremona and twenty manuscripts of this work have survived. The tradition of opposition and the grounds for it were, as will be seen, well known to Copernicus some three hundred years later.

It was also in the 12th century that *The Almagest* appeared in the West in Latin in two independent editions – one based on the Arabic translation, the other on a Greek text. The first appeared in 1160 in Sicily which, under Norman rule, enjoyed an intellectual renaissance. Together with Spain it was one of the great cultural and linguistic interfaces of Europe. At the court in Palermo, Byzantine Greeks, northern Christians, and Muslim traders and scholars were brought together and both Greek and Arabic were commonly spoken. By great good fortune, manuscripts which had been heard of but never seen by Western scholars for several hundred years were brought to light at the Norman court.

Sometime before 1160, Henricius Aristippus, archdeacon of Catania and translator of Plato's *Phaedo* and *Meno*, returned from a mission to Constantinople with some Greek manuscripts, gifts from the Emperor Manuel I to the Sicilian king, William I. One was a handsome codex of *The Almagest*. A letter from the anonymous translator conveys the excitement that such news caused to those desperately seeking to recover the learning of the past. Upon hearing of the arrival of the codex while studying medicine in Salerno, the future translator dropped his work and raced to Sicily where he found Aristippus at Pergusa. In order to be able to translate *The Almagest*, he not only willingly braved the terrors of Scylla and Charybdis as well as the perils of Etna ("Scilleos latratus non exhorrui, Caribdam permeavi, ignea Ethne fluente circuivi" – Haskins:191), but also took up the study of astronomy. With the help of "Eugene the Emir," a "man most learned in Greek and Arabic and not ignorant of Latin" (Haskins:160), he felt competent to make the translation. It appears to have been little known and used. Four manuscripts are known.

Shortly afterward Gerard of Cremona produced in 1175 a second translation, from Arabic, and for the next three centuries this edition served as the standard Latin text; some thirty-two manuscripts are known. Other attempts were made. A translation, also from Arabic, is known in fragments from 13th-century Spain while the first four books are known in a single copy in a pre-1300 manuscript. There is also a translation from Greek, done by George of Trebizond (1395–1484) in 1451 for Pope Nicholas V.

As in the Muslim world, *The Almagest* in Latin translation inspired a number of texts designed to simplify for scholars some of the basic ideas of the Ptolemaic scheme. One of the most elementary and popular of these works was *Sphere* by Sacrobosco or, to use his original name, John Holywood. Born in Yorkshire, Holywood taught in Paris where he died circa 1250. Known in many manuscript versions, it was first printed in Ferrara in 1472; by 1501 at least thirty editions had been issued while another two hundred appeared in the 16th century. An extremely elementary work, *Sphere* does little more than describe and name the main features of the celestial sphere with its various parts such as the ecliptic, celestial equator, and the zodiac. It is unhesitatingly Ptolemaic and within this framework sought to explain a number of puzzling astronomical features, such as eclipses and variations in the length of the day that occur at different latitudes. Also available from the 13th century was *Theorica planetarum*, anonymously produced, a work that attempted to do for planetary theory what Sacrobosco had done for the celestial sphere. It has survived in hundred of manuscripts.

At the same time, scholars began to question some of the basic assumptions of Ptolemaic cosmology. Long before Copernicus finally overthrew *The Almagest* in the 16th century, scholars who had become deeply suspicious of Aristotelian physics realized that, whatever the technical merits of classical astronomy, its foundations were profoundly shaky. In the 14th century Jean Buridan (1300–58) and Nicolas Oresme (1323–82) both wrote commentaries on Aristotle's *De caelo* and explored some of the subtleties of relative motion. Although both asserted that the earth did not rotate, they managed nonetheless to raise considerable doubt against such a view. Their work suggested that the universe would look the same to a terrestial observer whether he stood on a stationary earth around which the heavens moved or whether the stationary heavens were revealed to an observer standing on a rotating earth. The same arguments were brought forth by Copernicus.[16]

Reception – Modern Period

A new attitude is detectable in the 15th century. Western scholars were no longer prepared to accept the major works of classical science through the corrupting medium of indifferent Arabic and Latin translations. Thus, the ambition of Georg Peurbach (1423–61), working at the University of Vienna, was to discover the pure, undiluted text of the Greek *Almagest* and to make it available to the emerging class of astronomers and mathematicians of Renaissance Europe. Peurbach is best known for his *Theoricae novae planetarum*, a work intended to update the 13th-century *Theoricae planetarum;* it proved just as popular and, according to Koestler (1964:211), "was translated into Italian, Spanish, French and Hebrew" in fifty-six editions. He also published a table of sines and chords.

In search of a Greek *Almagest* Peurbach traveled to Italy where he met the remarkable Cardinal Bessarion (c. 1400-72). Born in Trebizond, he became bishop of Nicea and accompanied Emperor John Paleologus to the Council of Florence in 1439. There, it became evident in scholarly debate that Bessarion possessed a mastery of Greek documents whose existence let alone contents was unknown to Western scholars. It had long been suspected that much of classical wisdom could be found only in Greek manuscripts; Bessarion's stay in Florence clearly confirmed this view. Remaining in Italy, Bessarion joined the Roman church and was appointed a cardinal; twice he almost attained the papacy. Known to Italians as *Latinorum Graecissimus, Graecorum Latinissimus*, he brought with him to Rome an impressive collection of Greek books, as many as 500, which eventually formed the nucleus of the San Marco library in Venice. One of these must have been a fine manuscript copy of *The Almagest* for, on an official visit to Vienna, Bessarion approached Peurbach and requested his help in preparing a new edition and translation of Ptolemy's text. All Peurbach could achieve before his early death at the age of thirty-eight was a work entitled *Epitome*, a paraphrase of the first six books of *The Almagest*.

As he lay dying, Peurbach is reported, according to one romantic tradition, to have extracted from his pupil J. Muller (1436-76), known as Regiomontanus, a commitment to complete the task. Peurbach chose the wrong man. Although in many ways the most remarkable scientist of his century, Regiomontanus was not the kind of man likely to devote years to mastering and publishing a single text. He was a much traveled man, having spent several years in Italy searching for manuscripts and having also served for some time as librarian to Mathias Corvinus (1433–90), king of Hungary and founder of a great library in Budapest. (Its 50,000 manuscripts were unfortunately scattered by the Turks in the following century.) Regiomontanus was thus personally acquainted with the wealth of ancient science in the manuscripts of Archimedes, Euclid, Apollonius as well as the works of Ptolemy – works which existed in manuscript form in various areas of Italy and central Europe.

As a man of vision rather than scholarship, Regiomontanus saw that for science to develop it was necessary to make this material much more available than it had ever been. He realized that little would be gained by producing yet one more manuscript copy of *The Almagest*, however accurate it might be. Instead, he recognized the potential of the newly-invented printing press, which could make scores of copies available, and all of them identical. Tables of chords, sines, ephemerides, all notoriously difficult to produce by hand, could be printed accurately, cheaply, and in quantity to meet the demand of mariners, astronomers, mathematicians, astrologers and others. Wildly confident and ambitious, Regiomontanus found a patron in Berhard Walther, a wealthy citizen of Nuremberg, who enticed him to settle in Germany with offers of an observatory and a printing press. A catalogue was issued containing the names of the leading mathematicians and astronomers of antiquity among the twenty-two works Regiomontanus hoped to publish.

It was left to others to fulfill the vision of Regiomontanus for, like his teacher Peurbach, he died early at the age of forty in 1476. Called to Rome to advise on a new calendar, he died quite suddenly. Of his program, Regiomontanus only succeeded in printing a work entitled *Ephemerides* in which positions of the sun, moon, and planets were computed for the period 1475–1506; the *Theoricae* by Peurbach; various almanacs; and an edition of Manilius, a first-century A.D. Latin poet and author of *Astonomica*, a didactic poem on astrology and the zodiac. Ironically, his own works remained unpublished until long after his death. *Epitome*, Peurbach's work, which he completed, was not published until 1496; and his own *De triangulis*, an important work on trigonometry, was left unpublished until 1536. His efforts to produce an authentic *Almagest* were ignored by early printers. Thus, the first two editions of *The Almagest*, produced in 1515 and 1528, proved none other than the centuries-old Latin translations by Gerald of Cremona and George of Trebizond, respectively.

The early publishers were evidently more interested in Ptolemy's other works, and while there are no incunabula of *The Almagest*, there are four of his *Geography* (Vicenza: 1475; Ulm: 1482, 1486; Rome: 1490) and two of his *Tetrabiblos* (Venice: 1484, 1493).[17] After the early appearance of *The Almagest* in the 16th century, no more editions were published for almost three hundred years. The lack of interest in such an important work was due to the success as a text of *De revolutionibus* by Copernicus. Whether scholars accepted the geocentric or the heliocentric system, they still preferred to use as a practical astronomical textbook Copernicus's more up-to-date work, and it met what little demand there was for such texts. As a matter of fact, both books were too long, too similar, too technical, and too expensive to permit either one remaining in print.

When *The Almagest* was next published, in 1813, it formed part of a collected edition and was no longer a working astronomical text. It appealed primarily to historians of astronomy and mathematics, not to compilers of ephemerides. Attitudes, moreover, had changed and the eulogies and superlatives of medieval-period scholars was replaced by dismissive comments among the more critical astronomers of revolutionary France. J. B. Delambre (1749-1822), professor of astronomy at the College de France, raised in his *Histoire de l'astronomie ancienne* (1817) what became often-expressed doubts about the credibility and honesty of Ptolemy. These doubts have recently been revived by a modern astronomer, R. Newton, in *The Crime of Ptolemy* (1977). Newton somewhat extravagantly claims that "all of his own observations that Ptolemy uses in the *Syntaxis* are fraudulent, so far as we can test them. Many of the observations that he attributes to other astronomers are also frauds that he has committed" (pp. 378-79).

Two basic charges may be leveled against Ptolemy. The first is that the results presented in *The Almagest* as his own observations are in fact stolen from others. Thus the catalogue of 1,028 fixed stars is derived from the earlier work of Hipparchus, corrected for precessional effects. Since the catalogue of Hipparchus contained information on about 850 stars, it is nonetheless clear that Ptolemy made a small if considerably diminished contribution of his own to the list. But, further, if Ptolemy lifted the results of Hipparchus wholesale it should be the case that the difference between the values assigned by Hipparchus and his own would differ by a constant amount: that amount should be due to the movement of the equinoctial point in the two hundred and fifty years separating the two astronomers. This, however, is not the case, for where it is possible to make a comparison the differences are too variable to be simple adjustments.

The second charge, raised by Delambre, is that Ptolemy did not in fact carry out his own work, and Delambre asks, "Are not those he says he made but computations from his tables?" (Pannekoek: 149). The point is made more specifically by Newton who accuses Ptolemy of being an early practitioner of the Burt effect.[18] He asserts that Ptolemy's published results are far too accurate to have been made with instruments then available. Rather, he claims, Ptolemy "developed certain astronomical theories and discovered they were not consistent with observation. Instead of abandoning the theories, he deliberately fabricated observations from the theories so that he could claim that the observations prove the validity of his theories" (Newton 1977a: 80).

To date Newton's charges have found little support; rather, they have provoked countercharges that he little understands probability theory, the nature of ancient science, and the techniques of ancient astronomers. For those interested in this debate, a detailed and effective reply by Swerdlow (1979) may be consulted.

Publishing History

There are four 16th-century editions of *The Almagest*:

Venice, 1515

Edited and published by Peter Liechenstein, this is based on the Latin version of Gerard of Cremona, who translated it from an Arabic translation of an original Greek text. It was reprinted in Nuremberg (1537) and Paris (1546).

Venice, 1528

A reissue of George of Trebizond's Latin translation of the Greek text first published in 1451. The edition was revised by Luca Guarico, but Sarton still described it as "very imperfect." Giunti was the publisher. It was reprinted in Basel (1541, 1551).

Basel, 1538

The Greek *princeps* prepared by Simon Grynaeus for J. Walderus. It was based on the lost manuscript of Regiomontanus, the one claimed by Bessarion to be worth more than a province. A two-volume edition, the first volume consists of *The Almagest*; the second, the commentaries of Pappus and Theon.

Wittenberg, 1549

This joint Latin-Greek edition was prepared by E. Reinhold.

No other editions appeared for almost three hundred years. Two poor Latin translations, antiquated even by the standards of the time, and the Greek *princeps* remained for long the only available texts of "the most influential treatise on astronomy ever written." Nor was it among the libraries of later scientists – neither Newton, Locke, Hooke, or Flamsteed appear to have owned *The Almagest,* although most of them possessed other works by Ptolemy. It was not until the 19th century that it once more became available, this time in modern languages.

Paris, 1813-16

Nicolas Halma (1756-1828) intended to prepare a complete Greek-French edition. It remained unfinished at his death; he did, however, complete *The Almagest.* The French translation was reissued in 1927.

Leipzig, 1898-1903

J. L. Heiberg (1854-1928), who had prepared the definitive text of Euclid, began to prepare a complete edition of Ptolemy for Teubner. Like Halma's work, it remained unfinished, although he completed a two-volume text of *The Almagest.* Heiberg had access to thirty-six codices and based his edition on two ninth- and two tenth-century manuscripts. The text was translated into German by K. Manutius in 1912-13 and reissued in 1963.

Chicago, 1952

The first and only English translation was made by R. C. Taliafiero for the Encyclopedia Britannica's series, Great Books of the Western World.

Extracts

The substantive arguments on the basic structure of the universe, that is *Almagest* 1.2-1.8, can be found in many popular anthologies and readings: Hurd and Kipling (vol. 1); T. L. Heath, *Greek Astronomy* (1932); M. R. Cohen and I. E. Drabkin, *Source Book in Greek Science* (1958); and E. Grant, *A Source Book of Medieval Science* (1974).

Further Reading

A number of forceful, readable and not too technical accounts of the history of astronomy are available. Two of the best are Pannekoek (1963) and Ley (1961). For the background to Greek astronomy the refreshingly iconoclastic *Early Greek Astronomy to Aristotle* (1970) by D. R. Dicks is one of the few works to treat this period realistically and accurately. For those with a more technical capacity the works of Neugebauer (1969; 1975) are available and unsurpassed. The text of *The Almagest*, excluding Vesalius, is the most elusive of the classics and can be found in the translation by Taliaferro for the *Encyclopedia Britannica* (1952). These is however an excellent commentary by Ole Pedersen (1974) which, with his *Early Physics and Astronomy* (1974) constitute the best introduction for the serious student to the astronomy of Ptolemy. Much of value on Ptolemy and on the later medieval period can be found in Dreyer (1953).

Notes

1. How did a work originally entitled in Greek *Mathematike syntaxis,* or Mathematical Collection, become known throughout the world as *The Almagest?* Later Greeks referred to the work as *Megale syntaxis* (Greater Collection) and even later as *Megiste syntaxis* (Greatest Collection), although not, according to Neugebauer, before the 11th century. If this last title were common much before the 11th century it would explain why one of the earliest Arabic translators, Hajjaj ibn Yusuf ibn Matar, entitled his work *Kitab al-mijisti* (829-30), literally, "the book of the greatest." Under this name, it returned to the West in the 12th century as *The Almagest,* a latinized form of an Arabic form of the Greek superlative. There is a discussion of the question in the *Encyclopaedia of Islam* under the entry "Batlamiyus," the Arabic transliteration of the Greek Ptolemaeus.

2. The *synodic month* is the time taken by the moon to complete all its phases, from full moon to full moon or new moon to new moon; the *tropical year* is the time taken by the sun to complete one revolution with respect to the vernal equinox.

3. The verse continues: "The infidels are led into error by it. They allow it one year, and forbid it another, that they may make good the number of months which God hath hallowed, and they allow that which God hath prohibited. The evil of their deeds hath been prepared for them by Satan: for God guideth not the people who do not believe" (Sura 9:37).

4. For Aristotle it was clearly numerology, a consequence of the need to ensure that the heavens were occupied by ten and only ten objects: five planets, earth, moon, sun and the sphere of the fixed stars which with the counter-earth equalled ten, the sacred Pythagorean number. "They held that ten is a perfect number (1+2+3+4). . . . On this view they asserted that there must be ten heavenly bodies; and as only nine were visible they invented the 'counter-earth' to make a tenth" (Aristotle: 64).

5. By a heliacal rising is meant the star which on a particular day rises and is visible immediately before the sun. As the sidereal day, or star day, lasts 23 hours and 56 minutes, a new star will be visible on the horizon just before first light on most days. It was the heliacal rising of Sirius, the brightest star in the heavens, which signalled to the Egyptians the beginning of the year and which also at one time warned of the impending Nile floods.

6. The first assumption was given up by Hipparchus in the second century B.C.; the second was replaced by Copernicus in 1543; and the later two by J. Kepler only in 1609.

7. The first motion is the inevitable result of the earth's daily rotation around its axis; the second is a consequence of the earth moving in an annual orbit of the sun.

8. Different planetary configurations are conveniently named in fig. 4:10:

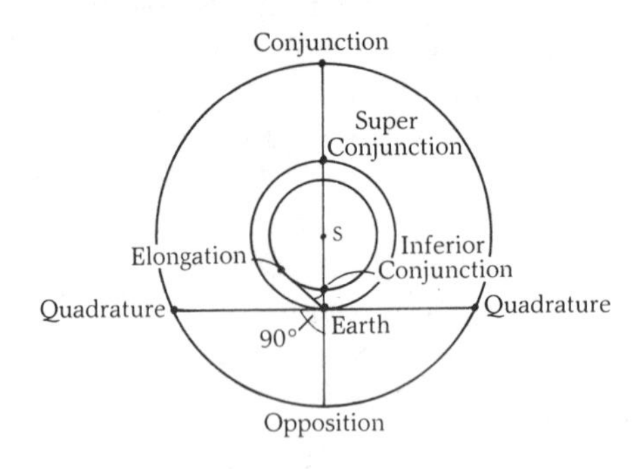

Fig. 4:10

Thus a planet at opposition will have a longitude of 180°, at quadrature of 90° and at conjunction it will have the same longitude as the sun; when a superior planet is in opposition, it is at its closest position to the earth (perigee) while in conjunction it is at its most distant (apogee).

9. Having once introduced the basic constructions of epicycle and eccentric, all kinds of complications and combinations of these two pure forms become possible. There can be, for example, an epicycle on an epicycle as in fig. 4:11; equally, the epicycle and eccentric model can be combined as in fig. 4:12; while, as a third complication, the actual center of an eccentric circle need not be fixed but can itself describe a circle (fig. 13). Within such constructions further variety can be introduced by adopting different radii, by allowing the orbits to move in the same or

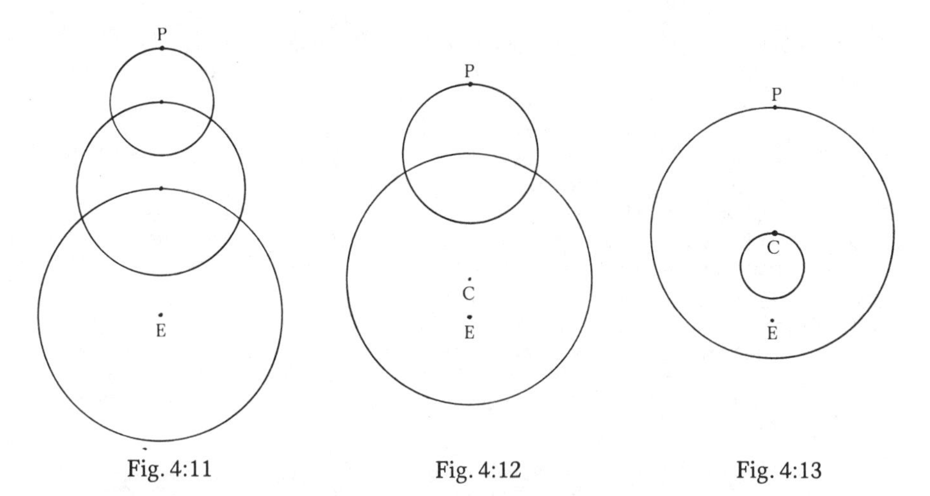

Fig. 4:11 Fig. 4:12 Fig. 4:13

opposite directions and by assigning different velocities. With such a wealth of variables at hand, it is clear that any motion, however bizarre, could be described in these terms. This is true in theory; in practice it proved to be rather difficult. It was first no easy matter to find a reasonably simple construction which did fit the known facts of a planetary orbit, and, secondly, once having been found the fit would be only temporary. What worked for Hipparchus had become inaccurate by the time of Ptolemy; while the repairs introduced in the second century A.D. were looking equally strange a few centuries later.

10. Using later observations Ptolemy rejected the figures of Hipparchus and claimed a precession of 36″ a year or 1° a century. The figure of Hipparchus was in fact more accurate. Connected with precession is the unfamilar notion of trepidation. Theon of Alexandria reported that some had held that the equinoctial point did not in fact precess completely around the ecliptic but merely oscillated 8° in the order of the signs for 640 years before spending another 640 years moving 8° back. The idea was not Ptolemaic; it was popular with a number of Islamic astronomers and is found in Copernicus.

11. Ptolemy's reasoning here can be sketched briefly as follows: He first proved the theorem, since known as Ptolemy's theorem, that in a quadrilateral (*ABCD*) inscribed in a circle with diagonals *AC* and *BD*, then:

(1) $AC.BD = AB.DC + AD.BC$. Thus, consider the quadrilateral in fig. 4:14 in which *AD* is the diameter and suppose *chord AC* and *chord AB*, referred to as *a* and *b* respectively, are known. The aim is to find *chord BC*:

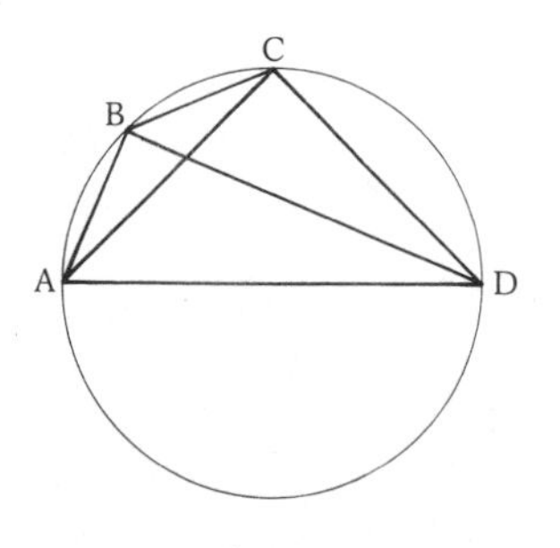

Fig. 4:14

(2) *DC* is *chord (180-a)* and is therefore known.

(3) *BD* is *chord (180-b)* and is known.

(4) *AD* is *chord 180* and is therefore known.

(5) *chord BC* is *chord (a-b)*.

If the appropriate substitutions are made in (1), the result is an equation in which all terms are known except *BC* or *chord (a-b)*, it is therefore a trivial matter to solve the equation. The result of this is that wherever we have the values for *chord a* and for *chord b* we can readily determine the value for *chord (a-b)*. In modern terminology this is equivalent to: (6) $\sin \theta - \phi = \sin \theta \cos \phi - \cos \theta \sin \phi$.

12. One work of Aristarchus has survived: *On the Sizes and Distances of the Sun and Moon.* His conclusion was that the sun is more than eighteen but less than twenty times further away from earth than is the moon.

13. The epicycle of Hipparchus worked reasonably well at conjunctions and at full moon but presented major difficulties when the moon was at quadrature on the epicycle. The moveable eccentric was very nearly equivalent to an epicycle when the moon was full or at conjunction but accounted for the discrepancies at other times. The cause of the problem was that the sidereal month, the time taken to retun to the same fixed stars, was slightly shorter than the anomalistic month, the time taken to return to the same velocity.

14. It is possible to dispense with the equant but, as Neugebauer has pointed out, it will involve the adoption of "extremely complicated combinations of circular motions" (Neugebauer 1969:199). Elsewhere, Neugebauer has claimed that the equant was "probably Ptolemy's most important discovery in the theory of planetary motion" (Neugebauer 1975: 155).

15. Ibn Aflah objected that Ptolemaic trigonometry was too complicated, the equant was an unacceptable construction, many of his parameters were wrong, and that Venus and Mercury were not between the Sun and Moon but beyond the Sun.

16. The arguments can be found in Toulmin and Goodfield (1963: 182-87).

17. *Tetrabiblos* is an astrology textbook, with a purer and clearer theory than can be found in today's textbooks. The early editions did not contain the later additions of houses, elections, and horary astrology. It is interesting, therefore, to note that at a time when Copernicus replaced Ptolemy's astronomy, there was a call by some astrologers for a return to the purity of *Tetrabiblos.* The key names here are Placidus de Titis, an Italian, and John Partridge, an Englishman. An English translation was published by John Whalley in 1701 under the title *Quadrapartite* (see Capp: 181-84).

18. The reference is to the British psychologist Cyril Burt (1883-1971). For decades Burt had apparently collected data on identical twins separated in early life that strongly suggested that intelligence was as much an inherited trait as eye color. After his death, the Princeton psychologist Leon Kamin conclusively demonstrated that Burt's data had been shaped by Burt himself to fit his own preconceptions. For a full account, see L. J. Kamin, *The Science and Politics of I.Q.* (1974).

CHAPTER 5

De revolutionibus orbium coelestium libri vi (1543)

by Nicolaus Copernicus

Mathematics is for mathematicians.

–Copernicus, *De revolutionibus*

While Bohr was summering in a cottage in the woods of Tisvilde, his neighbour, who was not a superstitious man, had a horseshoe put over his door. They asked him: "Do you honestly believe it will bring you luck?" "Of course not," the man answered, "but they say it helps even if you do not believe in it."

–Werner Heisenberg, *Physics and Beyond*

People give ear to an upstart astrologer who strove to show that the Earth revolves, not the heavens or the firmament, the Sun and Moon. . . . This fool wishes us to reverse the entire science of astronomy; but sacred Scripture tells us that Joshua commanded the Sun to stand still, and not the Earth.

–Martin Luther, *Table Talks*

Life and Career

The first classic work of modern science, if only by a few weeks, was undoubtedly *De revolutionibus orbium coelestium* by Copernicus, first published in 1543. Whereas Hippocrates, Euclid, and Ptolemy are figures of total anonymity, there is happily an abundance of biographical material on Copernicus. Consequently there is no difficulty in reconstructing much of his public and even some of his private life. His intellectual life, however, remains obscure. At several times in the course of his life, Copernicus behaved in a completely unpredictable manner, surprising contemporaries just as much as, over the centuries, he has continued to puzzle biographers.

Copernicus was born in 1473 in Torun, an area of great political complexity and instability. Then, as now, Poland was the reluctant victim of her geographical position, squeezed by both eastern and western powers. To the east lay the Mongols and pagan Lithuanians; to the south the powerful kingdom of Hungary. Poland itself was a loose collection of bickering principalities and bishoprics under a nominal king. Into this area of fluctuating political and dynastic power a new and powerful force, the Order of the Teutonic Knights, appeared in the 13th century. Originally an order of chivalry connected with the Crusades, the Knights had settled on the banks of the Vistula in 1230 at the invitation of Conrad of Mosovia, a Polish feudal lord, to lead a crusade against the Lithuanians. Torun, the birthplace of Copernicus, was initially a fortress-stronghold of the Knights and in his lifetime, it remained a center of intense political and military activity.[1]

By 1473, the year of Copernicus's birth, Torun belonged to Poland, now a stronger entity, thanks to a diplomatic marriage in the previous century which had done much to secure her eastern frontier. The Teutonic Knights who served as a vanguard of the German presence in traditional Polish territory nonetheless continued to cause inevitable and intermittent conflict. Suffering a defeat in 1462 at the hands of a Polish-Lithuanian force, its Grand Master became a Polish vassal, and Torun and the bishopric of Warmia became nominally Polish. The turbulent history of Torun placed Copernicus for much of his life in the center of military action. The picture of him as a "timid canon" and an isolated figure devoting himself to the solitary study of mathematics and the heavens is a complete fantasy. Copernicus, in fact, participated more in the affairs of state than have most modern astronomers.

This fact was directly related to the position of his uncle, Lucas Watzenrode (1447–1512), who was bishop of Warmia and his guardian. After Copernicus's studies at the universities of Cracow (1491–93), Bologna, Padua, and Ferrara, his uncle secured for him a position as canon of Frauenburg cathedral, which Copernicus took up upon his return from Italy about 1506. He also served as his uncle's secretary and perhaps as his physician.[2] When Lucas died in 1512, Copernicus's duties centered on his work as cathedral canon, although he remained free to offer astronomical and diplomatic help to the state when called upon.[3]

Indeed, he was often in the middle of hectic diplomatic negotiations, usually as a result of troubles fomented by the Teutonic Knights. While serving as administrator of the Bishop of Warmia's outlying estates at Allenstein and Mehlsack, Copernicus was probably one of the representatives who negotiated with the Knights after they invaded the estate in 1519. Peace talks broke down, Frauenburg was overrun, and in 1520 Allenstein itself, now under the command of Copernicus, was beseiged by the Order. An armistice was reached in 1521, bringing to a close a bloody interlude, and at the subsequent peace talks Copernicus was a leading figure. Despite these historical facts, Copernicus has been described as an academic recluse who, says Koestler (1964:121), lived for thirty years in "the north-west tower of the fortified wall surrounding the Cathedral hill of Frauenburg."

Among other matters of state, Copernicus was involved in finance and issued, in 1522, a memorandum, *De monete cudende ratio*, which dealt with the problem of Poland's debased currency. In it, Copernicus suggested consolidating the fiscal authority, and in a passage that bears resemblance to Gresham's law (bad money drives out good), he spelled out the consequences of a debased currency for the monarch and his subjects. "For such a ruler cheats not only his subjects but himself," the memorandum declares. "When he rejoices over a temporary gain . . . , he is like a stingy farmer who sows bad seed in order to save good ones, and then there will be even more bad produce than he sowed. This destroys the value of coins, just as weeds destroy the grain, when the former win the upper hand" (Kesten:219). Unless the custom of debasing the currency were stopped, Copernicus warned, "Prussia will soon have coins containing nothing but copper. Then all foreign trade will stop. For what merchant will sell his goods for copper coins?" (Kesten:222). The issue was so serious, and Copernicus felt so strongly about it, that he reissued the memorandum in 1527 and attended a number of the Prussian and Polish diets lobbying for his cause.[4]

"De revolutionibus" – Publication

Few books have been delayed so long and extracted so unwillingly from their author as *De revolutionibus*. In the dedicatory preface, Copernicus noted that the work had been kept "in store not for nine years only, but to a fourth period of nine years." This might lead to the assumption that he had waited thirty-six years before publishing the book and that it had been written in 1506. This is patently untrue, if only because the manuscript contains reports of observations made in 1529. Copernicus must have been referring, therefore, to the length of time he had been committed to the heliocentric hypothesis.

This interpretation would fit well with the earliest published account of his new theory. He had produced a brief summary of his views, known as *Commentariolus*, sometime before 1514. Untrumpeted, it was available in manuscript form and appears to have circulated by hand among European astronomers. Tycho Brahe (1546–1601), for example, had a copy, which had been handed to him at Ratisbon. He reported that "he sent the treatise to certain other mathematicians in Germany" (Rosen:6). Although all copies seemed to have disappeared in the 16th century, Curtze located a copy in a Viennese library and arranged for its first printing in 1873. As it uses heliocentric constructions quite different from those in *De revolutionibus*, it is obvious why Copernicus had no interest in preserving a work which already by the 1520s presented a misleading picture of his views.

From today's vantage point, three important points may be made about *Commentariolus*. First, it uncompromisingly rejected Ptolemaic cosmology and projected unambiguously heliocentric principles. It began with a statement of seven hypotheses, three of which are (Rosen:58):

> All the spheres revolve about the Sun as their mid point, and therefore the Sun is the centre of the Universe.
>
> The Earth . . . performs a complete rotation on its fixed poles in a daily motion.
>
> The apparent retrograde motion of the planets arises not from their motion but from the Earth's.

Second, Copernicus at this time had no reluctance in acknowledging such views as his own. The manuscript is entitled "Nicolai Coppernici de hypothesibus motum coelestium . . ." and thus clearly indicates that whatever inhibitions he may later have developed about publishing his work, Copernicus was free from them in 1514.

Last, Copernicus revealed in *Commentariolus* plans for the publication of a more substantial volume. For the sake of brevity, he declared, he had omitted all mathematical demonstrations; these he reserved for a "larger work" (Rosen:59). Nor was this the only reference made to a "larger work." In 1524 he took exception to certain astronomical views expressed in 1522 by the mathematician John Werner and consequently circulated manuscript copies of a "Letter against Werner," which ended, "What finally is my own opinion concerning the motion of the sphere of the fixed stars? Since I intend to set forth my views elsewhere, I have thought it unnecessary and improper to extend this communication further" (Rosen:106).

Why, then, did Copernicus delay so long? Why did it take some thirty years to publish the "larger work"—with its mathematical demonstrations—particularly since there is good evidence that the manuscript was completed long before it was finally published in 1543? Copernicus presumably felt that an explanation was needed, and mentions the subject in the remarkable and revealing preface of *De revolutionibus*, which is dedicated to Pope Paul III. It might be thought that Copernicus was restrained by a fear of what the Church might do, since his views were so clearly in conflict with Scripture, but Copernicus could have had no reason to hold such fears. If anything, the Church did its best to encourage publication. The main ideas of Copernicus seem to have reached Rome in the 1530s, where they were met with interest and excitement, in marked contrast to the disapproval Galileo received in the 17th century. No one objected, and a leading member of the Curia, Nikolaus Schonberg, archbishop of Capua, actually wrote to Copernicus in 1536 to "earnestly and repeatedly" beg him to send whatever he published. Schonberg went so far as to offer the services of a copyist.

The views stated by Copernicus in the preface to *De revolutionibus* are so clear as to leave no doubt of his position. Although the work is dedicated to the pope, Copernicus brashly denied the relevance of theology to astronomy and went on to proclaim the autonomy of his discipline in the stirring phrase, "mathematics is for mathematicians." Theologians who forgot this were likely to become figures of ridicule.

Those who might judge his work on the basis of Scripture he took no account of. "I consider their judgment rash and utterly despise it," he said. As a case in point Copernicus recalled the figure of Lactantius, a fourth-century father, who had quoted verses from Isaiah and St. Paul to demonstrate that the earth was not spherical. Copernicus dismissed him as "childish." In some cases, such as the ecclesiastical calendar, he said it was the Church which had to wait for astronomers to work out "the lengths of years and months and the motions of the Sun and Moon . . . with sufficient exactness." He concluded the preface by leaving his work "to the judgment of learned mathematicians," among them, presumably, Paul III.

Copernicus in fact made no secret of the cause of his delay, and in describing his dilemma he wrote, "I hesitated long whether, on the one hand, I should give to the light these my Commentaries written to prove the Earth's motion, or whether, on the other, it were better to follow the example of the Pythagoreans and others who are wont to impart their philosophic mysteries only to intimates and friends, and then not in writing but by word of mouth." If he did choose to publish he would, he realized, expose himself to "scorn . . . on account of the novelty and incongruity" of the theory and be despised "by such as either care not to study aught save for gain . . . or yet by reason of the dullness of their wits are in the company of philosophers as drones among bees."

Some independent corroboration confirms that Copernicus did indeed hold these views. Rheticus in *Narratio prima* noted that Copernicus was guided by the "Pythagorean principle . . . that philosophy must be pursued in such a way that its inner secrets are reserved for learned men, trained in mathematics" (Rosen:193). If such views are coupled with a genuine distaste of controversy, no further reasons need be sought to explain Copernicus's delay – and such delays, as we shall see, are by no means uncommon among major thinkers.

The solution of one problem raises another. After delaying thirty years, why did the sixty-eight-year-old Copernicus finally agree to publishing? Copernicus also addresses this question in the preface. "My misgivings [were overcome by friends who] urged that I should not any longer, on account of my fears, refuse to contribute the fruit of my labours to the common advantage of those interested in mathematics. They insisted that though my theory . . . might at first seem strange, yet it would appear admirable and acceptable when the publication of my elucidatory comments should dispel the mists of paradox. Yielding then to their persuasion I at last permitted my friends to publish that work which they have so long demanded." Most pressure came from Tiedemann Giese, bishop of Kulm and an old friend. The point is confirmed by Rheticus, who described in detail the pressure Giese put on him and some of the arguments put to him.[5]

Unfortunately, this is an insufficient explanation. If Giese had been pressuring Copernicus to publish for a number of years, why was the agreement that had been withheld for so long finally granted? The answer, no hint of which can be found in *De revolutionibus*, must lie with the 1539 appearance in Frauenberg of the striking figure of Rheticus (1514–76), a young mathematician from Wittenberg.[6]

Since 1536 Rheticus had been professor of mathematics at Wittenberg, the university of Luther and Melanchthon and at the heart of German Protestantism. He and a colleague, the mathematician Erasmus Reinhold (1511–53), compiler in 1551 of the first Copernican tables, had heard vague, unpublished rumors of the new heliocentric system. They agreed that the Protestant Rheticus would seek details in Catholic Frauenberg. In a manner characteristic of a Renaissance humanist, Rheticus arrived unheralded but bearing precious gifts: editions of Euclid and Ptolemy, works on trigonometry by Regiomontanus and Apianus, and *Optics* by Vitelio.

Copernicus must have found Rheticus acceptable, for he immediately made available to him the manuscript of *De revolutionibus*. Expecting perhaps to find a brief monograph, Rheticus examined a work comparable in scope, detail, and depth to *The Almagest*. Furthermore, its author allowed Rheticus to study it and to make known its main contents. Rheticus probably arrived in Frauenberg some time in May 1539; by 23 September he had mastered the manuscript sufficiently to have written a substantial monograph of his own, *Narratio prima de libris revolutionibus*, on its contents. The monograph was printed in Danzig in February 1540 and proved popular enough to require a second edition soon afterwards (Basel, 1541).[7]

Nowhere in *Narratio prima* is the name of Copernicus mentioned; throughout the text he is referred to simply as "my teacher." But little anonymity could have been gained by this ploy, for the seventy-six-page treatise carried a long title: "A First account of the Book of Revolutions by the most learned and most excellent mathematician, the Reverend Father, Dr Nicolas of Torun, Canon of Ermland." Not even this slight cover is retained in the Basel edition, where the name "Doctoris Nicolai Copernici Torunnaei Cannonici Vuarmaciensis" has replaced the "Doctor Nicolas" of Danzig. It is widely held that *Narratio prima* was a kite flown to convince Copernicus that ridicule and scorn would not necessarily result from the promulgation of his heliocentric views. The success of the minor work, it was hoped, would encourage him to publish his major work. While no hard documentary evidence favors this view, it is supported by the chronology of events.

After *Narratio prima* had been seen through the press in Danzig, Rheticus returned to Wittenberg to attend to his academic duties. By summer 1540, he was back in Frauenberg, and if a decision to publish *De revolutionibus* had not already been made, it must have occurred soon after his return. Rheticus remained with Copernicus until August 1541, during which period he made a fair copy of the manuscript. Academic duties called him to Wittenberg again, and not until May 1542 was he free to supervise the printing of *De revolutionibus*. It was appropriate that he chose Nuremberg, the city of Regiomontanus, in which to print the work, and Johan Petreius, who was experienced in mathematical printing, to undertake the task. Since Rheticus assumed a new post at Leipzig in about November 1542, he could not remain in Nuremberg until the job was finished, and instead recruited Andreas Osiander (1498–1552), a leading Nuremberg Lutheran who had corresponded in 1541 with both Rheticus and Copernicus on issues connected with astronomy, to replace him.

The task was completed in May 1543 by which time Copernicus, who had suffered a stroke in the latter part of 1542, was partially paralyzed and a dying man, confined to bed. There is a tradition that a copy of *De revolutionibus* reached Frauenberg on 24 May, the day of his death. Giese described the final scene in a letter to Rheticus: "For many days he had been deprived of his memory and mental vigour; he only saw his completed book at the last moment, on the day he died" (Koestler:189).

Examination of *De revolutionibus* (henceforth *DR*) reveals two particular features which shock and bemuse modern readers as much as they did Copernicus's contemporaries. The first is one of omission. In the preface, Copernicus mentioned those who had pressed him to publish. Two are singled out: Schonberg and Giese. Yet the name of Rheticus occurs nowhere, neither in the preface, nor in any other place in the book. Giese was embarassed by this omission and quickly wrote Rheticus, trying to pass it off as an oversight of a dying man. "Truly this was not due to indifference towards thee," Giese assured Rheticus, "but to his clumsiness and inattention; for his mind was already rather dulled, and paid . . . scant attention to anything not pertaining to philosophy. I know very well how highly he esteemed thy constant helpfulness and self-sacrifice" (Koestler:177). Rheticus was unlikely to have found Giese's explanation plausible, for he, or anyone who read the preface, could see that such a document could have come only from someone in full control of his mind.

Another suggestion attributes Copernicus's omission to political restraints. Koyré (1973:92) states "it is quite obvious that Copernicus could not mention colloboration with the Protestant Rheticus, when he was dedicating his book to Pope Paul III." If the point was so obvious why did not Giese offer it in explanation to Rheticus? Further, if contact with Protestants could not be acknowledged, why was a Protestant printer used and another Protestant, Andreas Osiander, chosen to see *DR* through the press? Koestler has pointed out just how free the humanist scholars were at that time. Just before Rheticus's arrival in Frauenberg a decree had banned all Protestants from the bishopric, and the following year a second decree banned all their writings. Yet Rheticus seems to have been free to come and go as he pleased (Koestler:158). In the following century Kepler, a Lutheran, found it equally possible to work and publish in Graz, a Catholic area – while other Protestants were being evicted – because the archduke was pleased with his discoveries (Koestler:282). In any case, Copernicus, whose preface claimed that "mathematics is for mathematicians" and was independent of theology, was free to mention any mathematician he liked.

If the reasons for Rheticus's omission from *DR* was neither politics nor an absence of mind, could it have been a deliberate snub? No hard evidence supports his view. Although Copernicus was not anxious to publish *DR* himself, he need not necessarily have felt gratitude to the amanuensis who undertook the task; he could well have felt that the attention bestowed on Rheticus for *Narratio prima* was compensation enough for his labor. Few scholars appreciate an amanuensis, however helpful and creative he may be. Doubts about an author's title to his property may emerge, and fears that others may overestimate the value of the help received. If the work is a success, the amanuensis may be seen as otiose, a mere messenger or clerk; if the work fails, the amanuensis may find himself blamed for, in some subtle way, having diluted the brilliance of the original. Few authors are as honest as Louis Leakey who, in the preface to his *Unveiling Man's Origins* (1969), noted that his coauthor, Mrs. Goodall, had been responsible for the "actual writing of Chapters I-XII." The book contains thirteen chapters. A more common response and the one perhaps adopted by Copernicus is simply to ignore the amanuensis.[8]

The second surprise of the published *DR* is one of addition. Left in charge of the work by Rheticus, Osiander felt free to add a brief one-page note to the text entitled "To the Reader Concerning the Hypotheses of This Work." He did not feel free enough, however, to indicate that the note came from him and not from the author. In it, he consoles those who may feel alarmed by the notion of the earth's motion with the thought that "these hypotheses need not be true nor even probable; if they provide a calculus consistent with the observations, that alone is sufficient." Constructions of the work are freely described as "absurdities" and inconsistent. But such judgments are misguided, for the point of the constructions is simply "to provide a correct basis for calculation." But, Osiander warned, "let no one expect anything certain from astronomy, which cannot furnish it, lest he accept as the truth ideas conceived for another purpose, and depart from this study a greater fool than when he entered it" (Rosen:25).

The doctrine of instrumentalism is not totally without merit and has been held at various times by the philosopher George Berkeley (1685–1753) and the modern physicist Ernst Mach (1838–1916). Indeed, much of the modern interpretation of quantum mechanics has been, at a very fundamental level, instrumentalist. A more important question is whether Copernicus knew of the note and agreed with it. He had corresponded with Osiander in 1541 and been told by him that astronomical hypotheses were not "articles of faith but bases of computation." Copernicus's side of the correspondence has not survived, but there is no reason to suppose he agreed with him. The heliocentrism of Copernicus was a firmly held belief. Giese clearly recognized this fact, and immediately wrote to the magistrates of Nuremberg to demand the removal of the fraudulent passage from *DR*. The demand was passed on to Petreius, who indifferently replied that it was not his responsibility. Osiander, at the insistence of Rheticus, publicly admitted his authorship, but no sign of this was carried in the second edition (Basel, 1566) of *DR*. Another who recognized the fraud was Kepler, who noted in 1609, "Do you wish to know the author of this fiction...? Andreas Osiander is named in my copy... Andreas regarded the Preface as most prudent... and placed it on the title page of the book when Copernicus was either already dead or certainly unaware of what Osiander was doing" (Koestler:173). Kepler's announcement seems to have gone largely unnoticed; in the 19th-century historians of astronomy such as Delambre still continued to write as if Osiander's contribution had come from Copernicus himself.

The misunderstanding was to have unfortunate consequences. To the Holy Office, a neutral instrumentalist account of science offered many advantages. Scientists could be granted complete freedom to work and write as they liked; as long as it was understood that their conclusions were "hypothetical" and not "absolute" they could never come into conflict with Scripture. The neat solution at a stroke guaranteed the integrity of Scripture from any future scientific advance. Cardinal Bellarmino made clear that the Church would stand firm against any threat to violate this doctrine. When such a threat came from Galileo, Bellarmino in 1615 issued a classic statement of what he considered to be the Copernican position: "It seems to me that your reverence and Signor Galileo act prudently when you content yourself with speaking hypothetically and not absolutely, as I have always understood that Copernicus spoke. To say that on the supposition of the Earth's movement and the Sun's quiescence all the celestial appearance are explained better than by the theory of eccentrics and epicycles is to speak with excellent good sense and to run no risk whatever. ... But to want to affirm that the Sun, in very truth, is at the center of the universe... is a very dangerous attitude and one calculated... to injure our holy faith by contradicting the Scriptures" (de Santillana:99). To the intense irritation of the Church this argument, supported by the apparent authority of none other than Copernicus, was to prove more attractive to theologians than to scientists. Galileo forced a major confrontation with the Church on this issue and was finally forced to give way but only under threat of the stake and not, as Bellarmino hoped, by scientists realizing the true status of their own theories.

One further addition to *DR*, presumably by the publisher, can be seen on the title page and somehow seems to represent all the confusions, hesitancies and inconsistencies in both the man and his work. Publishers in the 16th century had begun to insert at the front of their books a little advertising. The title page of *DR* carries the words: "In this newly produced and published work, learned reader, you have..." and goes on to draw attention to the "remarkable hypotheses" and "convenient tables" included. In a tone more suitable to a TV commercial than a work on mathematical astronomy it concludes by exhorting the customer, *"Igitur eme, lege, fruere"* ("Therefore buy, read, enjoy"). The last word, inevitably, is with the elusive presence of Copernicus. Directly under the Latin blurb is the starker injunction of Plato, in direct conflict with the happy invitation of the publisher: "Let none enter who knows no geometry."

"De revolutionibus" – Content and Argument

At the heart of *DR* and the Copernican revolution lie the two assumptions that the sun, not the earth, is at the center of the universe, and that the earth revolves daily about its axis and orbits annually around the central sun. Why did Copernicus feel compelled to assert such apparent absurdities against the received wisdom of the centuries?

We find it surprising that one reason *not* advanced by Copernicus was the failure of the Ptolemaic system correctly to predict planetary positions. The modern mind would consider this the first quality of an astronomical system to be evaluated. Copernicus was well aware that Ptolemaic predictions in the early 16th century were highly inaccurate. His own copy of the Alfonsine tables for 1492 carry the annotation "Mars superat numerationem plus quam gr. ij / Saturnus superatur a numeratione gr. 1½" ("Mars surpasses the numbers by more than 2 degrees; Saturn is surpassed by 1½ degrees"). Copernicus was referring to a conjunction of all the planets expected in 1504 in the sign of Cancer. The conjunction of Saturn and Mars – the time they would have the same longitude – had been predicted for 18 March. Copernicus noted that on that day Saturn was 1½ degrees behind the point of conjunction, while Mars had advanced 2 degrees beyond. Modern calculations show this was not a minor error, detectable only by computers working to several decimal places. The conjunction had actually taken place some *ten days* before the predicted event (Gingerich 1972:89).

Copernicus was perhaps wise not to pay too much attention to the accuracy of astronomical systems. The margins of error were too great and too numerous to be upset by inaccuracies or influenced by successes. In actual fact, the Prutenic tables in 1551, based on the Copernican system, were soon to be shown no more reliable than the Alfonsine tables. When Tycho Brahe looked for the conjunction of Saturn and Jupiter predicted for 1563 he found the Prutenic tables already inaccurate by a day or two. This shock prompted the sixteen-year-old Tycho to devote his career to the improvement of astronomical observation.

In reality Copernicus was less impressed by astronomical observation than by arguments of another kind. In all his writings and notes only about seventy of his own observations are recorded, and some of these are merely scribbled in the margins of books. *DR* contains only twenty-seven of his own observations. Copernicus saw himself as a natural philosopher and mathematician and was consequently swayed by arguments based on philosophical and mathematical principles. Observation had its use, but a secondary one. Serious issues like the structure of the universe were first worked out in terms of basic philosophical and mathematical principles; observation was employed afterward to provide the parameters for a structure created independently on other grounds. As such, these observations were relatively unimportant; any errors detected would lead to adjustments in the parameters, rather than to a rejection of the structure.

The starting point for Copernicus was not the heavens but *The Almagest* and its successors. His objection to the pure Ptolemaic system was that it had failed to abide by its own principles and reduce all heavenly motions to uniform, circular motion. As early as *Commentariolus* Copernicus had pointed out that Ptolemy allowed planets to move "with uniform velocity neither on its deferent nor about the center of its epicycle" (Rosen:57). He was of course referring to the equant. The point was repeated in the preface of *DR*, where it was noted that many astronomers had made "admissions which seem to violate the first principle of uniformity in motion." Far from being a revolutionary, Copernicus was calling for a return to a tradition already compromised by Ptolemy. It was for the purity of a system more ancient than Ptolemy's that Copernicus aspired and his willingness to innovate was controlled by this aim.

He could point to those who had sought to achieve this aim within the Ptolemaic framework and could note their failure. No acceptable geocentric system commanded broad agreement; instead, a variety used "neither the same principles and hypotheses nor the same demonstrations of the apparent motions and revolutions."[9] He noted that in any one particular work the structure was never presented as a harmonious whole. Acceptable constructions were proposed for each planet in turn, yet because of their dimensions and orbits it would be impossible to place them together in a single system. It was as if, Copernicus declared in a striking image, "an artist were to gather the hands, feet, head and other members for his images from diverse models, each part excellently drawn, but not related to a single body . . . The result would be a monster rather than a man."

Confronted by a multitude of incompatible geocentric "systems," none of which fully accepted the principle of uniform circular motion, Copernicus took the obvious step for a humanist scholar: he turned to ancient texts to see if it had ever been supposed "that the motions of the spheres were other than those demanded by the mathematical schools." He then discovered several scientists of antiquity – Hicetas of Syracuse and Philolaus the Pythagorean of the fifth century B.C., and Heraclides of Pontus in the fourth century B.C. – who had apparently spoken of a moving earth. To his surprise he found their "absurd" assumption made the whole system work, that "the heavens themselves became so bound together that nothing in any part thereof could be moved from its place without producing confusion of all the other parts and of the Universe as a whole." It remained to work out the details.

First it was necessary to persuade people of the possibility of terrestial motion. It is natural to suppose that the earth is stationary and that the heavens revolve around it. One of the most learned scholars of the 16th century, Jean Bodin (1530–96), who was referred to by his contemporaries as the "modern Artistotle," stated the arguments thus: "No one in his senses . . . will ever think that the Earth, heavy and unwieldy from its own weight and mass, staggers up and down around its own center and that of the Sun; for at the slightest jar of the Earth, we would see cities and fortresses, towns and mountains thrown down . . . also if the Earth were to be moved, neither an arrow shot straight up nor a stone dropped from the top of a tower would fall perpendicularly, but either ahead or behind" (Kuhn:190). Other, more sophisticated objections were expresed in terms of Aristotelian physics, but Copernicus knew that to convince people he must first dispose of the simplest of all objections proclaimed by countless scholars from Ptolemy himself to Bodin.

The first objection Copernicus used against itself. If the buildings of the earth are in danger of collapse from any supposed daily rotation, then should "we not fear even more for the Universe, whose motion must be as much more rapid as the Heavens are greater than the Earth?" (*DR* I:8). To argue against projectiles and falling bodies, Copernicus deployed replies developed by philosophers in the previous century. It was only necessary to suppose a "double motion of objects . . . rectilinear and circular." As the earth moves it carries around with it not only the oceans and the atmosphere, but "whatever else is thus joined with the Earth." A stone thrown in the air will fall down to earth because of its weight; it will also be carried with the earth on its circular path while falling to the ground.

Such arguments do not demonstrate that the earth does move; they merely dispose of a number of common objections against the proposal. Terrestial motion was no easy matter to prove before the age of rockets. Copernicus failed to do so, as did Galileo a century later. All Copernicus could do was to answer all objections against its inherent plausibility and to show the usefulness, if not the truth, of the assumption.

There remained the independent issue of the heliocentric nature of the universe. Arguments concerning the earth's motion were rare, but those capable of proving heliocentricity were nonexistent. Aristarchus of Samos, in the third century B.C., proposed such a structure, but his reasons have not survived. When Copernicus writes of the central position of the sun, with no arguments to advance, he resorts to rhetoric of a peculiar kind: "In the middle of everything stands the Sun. For in this most beautiful temple, who could place this Lamp in any other better place than one from which it can illuminate all other things at the same time? This Sun some people call appropriately the Light of the World, others its Soul or its Ruler. Trismegistus calls it the Visible God, Sophocles' Electra calls it the All-Seeing. Thus the Sun, sitting on its Royal Throne, guides the revolving family of the stars" (*DR* I:10).

Nor was the explanation any more sound in *Narratio prima*, in which Rheticus revealed as a basis for his master's cosmology the analogy that "in human affairs the emperor need not himself hurry from city to city in order to perform the duty imposed upon him by Good . . . the heart does not move to head or feet or other parts of the body to sustain" (Rosen:139). The sun, too, as the grandest body of the universe, had no need to move around its realm to assert its power.

Some see in these passages a connection between the Hermetic tradition of occultism, the Cabala and neo-Platonism so evident in Renaissance thought and the birth of modern science. The Trismegistus referred to by Copernicus was Hermes Trismegistus, believed to be an Egyptian contemporary of Moses who received from God revelations about the physical world comparable with the moral insights God granted to Moses. Hermes recorded the divine truths in a number of treatises known to scholars as *Corpus Hermeticum*. One of the great moments of Renaissance scholarhip occurred in 1463 when a Greek manuscript of fourteen of the fifteen Hermetic texts arrived in Florence. Cosimo de Medici (1389–1464) instructed Marsilio Ficino (1433–99), who was already working for him on a translation of the Platonic corpus, to drop everything and produce a Latin edition of the Hermetic texts as soon as possible. It is to be hoped that Cosimo was suitably impressed before he died shortly afterward, for *Corpus* stirs little excitement in many of its modern readers. Some find stimulating its implicit suggestion that within it lie truths and powers available only to the initiated; others consider it coy, pretentious and irritating. Nothing is ever revealed, although much is promised; rich in generalities, it is almost totally lacking in detail. It is not surprising that others filled the gaps and proposed techniques the exercise of which would attain the truths and powers described in *Corpus Hermiticum.*[10] While the connection between *Corpus* and the rise and spread of ritual magic in Europe was undoubtedly close, did it have a comparable effect on the emergence of science?

It is difficult to find any significant influence of the Hermetic tradition on Copernicus's astronomy. He would undoubtedly have considered Hermes an important thinker and would have further recognized *Corpus* as a repository of much wisdom. It is possible to see the reference to Trismegistus in *DR* as simply a trivial piece of rhetoric. Unable to defend his heliocentric position with astronomical arguments and unwilling to propose it as a purely speculative hypotheses, Copernicus resorted to a little intellectual padding. Certainly nothing but disappointment awaits those who look for Hermetic secrets in *DR*. They will find only trigonometry, geometry, and mathematical astronomy.

The interesting point about these early sections of *DR* is how little evidence there really was for the Copernican position, and that little had been available for centuries. Copernicus was in no better position than Ptolemy had been to construct a heliocentric system; the constructions used by him were basically Ptolemaic; and so was the mathematical framework. In principle the Copernican position could have been developed any time in the previous millenium. Copernicus differed from most other astronomers in his exceptional dislike of constructions that violated the original inspiration of uniform circular motion. Others were prepared to compromise; Copernicus refused and created modern astronomy in the process.

A great virtue of the Copernican system is its considerable explanatory power. Ptolemy can indeed explain much by providing appropriate constructions, but they are only constructions imported to deal with anomalies. With Copernicus the anomalies no longer seem anomalous and flow as necessary consequences of the original assumptions. Consider the well-known fact that, unlike the other planets, Venus and Mercury never move far from the Sun: Mercury is never more than 28°, and Venus's elongation – to use the technical term – does not exceed 45°. The other planets can achieve a quadrature of a full 90° away from the sun. An examination of the Ptolemaic diagram gives no hint why Venus and Mercury should be so restricted. In the Copernican diagram, the facts of elongation and quadrature fall automatically into place.

In fig. 5:1, using a Copernican diagram, an observer on Earth *(E)* moving around the central Sun *(S)* will, at the appropriate times, see the superior planet Jupiter *(J)* some 90° from the Sun; the inferior planet Venus *(V)* can never be seen further from the Sun than the angle *VES*.

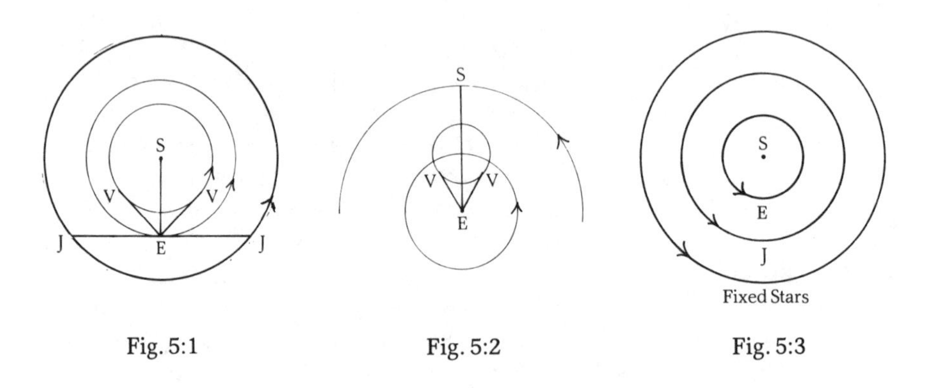

Fig. 5:1 Fig. 5:2 Fig. 5:3

To obtain similar results in Ptolemaic astronomy, an epicycle must be especially introduced, as in fig. 5:2. Here, if the deferent rotates about Earth just once a year, then the center of the Sun and the center of the epicycle will remain aligned with Earth in the manner represented in fig. 5:2. In this alignment, Venus can never stray further from the Sun than the angle *VES*.

While the same data can be represented in the two systems, as above, it requires little imagination to appreciate the attraction of the diagram in fig. 5:1 to that in fig. 5:2 and the power of the Copernican system, even at a distance of more than four hundred years. Thus the retrograde movement of the planets that caused so much trouble to Ptolemy and appears totally perverse on his assumptions emerges in a quite natural manner in the simple Copernican diagram, seen as fig. 5:3. Here the orbits of Earth, Jupiter, and the fixed stars are represented as they circle the Sun. It is obvious from the diagram that Earth in its smaller orbit will from time to time overtake the outer planet Jupiter. The effect will be to make Jupiter appear to a terrestial observer regularly to stand still and move backwards before continuing on its normal orbit. To a generation familiar with the illusory effects produced from racing cars passing and being passed by each other, the notion is elementary. From the diagram it is also obvious why an outer planet like Jupiter must be at opposition, closest to Earth, before it can be observed to retrogress. As has been seen the same planetary motions can be represented by Ptolemy with the use of an epicycle of appropriate dimensions and velocities. It is difficult, however, not to see the Copernican construction as an astronomical reality and that of Ptolemy as merely geometry.

Another example of how the assumptions of Copernicus could yield elementary insights into the constitution of the universe denied to Ptolemy is provided by discussions in *DR* on planetary distances. Astronomers of antiquity and the medieval period found it impossible to agree on such a simple detail as the sequence of planets. No one disagreed over the thirty-, twelve-, and two-year orbits of Saturn, Jupiter, and Mars respectively. The longer the orbital period, the further the planet from the center of the universe. The relative positions of Mercury, Venus and the Sun, however, led to prolonged disagreement. Some astronomers placed Mercury and Venus, in the jargon of the period, "below" the Sun or between the Sun and the Moon; others took the two planets to be "above" the Sun, that is, between the Sun and Mars.[11] It is quite extraordinary that the skillful astronomers of antiquity who detected such subtleties as the precession of the equinoxes could not agree on whether Venus or the Sun was nearest to Earth. For Ptolemy the sequence was Earth, Moon, Mercury, Venus, Sun; for Plato, Eudoxus, and Aristotle, the sequence was Earth, Moon, Sun, Venus, Mercury. No better reason for the Ptolemaic arrangement seems to have been available than that it was more harmonious, balancing the Sun with three planets on either side. The thought of so much empty space between the moon and the sun seemed to many to be absurd.

As Copernicus pointed out in *DR,* while all could agree about the superior planets, "Opinions differ as to Venus and Mercury which unlike the others, do not altogether leave the Sun" (I:10). But by accepting the principle "that the periodic times are proportional to the sizes of the Spheres" Copernicus was able to list the planetary order familiar to us today.

* * *

The contents of *DR* can now be briefly summarized. The main body of the work is divided into six books.

Book I consists of fourteen chapters which, for the most part, follow closely the design of *The Almagest*; indeed, for the first few chapters the casual reader could be forgiven for thinking he was reading *The Almagest.*

Chap. 1 argues that the universe is spherical on the grounds that the sphere is the most perfect shape of all and is the most capacious of shapes; the universe is likely, moreover, to be shaped on the pattern of its more perfect parts like the sun, which can readily be observed to be spherical.

Chap. 2 argues that the earth is also spherical, since "Italy does not see Canopus, which is visible in Egypt; and Egypt sees the last star of the River [Eridanus], which our region in a colder zone does not know."

Chap. 3 identifies the earth with its seas and oceans as a single sphere.

Chap. 4 continues the Ptolemaic theme and declares the motion of the heavenly bodies to be "uniform, circular, and perpetual, or composed of circular motions." The arguments are simple: rotation is natural for a sphere, and any observed irregularity in the heavens is apparent. Further examination will show it to be compounded of circular motions.

Chap. 5 for the first time introduces a point not found in Ptolemy. The earth rotates about its own axis and orbits the center of the universe. The point, Copernicus insisted, was far from new and could be found in the works of several ancient authors.

Chap. 6 declares, as with Ptolemy, the immensity of the heavens compared with the size of the earth.

Chap. 7 raises the question of why the ancients insisted on the immobility and central position of earth. The two main arguments were well-known: falling objects could not "reach their appointed place vertically beneath since the Earth would have moved swiftly away from under them"; and the speed of the earth rotating once in twenty-four hours would be vast enough to produce its own disintegration.

Chap. 8 argues for the insufficiency of these arguments. To the argument from falling bodies was added the claim that objects could possess a double motion, a circular aspect belonging to all bodies connected with the earth as they moved around with it, and a rectilinear aspect as they were dropped from a height. The second objection was neatly reversed, for if Ptolemy feared for the stability of an earth that rotated once in twenty-four hours, "should he not fear even more for the Universe, whose motion must be as much more rapid as the Heavens are greater than the Earth?"

Chap. 9 attributes further motions to the earth. Once one accepts that it can rotate on its own axis, why not also concede that it can orbit a particular center? If this were true, Copernicus points out, it would explain why the planets seem to vary in their distances from the earth.

Chap. 10 discusses the order of the heavenly spheres, and only here is the position of the sun at the center of the universe proposed. The argument is apparently Hermetic and has been discussed elsewhere.

Chap. 11 describes the orbital motion of the earth and shows how the sun will appear to a terrestrial observer.

Chaps. 12-14 introduce technical astronomy after some mathematics in which Copernicus shows how to derive the lengths of chords of a circle (Chap. 12), sides and angles of a plane triangle (Chap. 13), and of a spherical triangle (Chap. 14).

Book II continues the mathematical foundations and discusses problems in spherical astronomy. The circles of the sphere are dealt with, as are how to calculate tables of declination, right ascension, risings and settings. Here (II:14) Copernicus introduced his stellar catalogue. Taking the same stars as those used to illustrate Ptolemy's approach, Copernicus lists stars in the constellation Perseus as follows:

	Longitude	*Latitude*	*Magnitude*
The bright star in the left hand and in the head of Medusa	23° 0′	N 23° 0′	2
In the same head, the one to the east	22° 30′	N 21° 0′	4
In the same head, the one to the west	21° 0′	N 21° 0′	4
Still farther west of the foregoing	20° 10′	N 22° 15′	4

Copernicus had clearly failed to improve the clumsy means of identifying particular stars introduced by Ptolemy. An examination of his figures with those given by Ptolemy shows a complete agreement on latitude and magnitude; the figures for longitude differ by a constant 6° 40′ because, for reasons to be explained in the following section, Copernicus measured longitude not from the equinoctial point, as Ptolemy had done, but from the bright star "at the head of Aries," *υ Arietis* or *Mesarthim.*

Book III deals with equinoxes and solstices, precession, the length of the solar year, and the apparent motion of the sun. In some ways it is one of the most interesting sections of the entire work, for it shows how inaccurate and misguided Copernicus could be on quite fundamental issues. Classical astronomers had correctly described the phenomena of the precession of the equinoxes and the obliquity of the ecliptic, yet there was some uncertainty about the precise values to be assigned to them: Eratosthenes gave a value of 23° 51′ 20″ to the ecliptic's obliquity, and Hipparchus calculated the annual rate of precession to be about 45″. Compared with modern figures of 23° 26′ 54″ and 50″, respectively, their results were impressive. These were the first in a long line of estimates offered over the centuries by the astonomers of antiquity and of medieval Islam. Some were clear improvements, others were wildly inaccurate. Some of the more important published figures are listed on page 112.

Copernicus calculated from these figures that while the precessional rate before Ptolemy had been 1° per 100 years, it had increased to a rate of 1° per 65 in the centuries between Ptolemy and al-Battani and had then slowed down to a rate of 1° per 71 years between the ninth century and his own time. Similar conclusions were drawn about the angle of the ecliptic's obliquity. To explain these results Copernicus developed a complicated theory that neither the rate of precession nor the angle of obliquity were

Century	*Precession*		*Obliquity*
	Annual	*Time for 1°*	
Eratosthenes (3rd B.C.)			23° 51′
Hipparchus (2nd B.C.)	45″	80 years	
Ptolemy (2nd A.D.)	36″	100 years	23° 51′
al-Battani (9th A.D.)	54″	66 years	23° 36′
al-Zarkali (11th A.D.)	37″	97 years	23° 34′
Prophatius (12th A.D.)			23° 32′
Copernicus (16th A.D.)	50″	72 years	23° 28′

constant; that over a period of 3,434 years the obliquity of the ecliptic would vary between the limits of 23° 52′ and 23° 28′. At the same time the equinoctial point, Copernicus argued, did not move around the ecliptic at a uniform rate. His figure of 50″ per annum was a *mean* rate around which the actual rate varied throughout a full cycle lasting 1,717 years.

The price of such theorizing for Copernicus was high. It was a simple matter to state the essence of the theory in words, but to fit it into an already restricted geometrical scheme could only have the effect of throwing away all the gains obtained by assuming the earth moved. All the spheres saved by dropping the idea of a static earth were needed to explain the variable precession rate and changes in the obliquity. It is ironic, of course, that none of this complex theorizing was necessary, since it was caused by several centuries of errors of observation and calculation by others. It is a matter of some surprise that while Copernicus felt free to reject the theoretical part of *The Almagest,* he never seemed to consider that its observations could be just as unreliable. For Copernicus, a mathematical (as opposed to an observational) astronomer, observations were not doubted, carefully scrutinized, or checked; rather they were elements incorporated into geometrical patterns of ever increasing sophistication, additional variables in yet one more mathematical structure.

Books IV-VI deal with Copernicus's planetary theory. Book IV is devoted to the moon, and Books V and VI to the motions of the planets in latitude and longitude, respectively. The actual constructions used by Copernicus to describe planetary orbits are listed below, compared with those used by Ptolemy.

In concluding this section, did the Copernican adoption of a central sun and moving earth lead to any significant simplification? For many years it was claimed that the great virtue of the Copernican system was that it finally broke away from the endless circles and epicycles characteristic of the Ptolemaic tradition. At a stroke, so the argument went, Copernicus reduced the system of absurd complexity to one of elegant simplicity by his adopting the thesis that earth was neither immobile nor at the center of the universe. Koestler exploded this particular myth when he pointed out that "contrary to popular, and even academic belief, Copernicus did not reduce the number of circles, but increased them (from 40 to 48)" (1964:

Planet	*Copernicus*	*Ptolemy*
Earth	Movable eccentric	Stationary
Moon	Epicycle on epicycle	Epicycle and movable eccentric
Sun	Stationary	Fixed eccentric
Mercury	Epicycle and movable[12] eccentric	Epicycle and movable eccentric
Venus	Movable eccentric	Epicycle, fixed eccentric and equant
Mars	Epicycle and fixed eccentric	Ditto
Jupiter	Ditto	Ditto
Saturn	Ditto	Ditto

195). The eighty circles supposedly required by Ptolemy turned out on examination to be no more than forty. It was true that the assumption of the earth's annual and diurnal motion meant that fewer spheres would be needed; any saving here was, however, more than compensated for by the new need for circles to account for oscillations in precessional rate and angle of obliquity needlessly introduced by Copernicus. Nor did his refusal to use the equant lead to economy.[13]

"De Revolutionibus" – Reception

De revolutionibus was far from popular, since its complexity prevented it from competing with earlier texts like Sacrobosco's *Sphere* and later works in the same tradition. Koestler's judgment, that it was and is an "all-time worst seller" is somewhat extreme, however. It had only four reprints in three hundred years – although eleven more were issued in the last century. But the real measure of a work like *DR* is not how often it is published, but the extent and rapidity with which other authors took the material, reshaped it, and reissued it in a more comprehensible form, which is frequently the fate of an advanced, technical, and/or expensive treatise.[14] The same pattern occurred with the other major technical classic of 1543, Vesalius's *De fabrica,* which will be discussed in the next chapter.

A number of interested scholars have tried to determine the extent of the 16th-century readership of *DR*. Gingerich has done most of the work in this fascinating field. He speculated recently that "there are probably more people alive today who have read *De revolutionibus* carefully than there were in the entire sixteenth century." He then enumerated who they might have been: Rheticus; Rheinhold; Brahe; Kepler; Michael Maestlin (1550-1631), professor of astronomy at Tubingen and teacher of Kepler; Johann Schoner, Nuremberg astronomer and teacher of Rheticus; and

Christopher Clavius (1537-1612), a Jesuit mathematician at the College of Rome. Further research revealed two more 16th-century readers: Johannes Stadius, an ephemeris maker; and Caspar Peucer, son-in-law of Melanchthon and Reinhold's successor at Wittemberg. In addition to "careful" readers, capable of annotating the text, Gingerich also found "a fairly wide circle of casual readers, much larger than had been generally supposed" (Gingerich 1973: 99).

Gingerich extended his investigation, and by 1979 he had details of some two hundred and forty-five copies of the first edition of *DR*. He put his own estimate of the size of the edition at between four and five hundred copies, rather than the often-quoted figure of one thousand. "The great Copernicus chase," as Gingerich called his pursuit, produced some remarkable results. Many of the copies he examined were annotated, those of such workers as Rheticus and Kepler richly so. Others, more common, were minor, "rather uninspired, marginal tracks: catchwords that provide a running index. Usually such notes peter out after the first few chapters" (Gingerich 1979: 84). As Gingerich became familiar with more and more copies he discovered that the same sets of marginal notes reappeared in multiple copies. Reinhold's comments, for example, he found in more than a dozen copies. A unique network of dissemination for the difficult *DR* began to emerge. Gingerich thus supposed that "even if Copernicus's revolutionary new doctrine failed to find a place in the regular university curriculum, a network of University professors scrutinized the text and their protegés carefully copied out their remarks, setting the notes onto the margins of fresh copies of the books with a precision impossible by aural transmission alone" (Gingerich 1979: 84). The marginalia revealed several distinct networks of dissemination throughout Europe, which convinced Gingerich that his original suspicion of only a handful of 16th-century readers had to be rejected.[15]

Opposition to *DR* came from theological and the astronomical camps. The reception of the Protestant German establishment headed by Luther and Melanchthon is well-known; they even believed, remarkably inaccurately, that Copernicus must be the type of man who sought notoriety by arguing for absurd positions. Said Luther: "That is how things go nowadays. Anyone who wants to be clever must not let himself like what others do. He must produce his own product as this man does, who wishes to turn the whole of astronomy upside down." He continued: "But I believe in the Holy Scripture, since Joshua ordered the Sun, not the Earth, to stand still." Luther's authority, however, was extremely restricted, not even stretching into the center of Protestantism at the University of Wittemberg. There the two young professors of mathematics, Reinhold and Rheticus, not only accepted but actively campaigned for the new Copernican astronomy.

This rather literal interpretation of the Scriptures was also accepted by Rome. The Council of Trent (1545-63) had asserted that no interpretation of Scripture should be allowed that conflicted with the common opinions of the Church fathers. Galileo was later to argue that truths of Scripture and of nature could never come into conflict; apparent contradictions would be resolved by finding the "true sense" of the sacred passages which brought them into harmony "with those physical conclusions which have first been made certain and sure to us by manifest sense or necessary demonstrations" (Geymonat: 68). This view was eventually accepted by the Church in the 19th century.

But in the early 17th century, the Catholic Church took up Galileo's challenge and offered to legitimize Copernicanism "hypothetically" if not "absolutely"; it even conceded that it might rethink its position if "there was a real proof that the Sun is at the center of the Universe." The author of the Church's response, Cardinal Bellarmino, was aware – as was Galileo himself – that the Copernicans could not produce any such proof acceptable to the majority of their scientific colleagues. Consequently, in 1616, the Church could feel free to state its position. The Holy Office decreed "that the false Pythagorean theory of the motion of the Earth and the immobility of the Sun, which runs contrary to divine scripture and which Copernicus teaches in his book. . . has spread and been accepted by many. . . . Therefore, in order that such an opinion not continue to stalk the land and bring about the perdition of Catholic truth, the Congregation resolved that . . . *De revolutionibus* . . . is suspended until . . . corrected." This meant that no one was allowed to "print or publish" or "possess in any form or to read" *De revolutionibus* and other works favorable to Copernicus (Kesten: 375).

How effective was the Church's censorship? Its writ, of course, did not extend to Protestant countries, so it was possible to print a third edition of *DR* in Amsterdam in 1617. Nor did it stop Kepler from publishing an *Outline of the Copernican Astronomy* in 1618. Further proof of the restricted impact of the Church's censorship has been produced by Gingerich who examined some thirty copies of the second edition in Italian libraries. About sixty percent of them were lightly – almost formally – censored; for the most part, the copies were merely altered in about a dozen places, with such changes as the substitution of the term "hypothesis" where Copernicus had written "law." In reality there seems to have been no bar to the use of *DR* in Italy as long as the user was reasonably discreet and prepared to make a few minor changes in the text. What alarmed the Church were those individuals who, like Galileo, wanted not only to use the new astronomy but also to convert others to it.

After the issues raised by the trial of Galileo had died down, astronomers in the 18th century began to make representations to the authorities to withdraw the ban. In a series of apparently unrelated steps, the Church began the lengthy process of accepting the idea of the earth's motion. The first sign of a more tolerant attitude came in 1737 when Galileo's tomb in the Church of Santa Croce, Florence, was marked with a monument suitable to his status. This was followed by a decision in 1757 not to ban any further books simply because they supported Copernicus. Works already on the Index, however, remained there. The decision to remove them was taken in 1822, while the 1835 edition of *Index librorum prohibitorum* was the first not to contain the *DR* of Copernicus, the *Dialogues* of Galileo, the *Outline* of Kepler, and a host of minor texts. The first new edition of *DR* in more than two hundred years appeared in Warsaw in 1854; it was also the first to be translated into a modern language (Polish) and the first to appear in a Catholic country.

The response of astronomers was on the whole favorable, but it was not accorded with undue speed. It is difficult to generalize, for whenever substantial material was available astronomers seem to have adopted individual positions with numerous modifications to or reservations about the system of *DR*. In England, for example, the first reference was in *Castle of Knowledge* (1556) by Robert Recorde—the man who introduced the "=" sign into mathematics—who mentioned it in a dismissive reference rather than discussed it. The first discussion is found in a supplement to *Prognostication Everlasting* (1572) by Thomas Digges (1543-95) entitled "A Perfit Description. . . ," containing extensive extracts from Book I. Reprinted seven times before 1605, it must have been for many the only source for the actual words and arguments of Copernicus.

Where responses have been found, they were extremely variable. One of the first to support Copernicus in England was John Dee (1527-1608), mathematician and magus. In a preface to a 1557 *Ephemerides* Dee noted its dependence on the new tables of Copernican theory which he judged to be accurate. Although Dee possessed a copy of *DR* and might be presumed to have been a supporter of a heliocentric universe, he never actually came out in public support of Copernicus. The title page of Billingsley's Euclid, for which Dee wrote a famous preface in 1570, contains the figure of Ptolemy, not Copernicus. The attitude of another major English scientist, William Gilbert (1540-1603), author of *De magnete* (1600), was equally individual. He explicitly and publicly accepted the diurnal rotation of the earth (VI: 4), an explanation he thought fitted well with his claim that the earth was a giant magnet; the annual motion of the earth Gilbert simply ignored. It is curious how selective was the early support afforded Copernicus. An element of *DR* that appeared useful or attractive to another thinker was unhesitatingly adopted; the remainder of the theory was sometimes dismissed, but more commonly ignored.

An examination by Johnson (1958) of the main textbooks of the 16th century revealed that Sacrobosco's *Sphere* was the most common, with two hundred known editions and a "true figure that may well be twice that number" (Johnson: 294); it was too elementary to have dealt with Copernicus. Of the more advanced works the most popular was the 1570 commentary on Sacrobosco by Clavius, which was issued in about twenty editions over the following seventy-five years. The main theoretical problem tackled by Clavius and his contemporaries was not whether the universe was sun- or earth-centered, but whether it contained eight, nine, ten or eleven spheres. Spheres were clearly needed for the fixed stars and the seven planets; one who argued that these eight spheres were adequate was

Ricus in his *De moto octavae spherae* (1513). Most scholars, however, like Sacrobosco, demanded a separate sphere for the primum mobile, and others also required spheres to represent the obliquity of the ecliptic and trepidation. Clavius supposed eleven spheres, but all numbers between eight and eleven had their adherents. Maestlin, who favored Copernicus in his *Epitome astronomiae* (1582), worked in a traditional ten-sphere system.

The main difficulty in accepting the totality of the Copernican system was one of belief. Despite the abstract arguments developed by physicists, people found it difficult to believe in a moving earth. Presented with a hypothesis that ran directly counter to the experience and intuitions of everyday life, the usual response was total skepticism. Pages of complex physical reasoning and mathematical computation are unlikely to convince anyone, least of all physicists. In the same way it is widely known that as a consequence of Einstein's relativity theory it is possible for a man to travel for years in space only to find, on his return to earth, that he has aged less than his identical twin. I doubt, though, that this is sincerely believed by those lacking an understanding of the fundamentals of Einstein's physics. Authority and much else are apt to collapse before the overwhelming appeal of common sense.

One way out of such an impasse is convincing proof, expressible and comprehensible in everyday language and experience. Unfortunately for Copernicus and his followers, such proofs were hard to find and impossible to demonstrate. Most of the literate population of the 16th century were little concerned with the rate of precession or the nature of the solar apogee, but the motion of the earth did impress and excite them, and this they could be tempted to believe if it could be shown in some familiar way. Surely the rotation of the earth could not take place so discreetly as to be indetectable?

The one event that would clearly show the motion of the earth would be parallax, a technical concept based on familiar experience. Thus, if the earth did move around the sun the apparent change in the position of a celestial body against the background of the fixed stars should be detectable, as can be seen in the figure below:

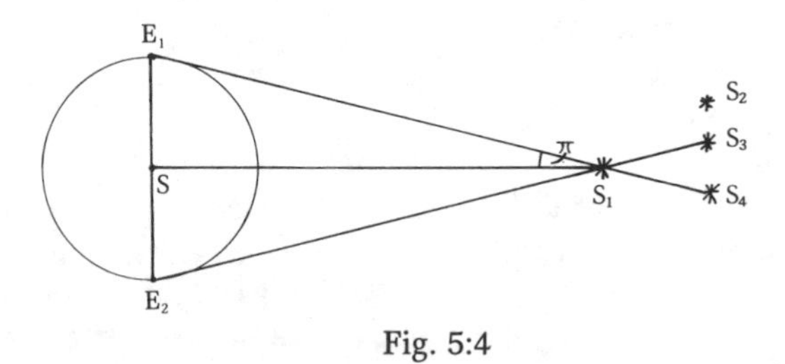

Fig. 5:4

Here an observer on earth at a particular time of the year when at E_2 would observe the star S_1 against the background of the stars as being in line with a particular star S_3. Six months later, when on the opposite side of earth's orbit, the observer at E_1 would observe S_1 to be in line with a different star, in this case S_4. S_1 would thus have been displaced by the angular distance π. The effect is familiar to anyone who has observed a particular point from opposite sides of a moving merry-go-round. In fact, from the Copernican point of view, the earth-sun relationship can be seen as a cosmic merry-go-round.

Yet observers as proficient as Tycho Brahe attempted to measure stellar parallax and could find no trace of it. In real terms this meant that the star S_1, when observed at six-monthly intervals, was seen with precisely the same stellar background. The absence of any parallactic displacement was sufficient to convince Tycho that Copernicus was wrong, and he consequently posited a third viewpoint, different from both *The Almagest* and *DR*. Tycho believed the universe to be geocentric, with the moon and sun rotating about earth in the normal manner and the remainder of the heavenly bodies orbiting about the sun. This way he was able to retain many of the advantages gained by Copernicus while avoiding the difficulty over parallax.

Nevertheless, an alternative explanation was possible. An examination of the preceding diagram shows that as the distance between earth at E_1 or E_2 and S_1 increases, the size of the angle (π) decreases. If the distance were great enough, the angle would be too small to detect. Is therefore the failure to detect stellar parallax due to the fact that there was no parallax or, perhaps, because it was too small to be detected? Copernicus took the position that the stars were such an immense distance away that in practical terms the parallactic displacement was too small to detect with the available instruments. To many such an argument was unattractive. That the universe was vaster than anyone could imagine was a possibility few were willing to consider. For one reason, there was a presupposition that the distances separating the orbits of various heavenly bodies were of approximately the same order. But if the stars proved to be a vast distance away this would be violated, and there would be for no apparent reason a great area of emptiness separating the orbit of the outermost planet, Saturn, and the fixed stars.

Although it was unknown in the 16th century, the angle actually was too small to be measured with the naked eye. Even with telescopes, the first such measurements would not be made until 1838.[16] Yet such an assumption was forced upon Copernicus. *Commentariolus* contains as its fourth assumption: "The distance from the Earth to the Sun is imperceptible in comparison with the height of the firmament" (Rosen: 58). Here is a striking demonstration of the power of a scientific theory to capture the imagination and command assent of generations of scholars long before it had been satisfactorily established. Proof of the validity of the Copernican system was decades away from the first publication of *DR*. And in this, *DR* fared no better nor worse than many classic works. If scientific theories could be falsified by producing conflicting evidence, then Copernicus's work would have been rejected upon publication. As it was, the power of the work was appreciated in advance of proof. Philosophers of science have long sought ways of characterizing the power of certain works to gather support readily and prematurely, and have suggested such factors as aesthetic appeal, simplicity, and comprehensiveness, but so far no clear-cut definition has emerged.

While the convincing physical demonstration of the new astronomy was lacking, considerable evidence emerged in the 17th century to prove the total inadequacy of the Ptolemaic system. Much of this new evidence depended on the invention of the telescope, and its first significant use in astronomy by Galileo led to the publication in 1610 of *Siderius nuncius.* Before the 1620s it had been observed that the moon was not the incorruptible body of antiquity, that Jupiter had moons and that Venus had phases, that comets seemed able to move through the heavens unhindered by restrictive spheres, and that new stars could appear and disappear. All these phenomena proved that the traditional cosmology of the Ptolemaic system was no longer viable. None of these features could actually prove that the universe was heliocentric, but with the removal of one competitor in a two-horse race, the effect was the same. Some strove for some time to promote the system of Tycho Brahe as a compromise, but with little real success.

Thus the Jesuit astronomer, G. B. Riccioli (1598-1671) in *Almagestum novum* (1651) mounted the last serious attack by a major astronomer on the plausibility of Copernicus. He deployed no fewer than seventy-seven arguments against the earth's motion. The striking title page shows Urania holding a balance, testing the Copernican and Tychonic systems against each other. The Ptolemaic system is shown discarded at the feet of a goddess, watched by a rueful Ptolemy who comments, "Errigor dum corrigor" ("I erred so that I could be corrected"). The scales of the goddess-judge are in favor of Tycho. In actual fact, the system of Riccioli was a modification of the Tychonic, with only Mercury, Venus, and Mars orbiting the Sun; Jupiter and Saturn, like the Moon and the Sun, were held to orbit the Earth. "Wonderfully complicated" is one modern judgment of this system (Dreyer: 419).

By this time it was not the pure system of Copernicus which Riccioli opposed, but that modified by the contributions of Galileo and Kepler. In a series of works, but particularly in *Astronomia nova* (1609) and *Harmonice mundi* (1619), astronomy under Kepler finally adopted its modern form. The obsession of centuries – the demand that all celestial orbits be circular – was finally abandoned and, speaking of Mars, Kepler declared that "its orbit is not a circle, but an oval." Further work led him to the conclusion that the orbit of Mars and all other heavenly bodies was an ellipse (Koestler: 333-38).

In terms of its spread beyond Europe, the Copernican system was "well established at Harvard" by 1659 (Jones and Boyd: 1). The evidence comes from almanacs: *The New England Almanac for 1659* contained an exposition of the Copernican system by Zachariah Brigdon. On its spread eastward, Needham commented, "Chinese books between 1615 and 1635 described the telescopic discoveries, but did not mention Copernicanism; then for a short time the heliocentric theory was described, but after news of the condemnation reached China, a curtain descended and a return to the Ptolemaic view took place (1959: III, 445-46). Thereafter the Ptolemaic system was deliberately taught by the Jesuits until their expulsion in 1773 on the grounds that the adoption of Copernicanism would give the impression of censuring "what our predecessors had so much trouble to establish" (Needham: 450).

As for Copernicus's fate among astrologers, or those who worked on the fringes of science, Capp's recent study of English almanacs shows that astrologers moved away from Ptolemaic astronomy as early as the beginning of the 17th century. They preferred the Tychonic system[17] over the Copernican in the beginning, but by mid-century and certainly by 1683, most astrologers had accepted the system of *DR* (Capp: 191-94). He found no evidence for the interesting claim, first made by Marjorie Nicolson, that the English Civil War was fought between "royalist Ptolemaics" and "parliamentarian heliocentrics." Rather, Capp noted, anti-Copernicans could be found among the parliamentary astrologers as commonly as Copernicans could be seen among the royalists.[18].

Publishing History

More interest has been shown in *DR* over the last fifty years than in the previous four hundred. Its editions are as follows:

Nuremberg, 1543

Johannes Petreius printed about 1,000 copies (but see above, page 114) which are now extremely rare and valuable. An indifferent copy from the 1979 Honeyman sale was sold for £22,000, while a copy belonging to Andreas Goldschmidt (1512-59), known as Aurifaber, which Rheticus and presented to him, sold for £44,000. Yet even this was not the highest price ever paid for one of Copernicus's books, for the thirty-eight leaves of *Narratio prima,* the rarest of all his works, was sold at the Honeyman sale for £75,000.

Basel, 1566

This is an uncorrected reissue of the Nuremberg edition, published with *Narratio prima.*

Amsterdam, 1617

Under the title *Astronomia instaurata, DR* was reissued with some minor errors corrected and explanatory notes provided by Nicolaus Mullerus of the University of Groningen. At this time the first edition was not yet exhausted but, with the exception of a reissue in 1640, it would remain the last issue of *DR* for nearly two hundred and fifty years.

Warsaw, 1854

The first edition in two and a half centuries, a Polish translation by Jan Baranowski, was also the first edition to appear in a modern language.

Torun, 1873

All previous editions of *DR* had been based on a copy made by Rheticus. The original manuscript survived, however, and was found in the 1830s in the library of Count Nostitz in Prague. It has been described as "a heterogeneous document, embodying numerous and extensive alterations, insertions and cancellations, evidently made at various dates" (Armitage: 64). In connection with the four-hundredth anniversary of Copernicus's birth, a critical edition was prepared by Maximilian Curtze, using Copernicus's own copy. Curtze believed the manuscript had been revised twice, and that it could not have acquired its final form before 1529, since the manuscript includes a celestial observation for that year.

Munich and Berlin, 1944-

In the middle of World War II, Germany surprisingly decided to produce the long-overdue *Gesamtausgabe,* or collected works, in nine volumes. Vol. 1, produced in 1944, consisted of a photographic facsimile of the Copernicus manuscript, edited by Fritz Kubach; Vol. 2, published in Munich in 1949, was a critical text, edited by F. and C. Zeller. No further volumes have since been published.

Translations

Torun, 1879: The first German translation, based on the Copernicus manuscript, was made by C. L. Menzer. A facsimile was issued in Leipzig in 1939.

Paris, 1934: The first French translation of Book I was made by A. Koyré; nothing more comprehensive is available in French.

London, 1947: An English translation of Book I was made by J. Dobson and S. Brodetsky and published in the *Occasional Notes of the Royal Astronomical Society* (2: 10). It was reissued in 1955.

Chicago, 1952: Four hundred and nine years after its first publication, an English translation of the entire text was published as Vol. 16 of the Encyclopedia Britannica's series, Great Books of the Western World. The translator was C. G. Wallis, Neugebauer has dismissed it as "not much more than a simple replacement of Latin words by English words. . . . It often requires re-translations to be intelligible. . . . A truly dilettante attempt" (1975: 69-71).

Moscow, 1964: A. A. Michajlova was the translator of the first Russian translation.

London, 1976: A new English translation by A. M. Duncan.

London, 1978: Another English translation by E. Rosen and edited by Jerzy Dobryzcki. It is part of an ambitious program of publications initiated by the Polish Academy of Sciences in 1973 to provide Copernicus's original text in a facsimile edition along with a critical edition, as well as a volume of Copernicus's shorter astronomical treatises and writings on other subjects. Envisioned as a three-volume set, the Academy hoped to provide translations in six modern languages. This is the English translation of the critical edition (Vol. 2 of the three-volume set).

Facsimiles

Facsimiles of the first edition have been produced in Paris (1927: by Herman); Turin (1943: by V. Bona to mark the four-hundredth anniversary of the publication of *DR*); and New York (1945).

Further Reading

De revolutionibus is now available in a number of editions of which the 1978 translation by Rosen is preferred. Recent studies by Armitage (1957), Kuhn (1957) and Fred Hoyle's *Nicolaus Copernicus* (1973) are all valuable but the work which above all others conveys the excitement of his times, the fundamental interest of his work and the enigmatic quality of the man remains Koestler's unquestioned masterpiece, *The Sleepwalkers.* Koestler is somewhat opinionated and his book is a little vague on the actual astronomy of Copernicus. It should therefore be used in conjunction with works like Dreyer (1953), the first part of Koyré (1973) and Rosen (1959). The spread of Copernicanism is traced in Kesten (1945) and in the detailed study by F. R. Johnson, *Astronomical Thought in Renaissance England* (1937). Recent work on Copernicus is surveyed in *The Copernican Achievement* (1975), edited by R. S. Westman.

Notes

1. What was his nationality, Polish or German? The question would have had little meaning for Copernicus, who wrote in Latin and spoke Polish and German depending on whether he was in a Polish-speaking or German-speaking town. He owed allegiance to the king of Poland and his father, Nikolaj Kopernik, came from Cracow, an ancient Polish capital.
2. During this period Copernicus published his first book, a translation into Latin of *The Epistles* of Simocatta (1509), an obscure seventh-century Byzantine historian. It consists of eighty-five fictitious letters between real and imaginary characters, written on pastoral, moral and amorous subjects. The eighty-fifth epistle, from Plato to Dionysius, is an example of the work: "If you would learn to master your pains, walk among the graves. . .and you will realise that beyond the grave the greatest bliss of mankind counts for naught" (Kesten: 125). Apart from a "comprehensive brevity" mentioned in the dedication of the work to his uncle Lucas, it is not known why Copernicus chose to devote time to this trivial and forced work of rhetoric. Perhaps the manuscript happened to be available and was the right length to tempt him: whatever, as the first original publication of a Greek work in Poland, it is worthy of note.
3. In 1514 Pope Leo X invited a number of scholars, including Copernicus, to Rome to advise him on the revision of the Julian calendar. Copernicus declined on the grounds that the courses of the sun and the moon were known with insufficient

precision. As a physician, Copernicus seems to have been summoned on a number of occasions to treat friends and the eminent. In 1539 he was summoned by his friend Giese to treat an attack of tertian ague, and in 1541 he was summoned to Königsberg by Duke Albrecht to treat a friend of his.

4. Sixteen letters of Copernicus have survived, ten of them to Bishop Dantiscus, who was appointed in 1523. The bulk concerns the appropriateness of Copernicus's living alone with Anna Schillings, his housekeeper. One letter mentions the matter of one hundred marks which Copernicus lent Canon Snellenburg who had, Copernicus complains, repaid only ninety.

5. They are found in *Narratio prima*. Although sensible, it is difficult to see why they should suddenly overcome the inhibitions of three decades. All Giese seems to have said was that better astronomical tables were needed, but publication of tables alone with no account of the system they were based on would be "an incomplete gift to the world" (Rosen: 192-93). Why Rheticus thought such views could persuade a stubborn Copernicus is far from clear.

6. In Rheticus we see a figure of Renaissance science at its most dramatic. His father was a physician who was beheaded for sorcery in 1528; he himself seems to have been a homosexual who cruised frequently and was forced to depart abruptly from towns and jobs. After his work with Copernicus he studied medicine in Zurich; and after practicing in Leipzig for several years he fled to Cracow in 1554 as a result of yet one more sexual scandal. His own most significant contribution to science was his work in trigonometry. With the aid of a team of computers he produced a ten-place table of all six functions for every 10" of arc, and a fifteen-place table of sines for every 10" of arc. His tables were published posthumously as the *Opus palatinum* in 1596.

7. The *Narratio* is a very strange work containing much more than a précis of Copernican astronomy. Not only does it end with an extravagant section "In Praise of Prussia," but it also includes a gratuitous account of some curious astrological views of Rheticus. He claimed changes in the sun's eccentricity from maximum to mean values to be linked with the rise and fall of Rome and the rise of Islam. When the sun reaches its minimum value the empires of Islam "will fall with a mighty crash." But when the other mean value is reached, we will have returned to the position it was in at the creation of the world and Christ will return (Rosen: 121-22).

8. Cases comparable to Copernicus's treatment of Rheticus are not unknown in science. In 1923 F. G. Banting (1891-1941) was awarded the Nobel Prize for the discovery and isolation of insulin. His collaborator, Charles Best (1899-1978), a graduate student, was completely ignored. This did not prevent the Nobel authorities sharing the prize with the head of the laboratory, J. Macleod (1876-1936), despite the fact that while Banting and Best carried out their work in Toronto, Macleod was in Europe. In fairness to Banting it should be pointed out that he was so appalled by the injustice of the award that he wished to refuse the prize; he was persuaded to accept it and shared the money with Best. Many other examples could be listed.

9. As well as Copernicus, the 16th century also saw a number of geocentric works. One was the *Homocentrica* (1538) of Fracastoro (1483-1553), the man who coined the term syphilis. He shunned epicycles and eccentrics and returned to the original homocentric idea of antiquity; the price to be paid for his purity was a system of 79 spheres. The work, according to Dreyer (p. 297), had a "total want of success." Even more spheres were required by G. B. Amici in his *De motibus* (1536). Their constructions also differ.

10. Cornelius Agrippa (1486-1535), for example, in *De occulta philosophia* (1533).

11. The simple observational test of a transit of Venus is quite rare. The first record of one in the West was made by J. Horrocks (1617-41) in 1639; since then there have been only four others: 1761, 1769, 1874, and 1882. The next one is due in 2004.

12. There is in fact more to the orbit of Mercury. Because of its highly eccentric orbit, the actual epicycle/eccentric described here did not suffice. Consequently Copernicus introduced a special device to account for the fact that "Mercury by its proper motion does not always describe the same circle, but very different ones" (V:25). Mercury is thus "made to move on a straight line segment such that its distance from the centre of its orbit varies with the proper period" (Neugebauer 1969:203). To the unsophisticated this looks very like a clear deviation from circular motion. Rosen (p. 88) describes the construction as "bieccentrepicyclic rather than eccentreccentric."

13. "Copernicus, as he loudly proclaimed, got rid of the equant circles... but this very refusal ... to allow the planets to move on their deferents with non-uniform orbital velocity forced Copernicus to replace it by a supplementary epicycle" (Koyré: 43).

14. It seems to have little impact at the textbook level. According to Johnson, "no textbook widely used in Europe in the sixteenth century expounded the Copernican theory, and few even noticed it" (Johnson: I, 285). It has been suggested that this was not because of disagreement with Copernicus, but because it was too technical. "Almost every university had at least one elementary astronomical text produced for local consumption.... The intricacies of the Copernican theory...were eschewed...as beyond the reach of the beginning students.... This dead weight of pedagogical tradition did far more to delay the spread and general acceptance of the Copernican hypothesis than any religious opposition to it" (Thorndike: VI, 6-7). In support of this view, Johnson contrasted it with the modern acceptance of Einstein's complicated theories. He examined eighteen physics textbooks available in the United States in 1943 and found that seven contained no reference at all to relativity theory, three offered only the merest allusion, and eight contained passages extending from one to three pages.

15. This mode of learning is not unknown elsewhere. Wilks, for example, has described how in the Dyula towns of West Africa "graduate" scholars would seek understanding of a difficult, essential, and important book from *ᶜulama* with the specialist knowledge of a particular text. In the same way students wishing to understand *DR* might have gone to Tubingen to study under Maestlin. See Wilks: 165-71.

16. In that year Friedrich Bessel (1784-1846) of the Königsberg Observatory detected a parallactic displacement of .3" in the position of 61 Cygni and attributed to it the massive distance of 10.3 light years.

17. In this, only Mercury and Venus orbit the Sun which, together with all other planets, orbits the Earth.

18. On her argument, see her "English Almanacs and the 'New Astronomy,'" in *Annals of Science* 4 (1939-40): 1-33.

CHAPTER 6

De humani corporis fabrica (1543)

by Andreas Vesalius

"Father," said Young Jerry, as they walked along. . . .
"What's a Resurrection-Man?"
Mr. Cruncher came to a stop on the pavement before he answered,
"How should I know?"
"I thought you knowed everything, father," said the artless boy.
"Hem. Well," returned Mr. Cruncher, going on again . . . , "he's a tradesman."
"What's his goods, father?" asked the brisk Young Jerry.
"His goods," said Mr. Cruncher, after turning it over in his mind,
"is a branch of Scientific goods."

– Charles Dickens, *A Tale of Two Cities*

Within a short time of each other, two works were produced in 1543 that changed the intellectual landscape of mankind. The first was *De revolutionibus,* published in Nuremberg in late May; the second was *De humani corporis fabrica* by Andreas Vesalius, a massive volume published in June in Basel. While Copernicus showed that man's traditional view of the heavens was wrong, Vesalius attempted to illuminate an even obscurer part of nature, the human body.

The human body and its interior has remained an almost total mystery to most of mankind. In recent years ethnographers have revealed just how knowledgeable many traditional societies are of the local flora and fauna, how acute their classifications often are, and how skillfully they exploit the environment for food, raw materials and drugs (see Levi-Strauss: Chap. 1). Others have described the knowledge of the heavens possessed by, for example, many Pacific island communities for whom the ability to sail over clear water was a considerable advantage (see, for example, David Lewis, *The Voyaging Stars,* 1978). Scholars have spent much time exploring the metallurgical, ceramic, agricultural and other productive skills possessed by many non-industrial communities with invariably impressive results. Noticeably missing, however, are studies describing the anatomical knowledge of traditional societies. Even the ancient Egyptians, whose knowledge of embalming is much celebrated, used only one word to refer to nerves, muscles, arteries, and veins (Glanville: 189).

Thus while the roots of astronomy were widely distributed, those for anatomy can be found in Greece and Greece alone.[1] About 300 B.C. Herophilus and Erasistratus began to dissect bodies. Arteries and veins were distinguished for the first time, the valves of the heart were described, motor and sensory nerves were distinguished, and the brain was recognized as the main organ of the central nervous system – to mention just a few of their numerous and important discoveries.

The scientific study of the human body thus commenced, and by Galen's time in the second century A.D. a body of anatomical learning had been assembled which proved as authoritative and long-lasting as the astronomy of Ptolemy, his near contemporary. The rudiments of anatomical terminology were developed, and the essential systems of the body – nervous, arterial, venous, muscular, and skeletal – had all been carefully isolated and described. Galen's achievements included the description of some three hundred muscles and the identification of seven of the twelve cranial nerves.

Galen explained the rationale for such knowledge: "What could be more useful to a physician for the treatment of war-wounds, for extraction of missiles, for excision of bones . . . than to know accurately all the parts of the arms and legs? . . . If a man is ignorant of the position of a vital nerve, muscle, artery or important vein, he is more likely to be responsible for the death, than the saving of his patients" (Singer 1956: 32-33). His argument would be repeated down the centuries, and in the 19th century, for example, Sir Astley Cooper (1768-1841) declared that "a half anatomist is a most dangerous surgeon."

But the difficulty with anatomical knowledge is that it can only really be obtained by the direct manipulation of a human corpse. In most communities there are social strictures against handling corpses, and dead bodies are disposed of as quickly as possible. Corpses are often imbued with powers and dangers, and it is believed that they will vent their rage on anyone near them. Special groups are given a dispensation to approach and dispose of corpses. One may bring up the example of the Nuer, a cattle-herding people of southern Sudan, who consider death "an evil thing." Only grave diggers are allowed to attend burials and once having dug a grave, they fill the shaft while "they sit or stand with their backs to it and scoop the earth back with their hands." The earth is beaten down to leave no sign of the grave (Evans-Pritchard: 144-45).

For the Nuer, corpses are just too potent to leave unburied any longer than is necessary. Other people revere the corpse and prohibit dissection by religious law.[2] Thus Muslims, who were predisposed to Greek science, were inhibited from pursuing anatomical studies because "Islamic law did not permit the dissection of the human body. . . . As a result, practically no dissections were carried out" (Nasr: 163).

Why the Greeks should be so different in this respect is far from clear.

Perhaps the influence of Plato for whom the body and the material world were of little value was of some significance here. Whatever the reason for its inception, the practice of human dissection seems not to have persisted. By the time of Rufus of Ephesus in the first century A.D. it was noted that while anatomy used to be taught from man, it had been discontinued. From Galen we know that it was rare to find usable bodies – those that might have been washed up on a river bank or uncovered by heavy rains and floods. Consequently, he recommended his students to work with "those apes that are most like man. . . In such apes you will find the other parts too arranged like man's" (Singer 1956: 3-4). Much of Galen's work was in fact based on the study of the barbary ape, *Macacus innus,* although other animals were also used, which was checked against his knowledge of human anatomy culled from his service as physician to the gladiators of Pergamum and with Emperor Marcus Aurelius during the German campaign. Remarkably, Galen constructed a most effective human anatomy.

Inevitably, an anatomy based on pigs, cows, cats, dogs and apes for some of the finer details could not be entirely accurate, and yet once incorporated into general anatomical tradition errors became so firmly entrenched that they are found in 17th-century textbooks. Running through the whole of the anatomical literature from Galen to Harvey are descriptions of animal features that are confidently and repeatedly ascribed to the human body. These include a five-lobed liver, a uterus with horns protruding on either side, a sternum divided into seven segments, a right kidney higher than the left, pores in the septum separating the two ventricles and the *rete mirabile.* This last feature, literally "wondrous network," is a vascular complex found at the base of a calf's or sheep's brain and is completely absent in man – as are all the other features mentioned above.

The practice of human dissection was revived on a very limited basis when new universities with medical faculties were founded in southern Europe. There are, for instance, 13th-century references to autopsies performed at Bologna, founded in 1156, and to the occasional dissection in the fourteenth.[3] The practice was officially recognized in university statutes dated 1405, and in 1442 the city of Bologna was instructed to provide two corpses each year for this purpose. The practice in fact spread to most centers of learning throughout the 15th century. Students could rely on a well-known dissecting manual, *Anothomia* (1316), by Mondino of Luzzi (c. 1275-1326).[4] It contained little new but reported the main details of the Galenian system, errors as well as truths. The author indicated, however, a direct acquaintance with both human and animal bodies, reporting, for example, on a "woman whom I had anatomised" who differed from an earlier "anatomised" body (Clendening: 123).

Illustrations from the period show that the jobs of the anatomist and the dissector were distinctly separate. This was an innovation. Galen had performed his own dissections. Not only did he frequently refer to what he himself had shown (which could, of course, have been a convention of the period), but he also specifically went out of the way to emphasize the value of even such an apparently trivial task as skinning specimens. At one time, Galen confessed, he allowed an assistant to "skin his apes, avoiding the task as beneath my dignity." But as he once found "by the armpit, resting on and united to the muscles, a small piece of flesh which I could not attach to any of them, I decided to skin the next ape carefully myself." As a result, Galen thereafter dispensed with the services of a skinner (Singer 1956: 8).

By the early Renaissance such scruples had disappeared. The anatomist then sat elevated on a chair, surveying the body and reading from a text, Galen or Mondino perhaps, what should be demonstrated. The dissection was performed by a barber, surgeon or butcher who were, no doubt, incapable of understanding the Latin instructions of the anatomist. Hence there arose the need for an ostensor, a kind of assistant professor, familiar with Latin who stood by the dissector and translated the anatomist's instructions; he also showed with a pointer the significant features exposed by the dissection.[5] Although illustrations and reports of the period all describe *public* dissections, in fact they rarely took place more than twice a year given the availability of corpses; the bulk of the anatomist's dissections was presumably carried out in private.

It is sometimes argued that the failure of medieval anatomists to expose the five-lobed liver and other gross errors of antiquity was due to this new division of labor: the anatomist took Galen's text as authoritative because he never compared it with an actual body; the dissector, who saw what was in the body, had no comprehension of Galen. Considering that even from his chair the anatomist could see what the body contained and, in any case, performed dissections, the question remains why did he continue reporting the anatomical fictions of Galen and antiquity? The history of science is full of examples in which facts and theories are unquestionably accepted for long periods of time as part of nature and then are suddenly declared fictions. Likewise, objects never observed are confidently declared as much a part of nature as the Alps. How is it that generations of anatomists described and drew the non-existent *rete mirabile* in the human brain while their counterparts in the astronomical sciences, when presented with a clear sight such as the Mountains of the Moon, failed to see them?

One answer to this subtle and complex issue is to consider how strongly the familiar serves as a model for rendering the new. What could be more familiar than the horse? Yet it was always drawn by classical artists with eyelashes on the lower lid, a human feature absent in horses (Gombrich: 70). What chance then would western artists have in portraying such unfamiliar animals as whales, rhinoceroses, and elephants?[6] Gombrich has pointed out that early 17th-century woodcuts of whales all depict the presence of an external ear, a feature lacking in whales, although they are said to have been drawn "accurately from nature" (Gombrich: 69). The observation of horses and whales seems relatively straightforward; the structure of the complex and possibly decomposing tissue of the vascular system of the brain is much more difficult to see. If a respected authority has asserted the presence of such a structure, then the neophyte will tend to dismiss any failure on his part not to see it as a result of his own inexperience. The more experienced anatomist might explain his failure to observe the *rete mirabile* by the knowledge that they are never easy to demonstrate and in some specimens are too ill-developed to be easily seen. In any case anatomists of the 14th-16th centuries were unable to obtain many cadavers. When Vesalius published *De fabrica*, he had examined only six female corpses.

Karl Rokitansky (1804-1878), Viennese pathologist, personally performed 30,000 autopsies; with experience of this kind, knowledge of the unusual, rare, common, abnormal, and never seen in the human body must be very strong. In contrast, it is little wonder that an experienced 15th-century anatomist who may have seen a mere dozen or so cadavers would have been reluctant to proclaim an inability to detect structures described by Galen and other authorities.

Even in modern times, little has changed since Vesalius's day. Medical students in Australia were reported in 1941 by Alan Gregg to have used for twenty years a manual on the English frog (*Rana temporaria*) as a guide in dissecting an anatomically distinct local species (*Rana australiensis*). Gregg also states that in that period only about a dozen students noticed a difference between the frogs they dissected and the frog described in their textbook: in other words, a generation of students had habitually identified anatomical structures which did not exist in the frogs they dissected. Of the dozen or so students who realized something was wrong, Gregg significantly pointed out, all of them objected that there was something *odd about their frog* (King: 279). Thus they resembled Renaissance anatomists who, when presented with a discrepancy between Galen and a real brain, sternum or liver, never doubted the authority of the book.

Life and Career

Vesalius was born in Brussels in 1514 at 5:45 A.M. on 31 December according to Cardano's horoscope.[7] No stars were needed to determine Vesalius's career and service: his family background was more than sufficient. His great-great-grandfather, great-grandfather and grandfather had all been physicians, the latter two serving the houses of Burgundy and Hapsburg. Vesalius's father, also named Andreas, followed the family tradition and served as apothecary to Margaret of Austria and later to her nephew, Emperor Charles V. Vesalius followed suit and after attending the University of Louvain (1528-33), he studied medicine at the universities of Paris (1533-36) and Padua where he was awarded his M.D. in 1537 and was immediately appointed professor of surgery and anatomy.[8]

Much is known about Vesalius's life through his writings, for even at their most austerely anatomical they contain many vivid autobiographical passages. Much of his work was polemical and frequently illustrated with dramatic events taken from his life, training and medical career. He presents a marvelously detailed picture of how difficult it was to be a serious anatomist in the early 16th century. He was scornful of the traditional way of teaching anatomy, describing anatomists as "jackdaws aloft in their high chair, with egregious arrogance croaking things they have never investigated but merely committed to memory from the books of others." As for dissectors, he writes they are "so ignorant of languages that they are unable to explain their dissection to the spectators and muddle what ought to be displayed according to the instructions of the physician who haughtily governs the ship from a manual since he has never applied his hand to the dissection of a body." He believed that a butcher displayed more information in his stall than an anatomist revealed in his theater (Clendening: 133).

The shortage of dissecting material presented a severe problem. The Paris anatomist Sylvius (1478-1553) was known to bring "a thigh" or "sometimes the arm of someone hanged," wrapped in his sleeve, to the lecture room for dissection. They stank so strongly that "some of his auditors would readily have thrown up if they had dared; but the cantankerous fellow. . . would have been so violently incensed. . . not to return for a week" (O'Malley: 49). Vesalius described how he overcame the shortage of material in *China Root Epistle* (1546)[9]. Apart from animals, he seems to have relied on the theft of bones and other parts from cemeteries or gallows. The Cemetery of the Innocents in Paris, where plague victims were buried, and Montfaucon, where executed bodies were taken and displayed suspended on beams, provided excellent sources of cadavers and usable remains. Competition came mainly from packs of savage dogs which, on one occasion, "gravely imperiled" his life.

Vesalius writes that in 1536, while out on a trip with his colleague Gemma Frisius looking for the bones of executed criminals, he came across a cadaver "partially burned and roasted over a fire of straw and then bound to a stake." He could scarcely believe his fortune, for the body was "dry and nowhere moist or rotten." He continued (O'Malley: 64):

> I climbed the stake and pulled the femur away from the hipbone. Upon my tugging, the scapula with the arms and hands also came away. . . . After I had surreptitiously brought the legs and arms home on successive trips. . ., I allowed myself to be shut out of the city in the evening so that I might obtain the thorax which was held securely by a chain. So great was my desire to possess these bones that in the middle of the night, alone and in the midst of the corpses, I climbed the stake with considerable effort and did not hesitate to snatch away that which I so desired. . . . I carried them some distance away and concealed them until the following day when I was able to fetch them home bit by bit through another gate in the city.

So expert did Vesalius and his colleague Mattheus Terminus become "that even blindfolded we could for the space of half an hour identify by touch any bone [our students] offered to us" (O'Malley: 60). Vesalius's talents were being noted, he reports, and in Paris he was invited by Sylvius to demonstrate his skill and to "perform in public . . . a dissection which dealt purely and simply with the viscera." Forced to leave Paris because of the outbreak of war, he returned to Louvain in 1536 where "the professors of medicine had not even dreamed of anatomy in eighteen years." He consequently "delivered lectures on the human fabric accompanied by dissections which I did a little more accurately than in Paris." A year later, in Padua, he regularly and personally performed dissections (Clendening: 134-35).

While at Padua, Vesalius produced his first anatomical work, *Tabulae sex.* Describing how this work came about, he writes he "made on a sheet of paper a hasty sketch of the veins. . . . That dissection proved so pleasing to all the professors and students of medicine that they strenuously urged me to supply a description of the arteries and nerves also" (Cushing: 12). The six *tabulae* were published in Venice in 1538: they dealt with the generative organs (horned uterus), the five-lobed liver of the ape, the heart and aorta showing the *rete mirabile* of cattle that he drew himself. The remaining three plates of a front, side, and rear view of the skeleton were made by van Kalkar, a Rhenish artist from the studio of Titian. With possible ulterior motives, he dedicated the work to Narcissus Parthenopeus, physician to Charles V. Judging by the number of times the plates were plagiarized – twelve editions were produced before 1557 in such places as Augsberg, Cologne, Strasbourg, Paris, Marburg, and Frankfurt – there was clearly a considerable demand for a work of this kind.

Five years later Vesalius wrote two works designed to satisfy for a century or more this demand for anatomical studies. They were his masterpiece, the massive *De fabrica,* and a student edition called *Epitome.* These two works run to over 800 pages and involved the preparation of over 250 blocks to provide the illustrations to the text. Nor were these the only works produced by Vesalius during this period. But at this point, his ambitions lay elsewhere. Though only twenty-eight, Vesalius aimed for imperial patronage. *De fabrica* was dedicated to "Divine Charles the Fifth, Greatest and Most Invincible Emperor," and in the preface are reminders of the service given by his father as an imperial pharmacist. He was not above including characteristic, elaborate, yet almost conventional, flattery in his remarks. Leaving nothing to chance, Vesalius made a special journey to the court at Speyer (Bavaria) to present a copy of *De fabrica* to the emperor. Vesalius achieved his aim: he was offered and accepted a post of physician to the imperial court.

The modern mind may not appreciate the appeal the imperial court held for Vesalius, especially since he had also received an offer of a position from Cosimo de Medici at the University of Pisa. Charles V was not the easiest of masters to serve. His constant movements around the widely-separated parts of his empire so wore him out that he eventually abdicated in favor of his son Philip II in 1555. From "mass to mess" is the phrase used to describe life at Charles's court. An insomniac and a glutton, he would normally retire at five A.M. and on waking would expect a twenty-course lunch to be followed throughout the day by regular helpings of pies, plates of anchovies, eels, and any number of fowl. Yet Vesalius's commitment to him was total, even to the extent of burning all his papers and declaring that "I should write nothing more." It was an act he later regretted.[10]

Vesalius occasionally emerged from the anonymity of his official position. In 1546 he published *China Root Epistle* in response to the import of an herb from China that supposedly cured most ills. Charles was eager to know if it could alleviate gout and other distressing consequences of his intemperance. On other occasions Vesalius advised on the treatment given Henry II of France who in 1559 was accidentally wounded in a joust by a splinter from a broken lance that entered just above his right eye. Traveling quickly from Flanders, Vesalius could do little beyond predicting that the king would not recover. In 1562 he attended the Infante, the gross Don Carlos, son of Vesalius's new patron Philip II, who had supposedly fallen down stairs while chasing a serving maid. Such episodes give an indication of the esteem Vesalius enjoyed among the royal families of Europe.

The last two years of Vesalius's life are known only through scarce fact and rumor. He made a pilgrimage to the Holy Land, traveling from Venice in 1564. It is hypothesized that he wished to return to academic life at Padua when the death in 1562 of Fallopius, professor of anatomy, opened up a suitable post. Since resigning from the service of a monarch was not possible, Vesalius may have thought that a pilgrimage would enable him to extricate himself from the Spanish court. Or, he may have wanted to expiate some unknown sin.[11] He died on 15 October 1564 on his return from Jerusalem. Only fifty years old, he was buried on the island of Zante (modern Zakinthos) in the Ionian Sea, a victim, it has been speculated, of plague, drowning, or murder.

Vesalian Anatomy

Vesalius and his early 16th-century contemporaries were much better placed to rewrite anatomy textbooks than any previous generation. To begin with, they belonged to the first generation of scholars with ready access to the complete works of Galen, still the most comprehensive and accurate texts available. The first Latin edition of Galen's *Opera* was published in Venice in 1490, the Greek *princeps* appeared a few years later in 1525. Vesalius himself worked on a later collection, the Giunta edition published in seven volumes in Venice in 1541, for which he edited three of Galen's anatomical works: one on the dissection of nerves, another on the dissection of blood vessels, and the most important of all ancient anatomical texts, *De anatomicis administrationibus (On Anatomical Procedures.)*

Vesalius knew the works of Galen as well as Copernicus knew the texts of Ptolemy. The advantages of having a fairly comprehensive treatment of a subject available to the research student are considerable and obvious, particularly in an area where the raw material of the discipline – corpses for dissection – were always in short supply. The Renaissance anatomist, however, could never be sure that the structures revealed by his scalpel were anatomically real or the product of his own clumsiness, the result of decomposition, or simply an abnormal growth. The presence of an authority in such a context was, therefore, invaluable, for if agreement with Galen was reached it could be presumed that the right techniques were being used; if the structures revealed differed from Galen's an interesting problem had emerged. A preexisting anatomical model provided an invaluable reference. The injunction to describe or draw what one sees, even for so-called "trained observers," is not so much difficult as impossible. Preconceptions also raise questions of objectivity and relevance. It is easier, to take a specific example, to compare an illustrated description of a human sternum with an actual sternum than to attempt to describe a sternum.[12]

Prior to Vesalius's time, the art of anatomical illustration did not exist. Earlier manuscripts contained figures, but they are better described as diagrams rather than illustrations. Singer's book gives examples: for instance, a drawing in a 14th-century manuscript in which the stomach is depicted as a perfect circle and the kidneys as a pair of connected ovals (Singer 1957: 80). The art of anatomical illustration had not much improved two hundred years later as the diagram of the heart from a 1513 edition of Mondino clearly shows (Singer 1967: 85). This was due, perhaps, to the lack of skill of early artists and engravers. The ability to represent the body's organs naturalistically depended on a mastery of artistic perspective, a 15th-century advance. The muscle *tabulae* of *De fabrica* where grotesquely posed bodies better reveal certain structures or the sections of the brain shown within an exposed skull from above could only have been drawn after artists had mastered the principles of perspective (Singer 1957: compare fig. 42 on p. 87 with the Vesalian woodcuts in pp. 133, 188-204). The anatomical sketches of Leonardo da Vinci (1452-1519), though confined to unpublished notebooks, brilliantly show these developments. The *Commentaria* on Mondino by Berengario da Carpi (c. 1530), published in 1521, has been described as the "first illustrated anatomical text in the modern sense" (Saunders and O'Malley: 23).

Vesalius made innovative use of the new skills of illustration.[13] That he attached considerable importance to diagrams is evidenced by the labor involved in producing some 250 blocks for *De fabrica.* He gave careful instructions to Oporinus, the publisher, on the placing of illustrations and how they should be related to his descriptive notes.[14] Moreover, he was annoyed by the sloppy ways illustrations were used in unofficial editions of his work. Thus he wrote in 1546, "Just now in England . . . , the illustrations of the *Epitome* have been copied so poorly and without artistic skill . . . that I should be ashamed were anyone to believe me responsible. . . . Everything has been shamefully reduced" (O'Malley: 224). The reference is presumably to Geminus's edition.

Curiously, his conservative colleagues surprisingly objected not so much to Vesalius's illustrations *per se* but to their presence. They objected, Vesalius reported, on the grounds that "however exquisite the drawings might be, these things had to be learned not from pictures but from diligent dissection and from observation of the things themselves" (Clendening: 138). Vesalius replied that he had no intention of discouraging dissection and hoped that the use of illustrations would stimulate rather than inhibit the practice.

Sylvius countered by claiming that illustrations were "fictive" and that they gave "false proportion." Some of the opposition may be laid to Platonism, which was so prevalent in Renaissance thought, for Plato believed that actual bodies were inferior copies of an ideal entity; consequently, illustrations of bodies were a further step away from reality and had no real worth. It was not as if early anatomical texts lacked illustrations: they simply lacked anatomical illustrations. They commonly depicted dissections or contained figures of men with the signs of the zodiac symbolically shown ruling over particular areas of the body: Scorpio over the genitals, Aries over the head, as, for example, in the well-known figures in the early 15th-century *Book of Hours* of the Duc de Berry. Symbolic truths could be illustrated, scenes could be depicted, diagrams shown, but not until the 16th century was it thought proper to produce a figure designed realistically and accurately to represent organs and parts of the body.

In Vesalius's work, there is no conflict between the use of illustrations and dissection. *De fabrica* is full of the language and imagery of the dissecting theater. Its remarkable title page shows a crowded theater with Vesalius himself positioned in the center, looking directly at the reader and standing over a cadaver with an opened abdomen. He is surrounded by assistants, dead animals ready for dissections and implements, while two disgruntled barber/surgeons are seen squabbling on the floor.[15] The book contains an illustration of a table with a full set of dissecting instruments and a number of historiated initials showing scenes from the life of a 16th-century anatomist. Naked *putti* are shown robbing graves, boiling skulls and, among many other scenes, dissecting dogs and pigs.

The importance of dissection is constantly emphasized throughout the text and in the illustrations. In Book I, for example, there are detailed instructions on how best to prepare a corpse to show the structure of the skeleton and attached ligaments. Traditionally, the body was placed in a box filled with lime, sprinkled with water, and holes were drilled allowing the "macerated remains of flesh" to drain away. Vesalius found this technique unsatisfactory, "troublesome, dirty, and difficult;" it destroyed those parts the anatomist wished to investigate. Instead, he proposed the use of a "large cauldron such as women used for the preparation of lye," which would allow the flesh to be boiled away from the bones (O'Malley: 328). Throughout the text there are frequent references to the handling of bodies, such as, "Now with a sharp knife free the clavicles from the sternum. . . ," and in any discussion of substantive matters Vesalius made a point of underlining what he had seen or failed to see at autopsy or at a dissection. When discussing the question of the number of segments in the sternum – seven according to Galen – Vesalius expressed the opinion, "I can state with certainty that I have never seen the occurrence of all seven bones" (Saunders and O'Malley: 68).

What kind of anatomy did Vesalius produce? The modern anatomist will be disappointed to find that Vesalius says nothing in *De fabrica* about the circulation of the blood, the lymphatic system, or had any appreciation of the function of the lungs, the various parts of the brain, or other features of the body now accepted as commonplace. Indeed, he hardly improved on Galen. The study of the human anatomy resembles in this way the progress of geographical exploration. The task required hundreds of workers and took centuries to accomplish. Early works did little more than conduct rapid surveys, noted major landmarks, corrected the most egregious of initial misconceptions, and laid down the principles under which further research could be directed. True detail and accurate descriptions came later from specialists and writers of detailed monographs. Early scholars are often more commonly judged by the number of errors they exposed rather than by the truths they established.

In *De fabrica,* an emphasis has been placed on how many fictions introduced by Galen into human anatomy were expelled by Vesalius. His attitude was transitional and cautiously sceptical; he was unsure as to how many of the structures proposed by Galen were really fictitious. An uneasy compromise was often evidenced, as Vesalius used illustrations to display traditional features while his accompanying text was more guarded, if not dismissive. As an example, Vesalius printed a detailed picture of the *rete mirabile* but added in the text: "How many things have been accepted on the word of Galen. . . . Among them is that blessed and wonderful reticular plexus. . . . In my devotion I never undertook the public dissection of a human head without having available that of a lamb or an ox to supply whatever I could not find in the human. . . . The internal carotid arteries wholly fail to produce such a reticular plexus as that described by Galen" (O'Malley: 179-80).[16]

While Vesalius's views on the *rete mirabile* could not have been clearer, his position on the existence of pores in the septum dividing the right and left ventricles of the heart was much more hedging. Certainly, he accepted the Galenic view that "just as the right ventricle draws blood from the *vena cava* into itself so also the left ventricle draws unto itself, each time the heart is dilated, air from the lungs through the vein-like artery and uses it for cooling the innate heat." But as to pores, Vesalius began by noting that the septum was the thickest part of the heart and that it was clearly pitted. Yet, of these pits, none, "so far at least as can be perceived by the senses, penetrates through from the right to the left ventricle, so that we are driven to wonder at the handiwork of God, by means of which the blood sweats from the right into the left ventricle through passages which escape human vision." Is this piety or sarcasm? In the second edition of *De fabrica* Vesalius suggests it was the latter. "However much the pits may be apparent," he writes, "yet none, as far as can be comprehended by the senses, passes through the heart from the right ventricle into the left" (O'Malley: 281).[17]

Vesalius was equally clear about the sternum with its seven segments and five-lobed liver and other features wrongly introduced by Galen. On the issue of whether nerves were hollow, as supposed by antiquity, he described "a man as yet warm and scarcely a quarter-hour after his decapitation. . . . I inspected the nerves carefully, treating them with warm water, but I was unable to discover a passage in the whole course of the nerves" (O'Malley: 171). This is a passage so typical of Vesalius: careful judgment, based on his own dissecting experience, and far from trivial in its implications.

Vesalius was more forceful when his observations confronted popular beliefs or superstitions. He dismissed the old belief that man had one less rib than woman. Both sexes, he declared, had twelve on each side, although he admitted he had occasionally seen cases with eleven or even thirteen. As for the so-called *os cordis*, a supposedly vestigial bone in the heart but in reality a cartilaginous plate found in the heart of young ruminants – which physicians supposed possessed especially potent therapeutic powers – he dismissed as "stupid." Nor did he find any evidence for the existence of "the bone of Luz," from which man's body was thought to regenerate on Judgment Day. This ancient idea derived from rabbinical laws and, according to the *Jewish Encyclopedia,* the bone was supposedly found at the base of the spine. One Joshua b. Hananiah, a rabbi, was reported to have demonstrated the existence of the bone to Emperor Hadrian: "They put it in water and it did not dissolve, in fire and it was not consumed, on a mill and it was not ground. They placed it on an anvil and struck it with a hammer. The anvil cracked and the hammer split."

Vesalius had also to deal with the confusing subject of anatomical terminology, a problem inherited from Galen who, among others, frequently described rather than named particular anatomical structures. In speaking of the cephalic vein, for example, Galen simply referred to it as the large shoulder vein, going on to speak of "another vein that is carried into the forearm" when referring to the basilic vein (Singer 1956: 75). Trouble partly arose because "cephalic" and "basilic" were so clearly derived from Greek yet had no specific connection with the "head" (*kephalos*) or a "prince" (*basileus*). Moreover, the terms were not actually used by Galen or other classical authors, but rather entered the vocabuluary via medieval Arabic translations of Galen. The Arabic terms, *al-qifal* and *al-basiliq* are as obscure in Arabic as they are in Greek (Singer and Rabin: 7).

Early Renaissance anatomists thus inherited a specialized vocabulary randomly derived from a variety of authorities in a variety of languages. Most anatomical structures were known by so many different names that even such a gross one as the aorta went by several names. Galen referred indifferently to it as "arteria megale" and "arteria orthe," and later authorities developed their own alternatives.[18] Later writers adopted an inelegant but presumably workable solution by using as many names as possible. Da Carpi, for example, when discussing the veins of the arm, wrote, "In these figures can be seen the course of the *salvatella* of Mondinus and Rhazes, the *sceylen* of Avicenna and the *salubris* of Haly and that branch of the *basilica* which ends between the little and ring finger which Rhazes calls also *salvatella*" (Singer and Rabin: lxxx). Vesalius also followed this usage. Referring to the liver in *Tabulae sex,* for example, he mentions the *vena porta* but found it necessary to add, "which is called *stelechiaia* by the Greeks and *weridh-ha-sho'er* by the Arabs" (Singer and Rabin: 3).

The forging of a new commonly-understood terminology was not the work of Vesalius, although he did introduce some terms such as "mitral valve," "corpus callosum" and "alveolus." A unified nomenclature was worked out by humanists and scholars more versed than Vesalius in the works of antiquity. They appeared in new editions of Galen, Celsus, and other ancient writers in the late 15th and early 16th centuries. Another source was *Onomasticon,* an early specialized dictionary, by Julius Pollax (A.D. 134-92). Printed in 1502 it contained in a handy form most anatomical terms from the classical period. Words taken from these sources by anatomists such as Gunther of Andernach (1487-1574) and Sylvius, both teachers of Vesalius, tended to replace the mixture of Greek, Arabic and Hebrew terms found in such profusion in most pre-Vesalian texts.

Vesalius's crucial role in this exercise was to provide the plates on which the new terminology could be stabilized. By becoming associated with a standard text, new terms could acquire broad acceptance and authority. Vesalius's plates, copied and pirated all over Europe during the next hundred years, provided this standard. Successive generations of medical students learned anatomy from the same text and examined the same plates. This was the particular debt owed to the publication in 1543 of *De fabrica* and *Epitome* and, more directly, to the hundreds of derivative editions these works inspired.

Reception

Although there was never serious doubt about the immediate success of *De fabrica,* opposition came from Sylvius who, in 1549, warned of a "certain, ridiculous madman, one utterly lacking in talent who curses and inveighs against his teachers" (O'Malley: 239). He later appealed to the emperor to defend Europe against this crude and confused "farrago of filth and sewage." Sylvius represented the extreme wing of traditional anatomy which believed that Galen was infallible and was prepared to go to extraordinary lengths to preserve his integrity against objections from Vesalius and other contemporary anatomists. Replying to the charge that the sternum of the ape and not the human body was divided into seven segments, Sylvius baldly countered that man's body had changed since ancient times in this and other respects. In this debate, Vesalius simply stated (in the preface to *De fabrica*) that "Galen corrects himself frequently, and in later works written when he became better informed he points out his own slips perpetrated in certain books and teaches the contrary" (Clendening: 136). Indeed, opposition to his views did not last long, nor was it particularly widespread.

Examining the publishing history of *De fabrica* by itself, it might be difficult to account for the enormous influence exercised by Vesalius. Four editions in the 16th century (one of which had no illustrations), one edition in the 17th, and the publication of the *Opera* in the 18th hardly suggests success in a discipline where a textbook is almost as indispensable as a scalpel. Rather, the dissemination of Vesalian anatomy occurred as a result of popular derivative works which drew heavily on useful parts of Vesalius's texts and illustrations. Reproduced and issued in a variety of forms and lanaguages, they met the needs of succeeding generations of medical students, artists and physiologists. In most cases additional material of varying quality was added.

Vesalius's work was introduced into England with remarkable speed in *Compendiosa totius anatomiae delineato,* published in 1545 by Thomas Geminus (c. 1510-62). Geminus, presumably a twin and like Vesalius a Belgian, moved to Britain about 1540. A skilled engraver, he produced in 1544 a number of copper-engraved anatomical sheets taken from *De fabrica* and *Epitome* published the previous year. One set was presented to Henry VIII who was sufficiently impressed to recommend publication, and in 1545 *Compendiosa,* with forty plates taken from Vesalius's work, was printed together with a somewhat abbreviated text of *Epitome.* This proved too difficult for its intended market, and an English version, *Compendious Anatomy,* replaced it in 1553; it was reprinted in 1559. Curiously, the text was not an English translation of *Epitome* but rather an adaptation of a 14th-century anatomical compilation enhanced with Vesalius's illustrations.

Geminus's work spawned a secondary wave of plagiarism. In England such elementary works as Bullein's *Government of Health* (1558) and Vicary's *Profitable Treatise of the Anatomie of Man's Body* (1577) prove to be little more than Vesalius filtered through Geminus. His plates were somehow obtained by André Wechel of Paris who, under the editorship of Jacques Grevin, published two Latin editions (1564; 1565) and a French translation (1569) of Geminus. Geminus had clearly found a successful publishing formula for meeting a new market, one which European publishers soon copied.

The first of these European derivatives appeared in Germany in an edition published by Baumann in Nuremberg in 1551. It contained all forty engravings of Germinus plus German indices. Another successful text was *Anatomia* by Juan Valverde (c. 1520-88), published in Spanish in Rome in 1556. It contained forty-two plates, all but four of which originated in *De fabrica.* It spawned another thirteen editions over the following century, appearing in addition to the original Spanish in Italian, Latin, and Dutch.[19] Other examples could be cited making it clear that the work, if not the books of Vesalius, was as widely known as any book in the 16th and 17th centuries.

Within this early tradition of anatomy, students were expected to learn by witnessing an expert dissection performed by an experienced professor of anatomy. There were too few available bodies to permit anything else. By statute of Queen Elizabeth I, for example, London physicians were allowed the bodies of up to six executed criminals; barber/surgeons were later granted up to four bodies a year on similar terms. But ten bodies a year were far from sufficient to satisfy the growing needs of new hospitals and academies that were so much a feature of 18th-century medical life.[20] William Hunter, writing of his student days at Guy's Hospital in London in the 1730s, noted that "the professor was obliged to demonstrate all the parts of the body upon one dead body" (Linebaugh: 70). By the end of the century it has been estimated that as many as 800 bodies a year were being supplied to British hospitals and academies.

The gallows could not be depended upon as a predictable source of corpses. Some condemned men were desperate enough to sell their bodies in advance to surgeons, but most resisted such extremes. As a matter of fact, the Murder Act of George II specifically decreed the consigning of bodies to surgeons as "a peculiar mark of Infamy added to the Punishment." It was consequently not unknown for a felon's kin and workmates to attend Tyburn in force and seize a corpse before the surgeon's agents could get their hands on it; sometimes they were so quick that a "condemned" man was actually revived. John Smith, for example, who supposedly was hanged in 1709, was subsequently revived and known ever after as "Half-hanged Smith."[21] Thomas Hill, hanged in 1743 for counterfeiting the duty-stamp on playing cards, came round on a surgeon's slab with an anatomist above him, scalpel in hand. "Twice-hanged Hill" was disposed of more professionally on the gallows in 1751 for committing another offense.

If the gallows and private purchase did not meet the demand, surgeons could fall back on "resurrection men." Astley Cooper told a House of Commons Committee that "there is no person, let his situation in life be what it may, whom, if I were disposed to dissect, I could not obtain." A body-snatcher reported to the same Committee that during the year 1809-10 he had provided "305 adults, 44 small subjects under three feet. . ., 37 for Edinburgh and 18. . . that were never used at all" to anatomical schools. Adults were sold for four guineas each, but "small ones were sold so much an inch" (Turner: 130). Two scandals in the late 1820s profoundly altered this situation. Burke and Hare in Edinburgh and Bishop and Williams in

London were discovered at about the same time to have provided surgeons with freshly-murdered specimens–even before they were buried.[22] Parliament passed in 1832 an anatomy act that empowered public institutions to dispose of cadavers to properly licensed medical schools. In due course Henry Gray (1827-61) produced in 1858 the first edition of a still continuing work, *Anatomy, Descriptive and Surgical,* and presumably based on a greater familiarity with human corpses. With 363 plates, it appears to have been only marginally more illustrated than the first edition of *De fabrica,* published some three hundred and fifteen years earlier.

Publishing History

Basel, 1543; First edition

Published by Oporinus, the first edition of *De fabrica* measured 42 by 29 cm and numbered 663 pages. It is divided into seven books: (1) Bones and Cartilage (72 woodcuts including three full-page illustrations of skeletons); (2) Ligaments and Muscles (36 woodcuts including 14 full-page); (3) Blood Vessels (22 woodcuts, with two full-page); (4) Nerves (9 woodcuts, two full-page); (5) Abdominal Organs (30 woodcuts); (6) Thoracic Organs (13 woodcuts); (7) Brain and Senses (21 woodcuts).

Lyons, 1552

Jean de Tournes published an unillustrated pocket edition in two volumes.

Basel, 1555; Second edition

Oporinus brought out an even more sumptious second edition, printed on better quality paper in larger typeface (49 lines per page rather than 57). It numbered 824 pages.

Venice, 1568; 1604

These posthumous editions were published by F. Sewense and sons.

De fabrica has never been translated in its entirety, nor has a modern edition ever been issued. Nonetheless and despite a sparse publishing history, its ramifications were considerable.

Fragments

"Some Observations on the Dissection of Living Animals," the last chapter of *De fabrica,* was translated by B. Farrington and appeared in *Transactions of the Royal Society of South Africa* (1931). He also translated the preface, which appeared in *Proceedings of Royal Society of Medicine* (1932). These fragments, together with other brief selections from *De fabrica*, are found in a number of readings and anthologies: Clendening (1960); Rook (1964); King (1971); and Runes (1962), and a dozen or more other volumes. A French translation of the preface, done by J. E. Vershaffelt, was published in Haarlem in 1924.

Facsimiles

A facsimile of the first edition was published in Brussels (1964), but an earlier attempt made in Leipzig in 1943 to celebrate the four-hundredth anniversary of *De fabrica*'s publication came to naught. The unsewn sheets appear to have been destroyed during the war. Curt Elze, who was planning to write an introduction to the edition, kept a set of the sheets which were later bound and can now be seen at the Yale University library.

Translations

Plans for a complete French translation were advanced in the 1920s, and the first six chapters were printed by Enschede of Haarlem. As the cost of producing these forty pages amounted to $1,800, it became evident that $40,000 would be needed to subsidize the complete translation and it was dropped. Cushing lamented, "Are there not, even now, forty people in the world who would be glad to give a thousand dollars to possess a French or English translation of this work?" (Cushing:95). A further attempt was begun by W. Wright in London in the late 1930s, but he died before making much progress. Ashley Montague in the 1940s was known to be preparing a translation and collation of the two editions of 1543 and 1555, but lack of funds also forced him to abandon the project. There appear to be no plans at the moment for what would be a very expensive project; nor is it likely that a translation will appear before the four hundred and fiftieth anniversary of *De fabrica*'s publication.

Epitome

Much of *De fabrica*'s impact came from the secondary literature it inspired. The most obvious example is Vesalius's own *Epitome*. Published in Basel in 1543 by Oporinus, it was printed in a 57 x 40 cm format, a size even larger than *De fabrica*. The work was intended for anatomy students and summarized the contents of the larger work. It contained nine plates and six brief chapters, and is no more than a sixth the length of *De fabrica*.

Epitome went through a number of editions. A German translation was published by Oporinus (Basel, 1543), and a Dutch translation appeared in Bruges (1569:105 pp.). Other editions: Basel, 1555 (second Latin); Paris, 1560 (without illustrations; 142 pp.); Wittenberg, 1582 (without illustrations, 108 pp.); and Wittenberg, 1603 (without illustrations, 110 pp.).

Collected Works

The *Opera omnia* was edited by H. Boerhaave and B.S. Albinus, published in two volumes in Leiden in 1724-25. No other edition has appeared.

The Blocks

Long after the death of Vesalius, the blocks of *De fabrica* and *Epitome* formed the basis for a number of anatomical works. The blocks were made in Venice and dispatched over the Alps to Oporinus in Basel. After the death of Oporinus they passed into the possession of Froben, another printing firm, which folded in 1603. A century later they turned up in Augsberg as the property of Ludwig Konig, a successor of Froben. Sold to Andreas Maschenbauer, a selection of the blocks were published in 1706 as works of Titian in his *Anatomie* (nineteen blocks) and reissued in 1723.

They disappeared again for three-quarters of a century, only to be found in Ingolstadt where in 1781 and 1783 they were again published, this time by H.P. Leveling, a Bavarian anatomist and surgeon, under the title *Anatomische Erklarung*. They disappeared again and remained hidden until 227 of the original 277 blocks were found in a cupboard of the library of the University of Munich in 1893. Remarkably, they were still serviceable despite being some three hundred and fifty years old and despite much use and considerable travel. An impressive limited edition was finally published as *Icones anatomicae* in New York and Munich (1934).

Writing of the blocks in 1943, four hundred years after they were first published, Harvey Cushing commented, "Time alone can tell what in another four centuries will happen to these historic wood blocks" (Cushing:108). Two years later, having survived the Thirty Years War, the French Revolution, and countless other upheavals, deaths, and bankruptcies, the blocks were finally destroyed in the Allied bombing of Munich.

Who was the artist who designed the blocks? It should be pointed out that *De fabrica* is one of the few books that gains immeasurably by being inspected in its original form. Although Vesalius's woodcuts will be familiar to anyone who has looked through the pages of Singer (1957) or any of the innumerable illustrated histories of medicine or science, the illustrations are considerably reduced, often to a quarter or less of their original size. When the muscle tabulae from *De fabrica* or even the larger format of *Epitome* are seen in their intended dimensions, they take on a magnificence undreamt of when thumbing through the pages of Singer. In the curious and often bizarre poses of the figures, the anatomical detail and clarity of the woodcuts lie evidence of an informed and powerful artistic imagination. Whose?

Some have attributed the work to Titian, the leading artist of Europe in the 1540s, but it has been objected that men of such eminence do not normally produce work on this scale – well over 200 blocks – anonymously. It was Maschenbauer who in 1706 first declared that the figures published in his *Anatomiae* were executed by Titian whom he favored over Kalkar because of the quality of the work. Vasari in *Lives of the Artists* had championed Kalkar despite the rather poor craftsman ship evidenced in his production of three blocks, produced in 1538, for *Tabulae sex*. Moreover, Vasari's knowledge of the issue has been called into question because he referred to the woodcuts in *De fabrica* as copper engravings.

The remaining candidate is Vesalius himself, already known as the designer of three of the plates in *Tabulae sex*. How he found time to write as well as to prepare over 200 detailed designs and fulfill his other obligations at Padua remains a mystery. The blocks and texts were completed by August 1542 and were unlikely to have been begun much before 1540, indicating just how efficient Venetian craftsmen could be. Two scholars who have made a special study of all Vesalius's illustrations concluded that the woodcuts "emanated from the atelier of Titian. . . . Kalkar and Campagnola and doubtless other artists participated in the work under the supervision of the master, but some of the plates were certainly the work of Vesalius himself" (Saunders and O'Malley: 29). Students of Vesalius may find this opinion unsatisfactorily vague and bland, if true.

De fabrica used not to be a particularly rare work. Osler notes earlier in the century that "copies are numerous and very often appear at prices ranging from £10 to £20 varying with condition." He himself had "distributed six copies to libraries." Indeed, on one occasion, he writes,

> Forgetting what I had done, I took out a copy in 1907 to McGill [University], and showed it with pride to Dr. Shepherd, the librarian, who pointed out in one of the show cases a very much better example presented by me some years before! Thinking it would be a very acceptable present to the Boston Library Association. . . I took the volume to Dr. Farlow who looked a bit puzzled and amused – "Come upstairs," he said; there in a case in the Holmes Room, spread open at the title page, was the 1543 edition and, on a card beneath it, "The gift of Dr. O."

Osler finally managed to donate it to the Library of the Academy of Medicine in New York. Such gestures were then open to wealthy physicians. Osler would have found it much more costly to indulge in such conspicuous generosity today. At the Honeyman sale in 1978-81 a copy of *De fabrica* in contemporary binding sold for £44,000 and in less prestigious salesrooms copies have gone in recent years for $24,000 while even a second edition has cost several thousand dollars.

Further Reading

Untranslated and available only in an expensive and limited facsimile reprint, *De frabrica* is likely to remain out of reach of most readers. Something of the richness of the illustrations can be gained from Singer (1957) and O'Malley and Saunders (1950). As if to compensate for the absence of his work, the life of Vesalius has been handled by O'Malley (1964) in one of the finest scientific biographies of recent years. It also contains extensive extracts from the Vesalian corpus. Works of Vesalius which are available and reveal the full power of his polemic are *The Bloodletting Letter* (1948) and extracts from the preface to *De fabrica* found in Clendening (1960), King (1971), Rook (1964) and O'Malley (1964). Also available is the *Tabulae anatomicae sex* which can be found in Singer and Raban (1946).

Notes

1. On Chinese anatomy Needham has commented: "I reflected on the strangeness of the fact that a similar rise and fall of ancient anatomy can be observed in China, beginning rather earlier with Pien Chhio, well attested in the time of Wang Mang (+9) and continuing a little later into the San Kuo period (c. +240) after which, as in Europe, it vanished until the late middle ages. Then the Sung anatomists precede Mondino de Luzzi by about a century but fail to go further" (Needham: 151). This however is from his Vol. I; we must await Vol. VI for further details; as of 1982, three substantial parts of Vol. V and all of Vol. VI remain to be published.

2. The early Church fathers were opposed to the practice of dissection. Tertullian in particular was especially strong in his condemnation; Augustine also wrote against it. In 1300 Boniface VIII issued a bull forbidding the practice of boiling flesh off bones, a custom apparently adopted by Crusaders in order to make possible the return of corpses for burial in Europe. I have never seen an explanation of why this practice was considered objectionable.

3. For Singer the beginning of "anatomical study" at Bologna and, therefore, modern Europe was in the decade 1266–1275.

4. One of the most popular books of the age. First printed in Padua in 1476, about forty further Latin editions were produced; vernacular editions are also known. Described as the "first work devoted to anatomy and not merely an appendage to a work on surgery" (Crombie: I,170). He wrote as if he had dissected in person. His work is little more than a dissecting manual even to the extent that it deals with the body in the order in which it would have to be dissected: he begins with the viscera and ends with the skeleton, a sequence reversed by Vesalius.

5. The clearest illustration of the tripartite division of labor can be seen in the *Fasciculo di medicina* (1493), reproduced in Singer (1957) p.77.

6 There is a fascinating account of the representation of the rhinoceros in western art by F.J.Cole, "The history of Durer's rhinoceros in zoological literature" in *Science, Medicine and History* (ed. by E.A. Underwood, 1953, vol. I). Cole showed how Durer who had never see a rhinoceros prepared in 1515 a woodcut of one. For various reasons he represented it not with two nasal horns but with one nasal and one much smaller spinal one. Once in the literature the image remained for nearly three hundred years despite the travelers, naturalists and engravers who must have noted the clear absence of a spinal horn from all specimens examined. It was only when James Bruce, explorer of Ethiopia, published his in his famous travelogue in 1790 that Durer's image finally disappeared from literature and art.

7. Cardano (1501-76) was a better mathematician – he was the first to solve cubic and quartic equations – than astrologer for in 1552 he predicted that the young 15-year-old king of England, Edward VI, would live to be at least fifty-five. Edward died the following year. For Vesalius, he predicted that the subject would be skillful with his hands. It is likely that Cardano, a Renaissance hustler, took the trouble to find out the identity of his subject in advance; even so, as in the case of Edward VI, he could be completely wrong.

8. Thus began the great tradition of Paduan anatomy which was to last well into the 17th century. Following Vesalius's tenure, there was Colombo (1542–59) whose *De re anatomica* (1559) demonstrated the pulmonary circulation; Fallopius (1559–62), author of *Observationes anatomicae* (1561) and discoverer of Fallopian tubes; Fabricius (1562–1604), teacher of Harvey and discoverer of valves in the veins; Casserio (1604–1616), author of *De vocis auditusque organis historia anatomica* (1601), a comparative study of the vocal organs and auditory anatomy of men, apes, sheep, pigs, and mice; and Spigelius (1616–27) who ended the tradition. In its place there arose a tendency to work on human anatomy and, with Harvey, a shift to physiological studies. The role of Padua in the scientic revolution was clearly a crucial one. In the 16th century Copernicus, Harvey, Vesalius and Galileo were all connected with it, the former two as students and the latter as teachers.

9. This important epistle is only partly about the china root and its properties; it is mainly concerned with a defense of his approach to anatomy against the intemperate and abusive attack of Sylvius, his former teacher.

10. At first he expressed relief that he would no longer have to "spend long hours in the Cemetary of the Innocents" or steal bones from gibbets. Another source of pleasure was his realization that he would no longer have to "put up with the bad temper of sculptors and painters." As for his unpublished writings he declared: "I burned everything with the idea that in the future I should write nothing more. However, I have regretted it several times, and I am sorry that I did not allow myself to be dissuaded by my friends who were present" (Cushing: 156).

11. One rumor was that Vesalius was conducting a post mortem on a Spanish nobleman when he discovered the heart was still beating. Accusations of murder and impiety were made against him and, sentenced to death by the Inquisition, he was allowed by Philip II to expiate his sins by a pilgrimage to the Holy Land. One reason for not taking the story too seriously is that similar charges of vivisection were made against all the leading anatomists of the century.

12. The sternum of the macaque monkey is divided into seven segments. The figure of the first skeleton from *De fabrica* (Singer 1957: 190) clearly shows the seven segments. Yet in man the sternum or breast bones, according to current anatomy textbooks, is a flat bone divided into three parts.

13. "Illustrated anatomical textbooks do not appear until the third decade of the 16th century" (Singer and Rabin:xxxi). Between the appearance of the first in the 1520s and *De fabrica* in 1543 less than a dozen illustrated such texts were published. None is remotely on the scale of *De fabrica*. John Dryander, for example, in *Anatomia capitis humanis* (1536), has eleven woodcuts while da Carpi in his *Commentary* (1521) had just one illustration of an internal organ. Contrast such work with the 200 or so woodcuts of Vesalius; this is an approach on a quite different scale from anything that had gone before.

14. Vesalius wrote to his printer, Oporinus, in Basel: "Greetings. Through the agency of the Milanese merchants, the Danoni, you will shortly receive, with this letter, the wood blocks for my books the *Fabrica* and the *Epitome*. . . . I carefully packed them with the help of the engraver. . . [and] I have written where they are to be placed.".

15. The frontispiece is a document rich in meaning and deserves careful study. It can be seen considerably reduced in Singer (1957). Above the title is the crest of Vesalius, three weasels *courant*—a play on the family name Wessels. Oporinus is shown in the gallery at the top, next to his monogram. Some have claimed to see on the book held by the central figure on the front row, with eyes acuter than mine, the letters S.C. for the artist Calcar or C.G. for the anatomist Galen.

16. Opinions in the early 16th century on the existence of the *rete mirabile* were varied. It was drawn by Leonardo and affirmed in the *Anatomiae* (1538) by Nicolo Masa (1480–1569) who wrote, "Some dare to say that the rete is a figment of Galen. . . but I have myself often seen the rete, and have demonstrated it to the bystanders. . . though sometimes I have found it very small" (Singer and Rabin: xxxi). In contrast da Carpi insisted that he never saw the *rete*. It was still being discussed in the late 17th century when in 1686 J. Peyer felt it worth insisting in his belief that the *rete* could not be found in man. References to the *rete* can also be found in the literature of the period. Thus in Rabelais (1494?–1553?), himself a well-known physician, there is an item in Xenomanes's anatomy of Lent: "His miraculous network like a head-stall" (Book IV, Chap. 13). The actual name *rete mirabile* appears in the 13th century as a mixture of the various terms used by Galen: he called it *thauma megiston* (greatest miracle) and *diktyoidies plegma* (net-like wreath). The two together become *rete mirabile*.

17. Vesalius was quite willing to contradict the views of Galen without reservation or comment: "The following chapter (Book I, Chap. 2) will teach the falsity of

Galen's statement in which he asserts that the capacity of the vena cava is very large when it is joined to the liver" (O'Malley: 167). It is much larger, Vesalius went on to argue, where it is joined to the heart.

18. Galen speaking about the aorta: "Some call it by that very name 'greatest' *(megiste)*, others simply 'the great' *(megale)*, other 'the thick' *(pacheia)*, and others 'the straight' *(orthe)*" (Singer 1956: 172).

19. There were in fact two works: the first, *Anatomia* or *Historia de la composicion del cuerpo humano* which first appeared in Rome (1556), was available in at least nine editions before 1682 and usually consisted of plates plus the text of *Epitome*; the second work, *Vivae imagines*, first published in Antwerp (1566), was known in at least five editions and normally contained just the plates. Vesalius expressed his opinion of the author: "Valverde who never put his hand to a dissection and is ignorant of medicine as well as of the primary disciplines, undertook to expound our art in the Spanish language only for the sake of shameful profit" (O'Malley: 294).

20. A few foundation dates of London hospital will give some idea of the expansion which took place in the 18th century: St. Thomas's (1693), Westminster (1719), Guy's (1725), St. George's (1734), St. Bart's (1739), London (1740), Middlesex (1745). Comparable American dates are: Charity Hospital, New Orleans (1736), Pennsylvania (1752), New York (1776), Bellevue (1794).

21. The practice and soubriquet appear to have been not uncommon. In 1724 in Edinburgh, for example, "Half-hangit Maggie Dickson" was revived by her family after her body had been snatched from the surgeons (Rae: 52). There is reason to believe that the hangman, for a fee, was prepared to adjust the noose in ways which would delay death.

22. Things were arranged better in Paris. The *chef des travaux anatomiques* saw to it that the two official anatomical schools, the École de la Médecine and the Hôpital de la Pitié, received an adequate supply of bodies from Parisian workhouses and hospitals.

CHAPTER 7

Siderius nuncius (1610)

by Galileo

I send herewith unto his Majesty the strangest piece of news (as I may justly call it) that he hath ever yet received from any part of the world; which is the annexed book (come abroad this very day) of the Mathematical Professor of Padua, who by the help of an optical instrument (which both enlargeth and approximateth the object) invented first in Flanders, and bettered by himself, hath discovered four new planets rolling about the sphere of Jupiter besides many other unknown fixed stars; likewise the true cause of the Via Lactae. . . and lastely, that the moon is not spherical, but endued with many prominences. . . . So, as upon the whole subject he hath first overthrown all former astronomy – for we must have a new sphere to save appearances – and next all astrology. For the first of the new planets must vary the judicial part, and why may there not yet be more?

– Letter of Sir Henry Wooton,
British ambassador to Venice, 13 March 1610

Life and Career

The name of Galileo has been familiar to many generations of students not because of his contributions to mechanics and astronomy but because of certain incidents in his career that are the stuff of legend – if not history. They are also incidents that have appealed to artists and dramatists over the centuries. In the first, Galileo stands atop the leaning Tower of Pisa with a 100-lb. weight in one hand and a 1-lb. weight in the other. Aristotle had claimed that the rate at which a body falls to earth is a function of its weight, but according to popular legend no one had actually bothered to test the belief by dropping bodies of different weights and observing the results. Galileo is thus represented as dispelling centuries of abstract nonsense at the drop of a cannonball and supposedly demonstrating the crucial role experiment was to play in the development of modern science.[1]

In a second incident the aged Galileo, threatened with torture and execution by the Inquisition, is finally forced to denounce the cosmology of Copernicus; on rising from his knees he stamps the floor and quietly, yet audibly, insists, "Eppur si muove" ("And yet it moves").[2] A third image, which was equally attractive to Victorian painters, is of Galileo in his confinement, lonely, isolated, and blind, visited by the poet John Milton (1608-74), who himself became blind in 1652. In several paintings Milton is shown looking through a telescope as the wonders of the heavens are described by the discoverer, a Galileo to whom they are no longer visible.[3]

There is still a fourth image, one more soundly based on fact that the others. It involves Galileo using the newly invented telescope to discover the features of the heavens unsuspected by the astronomers of antiquity. The failure of his predecessors and colleagues to observe the same phenomena has often been cited as an example of authority and bigotry succumbing to reason and evidence. In actual fact the response of the traditionalists was more complex and justified than is sometimes allowed. For Galileo it was nonetheless a timely discovery.

When Galileo first observed the moons of Jupiter in 1610 he was already forty-five, an age by which most scientists have completed their most important work. Galileo had previously failed to gain the public esteem or position he felt necessary to complete his work and desperately needed a discovery simple enough to present to the public and his patrons yet dramatic enough to catch their imagination. The sights in his telescope in late 1609 were, he must have realized, his last opportunity to establish a European reputation.

Galileo was born in 1564 in Pisa, the son of a musician and the eldest of seven children.[4] After studying medicine and philosophy at the University of Pisa he was appointed to the chair of mathematics there in 1589. The earliest lasting influence on Galileo was probably Archimedes, whose works were finally made available in fairly complete Latin translations in the 16th century.[5] What Archimedes had done for statics, Galileo resolved to do for motion. He had the mathematical ability, the physical intuition, and the experimental skills to work out for the first time an adequate mathematical treatment of the facts of motion and acceleration. Somehow all Galileo's natural talent failed to produce the comprehensive and definitive treatise that lay within his power. All he had produced before 1610 had been some minor work, among them *De motu* (1592), and, in 1606, an account of a new instrument he had designed some years earlier.

Galileo's private life helped to explain his comparative failure. His father died in 1591, leaving Galileo to provide dowries for his two sisters who married in 1591 and 1601. He also had to support his mistress, Marina Gamba, and their three children, two girls, born in 1600 and 1601, and a son, born in 1606. His post at Pisa had been full of conflict and was of small salary. When he moved to Padua in 1592 his salary there was only 180 florins a year (while his friend, Cesare Cremonini, professor of philosophy, was receiving 2,000 florins a year). The low wages and heavy domestic expenses forced Galileo to seek outside sources of income. He took large loans from the university, tutored many private pupils, and began to manufacture and sell mathematical instruments of his own design.[7] By 1604 he was seeking more lucrative employment, first with the Duke of Mantua and later with the Grand Duke of Tuscany. His reasons were the demands made upon his time in Padua. "Giving private lessons" and "taking scholars as boarders" did not leave him sufficient leisure to complete his works. If only he could have a 1,000 florins a year and sufficient free time, he wrote to Tuscan officials, he would at last be able to complete "two books on the system and constitution of the universe – an immense conception . . . three books on local motion, an entirely new science . . . three books on mechanics . . . also various little works on physical subjects" (Drake:160). To break away from the routine of Padua he first had to convince a patron that he was worth supporting. He was already well known as being disrespectful of authority, a disruptive influence, and extremely argumentative. A prospective patron could have been forgiven for choosing to overlook his many virtues and be warned off by his temperament.

In 1609, however, an instrument arrived in Venice that would revolutionize astronomy and fundamentally alter the fortunes of Galileo.[8] In Galileo's own words: "About ten months ago a report reached me that a Dutchman had constructed a telescope, by the aid of which visible objects, although at a great distance from the eye of the observer, were seen distinctly as if near" (Hurd and Kipling:139).[9] From his own "deep study of the theory of Refraction" he was able to work out the basic principles of the telescope and build his own instrument, in which he fitted "two glass lenses, both plane on one side, but on the other side one spherically convex and the other concave." Objects examined through it appeared "one-third of the distance off and nine times larger." Further work improved the instrument to such an extent that it magnified nearly a thousand times and brought objects more than thirty times nearer.

It is possible to be precise about the dates of the above sequence. The news of the instruments reached Galileo from Paris in July 1609 from a former pupil, Jacques Badovere. By 21 August Galileo was sufficiently adept to produce a model of his own and confident enough to demonstrate its power to the Senate of Venice from the campanile of St. Mark's. To the amazement of the Senators it proved possible to perceive the congregation of the Church of Saint Giacomo, Murano, some miles away. The instrument was presented to the Doge, and Galileo was suitably rewarded with a lifetime appointment and a salary of 1,000 florins a year. Padua had in fact treated Galileo quite handsomely. There was still no indication that the device would be other than a princely toy; at best, it was thought, it might find use at sea or in the hands of the military. Above all, there was no suggestion that it would reveal any secrets of the heavens, for to astronomers of both the Ptolemaic and Copernican systems there *were* no secrets in the heavens.

The assumption was still widely accepted that the heavens were incorruptible. Consequently, examining celestial bodies at close quarters would be pointless, simply revealing an homogeneous, flawless crystal structure of some kind. The battles of the craft of the 16th-century astronomer were fought on paper with calculations and geometrical constructions. As the heavens were changeless, the few observations needed were as acceptable from Ptolemy as they were from more recent observers. In such a tradition, the developments in trigonometry, the construction of tables of chords, and new calculating techniques offered hopes of fundamental advances rather than an instrument which allowed the view to see only what he could already see unaided. The strange phenomena in the sky – comets, the appearance of a new star in 1572, and yet another in 1604 – were dismissed as atmospheric rather than celestial events.[11]

Galileo was no astronomer; the drudgery of calculation and regular observation were alien to his personality, and the isolation of astronomical work did not suit his social nature. Drama, controversy, argument, and rhetoric were what fired Galileo. Yet when he did turn the telescope to the heavens in late 1609, seldom have major discoveries come so readily, so frequently and so fast. In little more than a few months Galileo was able to reveal the existence of at least three totally unexpected features of the heavens – all of which were to prove of fundamental importance. This was clearly no time for delay; and Galileo quickly (in January/February 1610) put together the twenty-four pages of *Siderius nuncius* – the last observation in it is dated 2 March – and rushed it to the printers. It was published on 12 March in Venice.

The Discoveries of "Siderius Nuncius"

The first object examined by Galileo was the moon. The traditional view of the heavens, it will be remembered, as expressed by the 13th-century Sacrobosco was: "The Heavens – around the elementary region revolves with continuous circular motion the ethereal, which is lucid and immune from all variation in its immutable essence" (Bagley and Rowley:140). Yet when Galileo turned his telescope to the moon he found that it was "not perfectly smooth, free from inequalities, and exactly spherical [but] on the contrary, it is full of inequalities, uneven, full of hollows and protuberances, just like the surface of the Earth itself, which is varied everywhere by lofty mountains and deep valleys" (Hurd and Kipling: 140).[11]

Galileo did not merely note the ravaged nature of the moon's topography; he went on to analyze it in some detail. The division between the dark and light parts of the moon was not perfectly elliptical, as it should have been if the moon were perfectly spherical. He also observed several bright areas in the dark part and many "blackish spots" in the part "flooded with the Sun's light." He noted that the spots have "dark parts toward the Sun's position, and on the side away from the Sun they have brighter boundaries." To explain the mystery Galileo sought a comparable experience on the earth's surface; around sunrise we see valleys still out of reach of the sun's path, but the mountains surrounding them opposite the sun are "already ablaze with the splendour of his beams."

The light spots on the dark part of the moon he explained in a similar way as being the peaks of the highest mountains catching the sun's rays: "when the sun has risen, do not the illuminated parts of the plains and hills join together?" (Hurd and Kipling: 141-42). Galileo had discovered a world undreamed of by previous astronomers; its mapping has occupied them ever since.[12] His own contribution was a rough calculation of the height of the mountains of the moon. Knowing the angle at which the sun's light struck them and the length of the shadow they cast, it was a simple problem to work out that some of them were about four miles high – a slight overestimate.

Observing the fixed stars, Galileo pointed out that, unlike planets, they did not present images of perfectly round discs, but looked pretty much the same in the telescope as seen with the naked eye – "blazes of light, shooting out beams on all sides and very sparkling." Nor did they increase in magnitude "in the same proportion as other objects." What did surprise Galileo and will give pause to anyone who first looks through a telescope was the "host of other stars, which escape the unassisted sight, so numerous as to be almost beyond belief." He added eighty stars to Orion's belt and forty to the Pleiades, while the Milky Way "is nothing else but a mass of innumerable stars plaited together in clusters. Upon whatever part of it you direct the telescope a vast crowd of stars presents itself to view; many of them are tolerably large and extremely bright, but the number of small ones is quite beyond determination" (Hurd and Kipling: 144).

So far Galileo had done little more than open his eyes; his next discovery required investigation. On 7 January 1610 he examined Jupiter through the telescope. He noted near the planet three small stars that attracted his attention because they seemed to be arranged in a straight line and were somewhat brighter than the rest of the stars.[13] Their relative positions were depicted by Galileo as follows:

East *x* *x* Jupiter *x* West

with *x* representing the position of each of the three stars relative to Jupiter. For no particular reason, "led by some fatality" was the phrase he used, on the following night he examined the same part of the sky. Much to Galileo's surprise and mounting excitement he found a very different state of affairs:

East Jupiter *x* *x* *x* West

"How," he asked, "could Jupiter one day be found to be the east of all the aforesaid fixed stars when the day before it had been to the west of two of them?" Could Jupiter have moved to the east? It was not supposed to do so at that time of year, but astronomers had been wrong often enough to make Galileo's original explanation the most likely. He waited eagerly for the next night to see if Jupiter had moved even farther east, only to find a sky "covered with clouds in every direction." On 10 January conditions were clear and he saw

East x x Jupiter West

with only two of the stars visible and still aligned with Jupiter. The third star he assumed to be hidden by Jupiter. The following night he saw only the two stars lying east of Jupiter, but now "the star furthest to the east was nearly twice as large as the other one; whereas on the previous night they had appeared of equal magnitude" (Hurd and Kipling:146). Such data could be explained not by the movement of Jupiter but only by the stars themselves "moving about Jupiter, as Venus and Mercury round the sun." Observations over the rest of January and February amply confirmed his original suppositions with a fourth "erratic sidereal body"[14] discovered later in the month. Kepler later suggested they be called satellites, from the word which originally referred to "an attendant upon a person of importance."[15]

Galileo in 1611 worked out the orbital period of the satellites as they accompany Jupiter in its 12-year journey around "the centre of the world". His results, with modern figures in brackets, were: *Io*–1.78 days (1.77); *Europa*–3.55 days (3.55); *Ganymede*–7.16 days (7.16); *Callisto*–16.75 days (16.69). Galileo's competence as an observer should not be allowed to obscure the magnitude of his achievement at this point. It does not belittle the skills of David Jewitt and G. Edward Daneielson of the California Institute of Technology to point out that in 1979 when they discovered a fourteenth satellite of Jupiter, only a few dozen kilometres in diameter, they were working with familiar concepts in a well-established tradition; they knew what they were looking for. In contrast, Galileo did not even have the concept of a satellite. If anything, the word would have suggested to him a flunky, a court attendant rather than anything to do with the heavens. It is hard for a generation that has seen satellites launched on T.V. with the regularity of a breakfast entertainment to realize that a major intellectual insight was called for to identify a few ill-defined stars in a primitive telescope as satellites.

In the Ptolemaic system there were no satellites; everything in the heavens simply went round the central earth. In the Copernican system with its central sun, the earth with its lunar satellite (the only satellite in the whole system) looked particularly anomalous. The temptation to note the existence of Jupiter's satellites and to pass on to something else must have been considerable; if Galileo had done so he would have been in good company. It is a well-established astronomical tradition that whenever anything new is found in the heavens, the observations of the past should be searched for any earlier trace of it. Invariably it appears to have been sighted several times before. (When the planet Uranus was first observed by Wiliam Herschel (1738–1822) in 1781, it was no exception.)

The point is not that others saw the Jovian moons before Galileo but that, in astronomy, it is no straightforward matter to see first what later becomes obvious to many.

One consequence of Galileo's discoveries was to make the moon's position as a satellite of the earth, in the Copernican system, more credible. Five planetary satellites in the heavens seemed more convincing that a solitary moon. Further, the mere fact that celestial bodies could be seen to orbit objects other than the earth was itself evidence against Ptolemaic cosmology. Galileo's *Siderius nuncius* contained even more reasons to doubt the traditional astronomy. Basic to the pre-Galilean position was the assumption that there was a fundamental distinction between heaven and earth, in physics as well as theology. The distinction was most apparent in two main areas: the constitution of objects in each domain, and their characteristic motion. Thus, heavenly bodies typically moved in circles, while bodies on the earth normally moved in straight lines or in some irregular manner. Earthly bodies were made from the normal elements of earth, air, fire, and water and were subject to change, decay, and corruption; in contrast, the celestial bodies were made from some special, literally quintessential substance, sometimes known as ether, which in the words of Dante was "lucid, dense, solid and polished" as well as incorruptible and changeless.

The effect of such distinctions was to require two kinds of physics: one developed to deal with the heavens, the other with the earth. Any approach or equations that failed to make the necessary distinction would be automatically dismissed as nonsense. In truth, of course, physics is physics, and bodies move, wherever they may be, in accordance with the same simple laws. This realization, rather than the change from the Ptolemaic to the Copernican cosmology, truly marked the great contrast between ancient and modern physics. The importance therefore of *Siderius nuncius* was that it was the first major work to begin to break down the assumption on which the traditional view rested. If the moon could be seen to have mountains and to be other than a perfectly smooth body, the heavens could no longer be seen as perfect and unblemished. Mountains mean change, a wearing away from an initially smooth surface. It was, Galileo claimed, just like the earth itself. His language at this point is revealing. Mountains are not just mountains or valleys simply valleys; instead they become "wrinkles," "excrescences," "small blackish spots"—terms which automatically assume any fall from smoothness and uniformity must be the result of decay.

The first step had been made. Sunspots in 1613 and the appearance of three comets in 1618 served to drive home the basic argument of *Siderius nuncius* that change was no stranger to the heavens.[16] The second step, the realization that the motions of heaven and the earth were basically the same, was beyond Galileo.[17] He made considerable steps in its direction, but the final insight was to come, as will be seen later, with Newton toward the end of the century.

Reception of "Siderius Nuncius"

Galileo was determined to gain maximum benefit from his *Siderius nuncius* (hereafter *SN*) and consequently sought the patronage of the Medici in his native Tuscany. He named the four new satellites the Medici planets (Medicea Sidera), dedicated the book to Cosimo II, the Grand Duke, and sent a telescope with a copy of the book to Cosimo in Florence. He also promised to visit him to demonstrate the newly discovered stars. At the same time he corresponded with Cosimo's secretary, B. Vinta (1542–1613), outlining the important works he could complete if only he could return to Pisa under the patronage of the Medici. Events moved quickly. *SN* was published in March; Galileo wrote Vinta on 7 May asking for 1,000 florins a year and unlimited free time; and on 22 May he was offered the post of chief mathematician at the University of Pisa with the salary and conditions he had proposed.

In September Galileo ended his eighteen-year stay in Padua, separated from his mistress, and returned to Pisa. It was to be a major mistake. His first problem, however, was to convince European scholars of the accuracy of his descriptions. It was surprisingly difficult in certain quarters, and the response of some astronomers to Galileo's work is in some respects the most interesting aspect of his story.

The first attempt to justify his work took place on 24–25 April 1610 in Bologna where, on his way to Pisa, he had stopped to show his results to Giovanni Magini (1555–1617), professor of mathematics there since 1588 when Magini had been chosen over the younger Galileo. The visit was a famous one and has since been much discussed, for Magini and his colleagues have since come to represent a particular kind of bigotry: bigots are so under the dominion of a particular theory that they are literally unable to see any conflicting evidence, however compelling and obvious it may be.

Also present on that occasion was Martin Horky (c.1590-c.1650/7), a pupil of Kepler and assistant to Magini, who sent an account of the meeting to his teacher dated 27 April. Although a far-from-objective account, it is a fascinating document. "Galileo Gallilei, Paduan mathematician, came to visit us at Bologna, bringing his telescope with which he saw four feigned planets. I never slept on the 24th or 25th of April, day or night, but I tested this instrument of Galileo's in a thousand ways, both on things here below and on those above. Below it works wonderfully; in the sky it deceives one, as some fixed stars are seen double. I have as witnesses most excellent men and noble doctors. . . and all have admitted the instrument to deceive. Galileo fell speechless, and on the 26th. . . departed sadly early in the morning. . . not even thanking Magini for his splendid meal" (Geymonat:45). Magini also wrote to Kepler at the same time affirming that Galileo "has achieved nothing, for more than twenty learned men were present; yet nobody has seen the new planets distinctly; he will hardly be able to keep them" (Feyerabend:123-24). A few months later, however, Magini's comments were a littled weaker, for he admitted, "Only some with sharp vision were convinced to some extent."

This is the reverse of the ease with which Renaissance anatomists could locate non-existent structures, such as the *rete mirabile* in the human body; here their astronomer colleagues seemed unable to see what did exist when it was pointed out to them. Horky rushed into print with his *Brevissima peregrinatio contra Nuncio Siderium* (1610), published in both Mantua and Bologna, in which he ridiculed the claims of Galileo. Oddly enough, support came first from Kepler, who was then in Prague where he was serving as Imperial Mathematician to the mad Hapsburg emperor, Rudolph II. Galileo had sent a copy of *SN* to the Tuscan ambassador in Prague, on 8 April it was passed to Kepler, who on 3 May published his reply, *Dissertatio cum Nuncio Sidero*—events certainly occurred rapidly in 17th-century science. The surprising feature of Kepler's work is that it was written before he had access to a telescope. It was nevertheless favorable: "I may perhaps appear rash in accepting your claims so readily with no support from my own experience," mused Kepler, but went on to still his doubts with the comment, "But why should I not believe a most learned mathematician whose very style asserts the soundness of his judgments?" (Rosen 1965:12-13).

Kepler's support seemed no better based than his general conviction that as Galileo would be unlikely to mislead his colleagues or to indulge in a hoax, his reports should be trusted. By August 1610 Kepler was feeling less confident. So many doubts had been raised by observers, those who unlike Kepler had actually looked through a telescope, that he wrote to Galileo asking for some favorable witnesses. Galileo's reply of 19 August probably did little to relieve Kepler, for after admitting that Italian astronomers were indeed hesitant to give their support he offered as a further witness "myself, who have been singled out by our University for a lifelong salary of 1,000 florins" (Koestler:381). Kepler must have been relieved when he had access to a telescope and was able to observe for himself the moons of Jupiter. He reported the results of his observations in a further work on the issue, *Narratio de observatis* (1611), which incidentally contained the first published use of the term "satellite."[18] (The first use in an English work was 1665.)

If Kepler could confirm the existence of the satellites without actually looking at them, others were equally confident, on no stronger grounds, of their non-existence. One was Cesare Cremonini, professor of philosophy at Padua and a good friend of Galileo who in 1608 had arranged for Galileo to receive a large personal loan from the university. Cremonini was no servile traditionalist; indeed, he was the kind of critical and independent Aristotelian the Church felt distinctly uncomfortable with. In 1611 he was investigated by Cardinal Bellarmino and the Holy Office but released; ominously, it was on this occasion that the authorities in Rome began to take an interest in Galileo. Their first step was to see if there were any link between Cremonini and Galileo.

In 1611 a friend of both scholars wrote an account of a conversation between himself and Cremonini. "Signor Galileo is anxious to know what is in your book," the friend, P. Gualdo, asked Cremonini, referring to the *Disputatio de caelo* published later in 1613. Cremonini replied, "He has no reason for anxiety, because I make no mention whatever of these observations of his. . . . I believe no one but Galileo has seen them; and besides, looking through these spectacles gives me a headache. Enough; I do not want to know any more about it." Cremonini concluded prophetically: "Oh, how much better Signor Galileo would have done not to enter into these capricious ideas, and not to leave the freedom of Padua" (Geymonat: 43-44).[19]

It was presumably this incident Galileo had in mind when he wrote to Kepler in 1611: "What would you say of the learned here, who, replete with the pertinacity of the asp, have steadfastly refused to cast a glance through the telescope? What shall we make of all this? Shall we laugh, or shall we cry?" (de Santillana: 9). Cremonini, who earlier refused to look through the telescope as a matter of principle, now seems to have been reluctant to accept results derived from some new instrument which might turn out to be worthless. His critical response would be very familiar to modern scientists. Today's scientific literature is full of claims made by operators of expensive and complex instruments for the discovery of a variety of exotic objects. Gravity waves, magnetic monopoles, quarks, decayed protons, black holes, and scores of other much-sought-after entities are regularly claimed as discoveries by enthusiastic workers in laboratories all over the world. No one takes them terribly seriously, even if they come from the most prestigious of institutions, for it is common knowledge how easy it is to be mistaken and how unreliable complex machinery can be; instead the emphasis is on repeatability. If results are so ephemeral that they can be obtained only be one group of workers, they are better ignored. Only data which can be established by a variety of workers, in a variety of institutions and with a variety of instruments, are worth discussing. There was something of this attitude in Cremonini.

The main problem Galileo faced was not Cremonini, but critics like Magini who did look through the telescope but saw none of the new worlds described by Galileo. Were they liars? Were they bigots? Too many observers failed to see Jupiter's moons to make such a simplistic answer convincing. Two general answers have attempted to account for this failure – the quality of the early telescopes, and the inexperience of the observers. Galileo reported to Kepler that many of the observers "are entirely unable to distinguish Jupiter, or Mars, or even the Moon as a planet" (Feyerabend: 124). Even today, the neophyte invariably experiences disappointment when first he looks through a telescope; the much-magnified photographs in books simply fail to appear. Even what he can see with the naked eye vanishes when he screws his eye up and peers through the lenses: unrelated blobs of light replace familar constellations, and even the moon can become an elusive object. With the early instruments such difficulties must have been even more acute. They had no fixed mounting, as refractors they must have suffered from severe chromatic aberration, and they were small and difficult to handle. It might also be wondered how much wine was drunk before the would-be observers failed to make the crucial sighting. Horky rebuked Galileo for failing to thank Magini for his "splendid meal;" the meal may well have been *too* splendid.

Such practical difficulties may have explained the failure of the various attempts to detect the moons of Jupiter, but another, more important reason was probably the observers' inexperience. Individuals must be taught how to read the plates produced with a spectrograph. The astronomer will glance casually over the spectrum clearly displayed and note the presence of hydrogen and the absence of sodium, and perhaps also make judgments about the temperature and distance of the stars involved. In contrast it is assumed that an observer need only pick up a telescope and, regardless of previous experience, be able to immediately *see* whatever lies in his field of vision. This may work if he is looking at a familiar object like a ship in the harbor or the porch of a local church; but presented with strange objects, an observer will either express his incomprehension or provide a variety of incompatible answers. Hence the report from Kepler that he saw square stars (?) or, to take a different example, the curious published descriptions of Saturn.

After telescopic observation in 1610, Galileo described Saturn as flanked on either side by a sphere, and compared it to two servants supporting an old man. Others thought it resembled a cup with handles. It was left to Christian Huyghens (1629-95) in 1659 to publish his view that observers had been looking at Saturn surrounded by an equatorial ring. It had taken nearly fifty years before astronomers could see an object that today is quickly picked up by inexperienced observers conditioned by photos and diagrams.

The critics of the Jovian satellites learned quickly; within a year most had publicly admitted they had been hasty in their rejection. One of the most distinguished of these was Chrisopther Clavius (1538-1612), who was well known for his work on the reform of the calendar and was Rome's leading mathematical astronomer. The first reaction of the seventy-two-year-old Clavius was scorn, but by December 1610 careful telescopic observation convinced him of the truth of Galileo's claims. A letter from Christopher Grienberger, colleague of and successor to Clavius, to Galileo in January 1611 expressed the view of Clavius and, in passing, managed to make the most sensible comment on the whole affair: "Things so hard to believe. . . Neither can be nor should be believed lightly; I know how difficult it is to dismiss opinions sustained for many centuries by the authority of so many scholars. And surely, if I had not seen, so far as the instruments allowed, these wonders with my own eyes. . . I do not know whether I would have consented to your arguments" (Pannekoek: 230). The Vatican went to Clavius to discover how reliable Galileo's results were. In consequence of Clavius's advice, when Galileo visited Rome in

March 1611 he was shown much respect and treated as an important guest. He took with him a telescope and met Pope Paul V and Cardinal Bellarmino; all appeared to be impressed.

Through these Jesuits the telescope and Galileo's discoveries spread far beyond the confines of Europe. By 1612 news of it had reached India. Anthony Rubino (1578-1643) wrote in late 1612 to Grienberger in Rome, saying that he had heard of a certain *occhiali* "by means of which . . . many discoveries have been made in the heavens." Would Rome send him one? If not, would they send him "the manner of their construction, so that I may have them made in this land of many officials and abundance of crystals" (D'Elia: 17).

In 1618 Terentius, a former student of Galileo and a Jesuit missionary, arrived in China with a telescope. Even earlier, Emmanuel Diaz, another Jesuit missionary in China, had published in Chinese his *Problems of Astronomy* (1615), in which he reported to his congregation: "Lately a Western scientist versed in astronomy has undertaken to observe the mysteries of the Sun, Moon and stars. But grieved at the weakness of his eyes, he constructed a marvellous instrument to aid them . . . Viewed with the instrument, the Moon appears a 1000 times larger . . . This instrument shows Jupiter always to be attended with 4 small stars moving around it very rapidly" (D'Elia: 18). The first treatise on the telescope, by A. Schall von Bell, was published in Chinese in 1626 and contained a plate of what was called the "far seeing optic glass." Yet not until 1640 was the name of Galileo – Chia Li Lueh – introduced to China; previously he was referred to as "a wise scholar from the west" (Needham: 444-5).

From China news of the telescope had reached Korea by 1631. It is heard of in 1638 in Nagasaki where, on a hill outside the town, it was placed to give early warning of the arrival of foreigners. In astronomy its first recorded use is in a work of K. Yoshinobu (1601-84), who used it to observe the Milky Way (Nakayama: 100). Westward the spread of knowledge was just as rapid. By 1672 John Winthrop Jr. (1606-76) was in a position to donate to Harvard a 3.5-ft. telescope. With this instrument Thomas Brattle observed the comet of 1680, the results of which observation were used and acknowledged by Newton in his *Principia* (1687) – perhaps the first recognition by the Old World of New World science.

Thus to the 45-year-old Galileo in 1610, life at last was beginning to look secure. Soon he would have a worldwide reputation, from Massachussets to Peking, and be one of the first scientists to be so noted. He had his job and position in Pisa, and approval for his work had come from the Vatican. Financially secure and honored, it was time for Galileo to turn to the ambitious scientific projects he had announced so eagerly to Vinta just a year before.

Galileo's Later Work

The wealth, fame, and freedom won by Galileo with *SN* did not in fact permit him to produce the major works about which he had written to Vinta so confidently. Much of his first ten or so years in Pisa was taken up with a long series of controversies. He first became involved wtih Christopher Scheiner (1575-1650), a Jesuit astronomer, in a priority dispute over the discovery of sunspots.[20] At the same time, in 1612, he was quarreling with the Aristotelians on why bodies float.[21] The disputes were drawn out, lasting at their height for several years, and involved more than one publication. The appearance of three comets in 1618 instigated a prolonged dispute into their nature – whether they were atmospheric phenomena or truly celestial – culminating for Galileo with his *Il saggiatore* (1623).[22]

The effects of such labors were two. First, they took up time: while Galileo was fashioning his polemical works he could not write the major works he had promised Vinta. Galileo's minor polemical tracts may well be of more interest than the major scientific works of almost any of his contemporaries; nevertheless, by the exacting standards by which major figures are judged, Galileo's work was minor and distracted him from the completion of the major works of which he was capable.

A second effect was to create enemies. In many quarters of Italy Galileo was not a popular man. His relish for debate and his ease with language frequently led him to make judgments and remarks which wounded those they were leveled against. On the death of Libri, a critic who failed to see the moons of Jupiter, Galileo commented: "Libri did not choose to see my celestial trifles while he was on earth; perhaps he will do so now he is in heaven" (Koestler: 374). As an example of the Galilean style and the ease with which he could reduce an opponent's position to nonsense, consider his objection to the traditional claim that the sphere is the most perfect of all forms: "I do not see how it is possible to assert that the spherical form is absolutely more or less perfect than the others. One can say this only with respect to some particular thing . . . for a body that must be able to rotate in every direction, the spherical form is most perfect; and for that reason the eyes and end of the thigh bones have been made by nature perfectly spherical. On the other hand . . . anyone who in the construction of a wall should make use of spherical stones would do very badly" (Geymonat: 52). Again there is a tendency to make the traditionalist appear not only misguided but ridiculous.

The main work of Galileo for some twenty years after *SN* was as a propagandist for Copernicus. His efforts culminated in his *Dialogue on the Great World Systems* (1632), an ostensibly impartial work but with an impartiality that fooled no one – certainly not the Holy Office nor the many who over the years felt they had been treated by Galileo with less respect than their due. Many old scores were settled when Galileo was brought before the Inquisition in 1633 and forced to affirm: "I do not hold and have not held this opinion of Copernicus since the command was intimated to me that I must abandon it" (de Santillana: 303). He was sentenced to imprisonment "during the pleasure" of the Inquisition, in effect for the rest of his life, but he was allowed to spend his confinement within the comfort of his own villa at Arcetri in the hills overlooking Florence.

It is ironic that at Arcetri, in the last years of his life, cared for by his daughter and attended by a few faithful pupils, blind and isolated, with all his works banned, Galileo finally produced the major work on physics he had promised his patrons for so many years. *Discourses and Mathematical Demonstrations Concerning Two New Sciences* (1638), published four years before Galileo's death, dealt with problems of engineering, the strength of materials, the first new science, and tackled the more fundamental problem of bodies in motion.[23] He developed the latter subject to its highest level before it received, later in the century, its definitive treatment from Newton.

Even after 1638 Galileo continued to work on scientific problems and the design of instruments he had first contemplated in earlier days. On his death in 1642, the Church informed the Florentine authorities that "it is not good to raise a mausoleum over the corpse of one who has been punished in the Tribunal of the Holy Inquisition and has died under that punishment" (Geymonat: 202). Consequently, a suitable monument was not allowed by the Holy Office to mark the tomb of Galileo in the Church of Santa Croce, Florence, until 1734.

Publishing History

Venice, 1610: First edition

Written in January and February, *Siderius nuncius* was published on 12 March. Five hundred and fifty copies were printed and sold out within a week. Written in Latin, it is dedicated to Cosimo II and consists of just twenty-four leaves.

Frankfurt, 1610: Second edition

Although the second edition was fifty-five pages, it was a simple reprint of the first. Galileo spoke vaguely about producing an Italian version, but, as with so many of his projects, nothing came of it. These are the only two editions published in Galileo's lifetime.

London, 1653

Pierre Gassendi (1592-1655) published the second edition in his *Institutio astronomica.* It was reissued in various forms in London (1675; 1683), Amsterdam (1682), and Cambridge (1702).

Opera

Siderius nuncius was repeatedly published as part of several editions of Galileo's supposedly collected works. Editions have been brought out in

Bologna (1655-56, 2 vols.), Florence, (1718, 3 vols.), Padua (1744, 4 vols.), Florence (1768, 3 vols.) Milan (1808-11, 13 vols.), Milan (1832, 2 vols.), and Florence (1842-56, 16 vols.) This latter edition, edited by E. Alberti, is the first to be serious in its claim to completeness. It was superseded, however, by the definitive Edizione nationale edited by A. Favaro and published in Florence (1890-1909, 20 vols.). It has since been reissued (1929-39).

Translations

Siderius nuncius was first translated into English by E. S. Carlos (London, 1880; reissued, 1960). Substantial portions of this translation have been used in anthologies and collections, such as Hurd and Kipling (1964, vol. 1); Shapley and Howarth (1929), and Stillman Drake (1957). An Italian translation was done by Maria Cardini (Forence, 1948), and a German translation was published in Frankfurt (1965). A facsimile of the first edition was issued in Pisa in 1964 to mark the four-hundredth anniversary of Galileo's birth.

Despite its brevity, *SN* regularly outsells Galileo's more substantial and bulkier works. The original 550 copies are much sought after. Whereas *Dialogues* could be bought in 1979 for less than £3,000, a first edition of *SN* was sold at auction the previous year for £17,000; another copy, in 1977, sold for $39,000.

It is worth considering why *SN* was published so rarely. Once it is removed from various collected works and its publication with works of other authors, as in *Institutio astronomica,* only the two editions of 1610 existed for more than two hundred years. One factor was size: publishers do not like to issue such small works because they are difficult to price and sell. Consequently, they try to find some other work to combine into a harmonious and more substantial volume.

There is another reason. *Siderius nuncius* was too successful; its results were simple enough to pass by word of mouth and, in any case, were rapidly incorporated into textbooks. What need then was there for it?

Further Reading

Pannekoek, Ley and Koestler should again be consulted although the reader should be warned to bear in mind Koestler's unconcealed bias against Galileo. The politico-religious aspects of Galileo's career are dealt with at length by de Santillana (1961) while two recent biographies, Ronan (1974) and Geymonat (1965), provide the basic chronology of his life and the events of 1609-10. Extracts from *Siderius nuncius* can be found in Hurd and Kipling (1964) and Shapley and Howarth (1929). Much relevant critical data can be found in McMullin (1967) and Drake (1980).

Notes

1. Did this event actually take place? Galileo himself made no reference to it. The story first appeared in the writings of Vincenzo Viviani (1622-1703), an early biographer of Galileo. His biography was completed in 1654 but remained unpublished until 1717. Viviani was writing of events he himself did not witness, but noted that Galileo disproved Aristotle's view that the velocity of falling bodies was proportional to their weight; "with repeated experiments from the height of the Campanile of Pisa in the presence of the other teachers . . . and the whole assembly of students" (Ley:109). Doubt was first cast on the truth of Viviani's account by a German scholar, Emil Wohwill, in 1909, and the whole issue was reviewed by Lane Cooper in *Aristotle, Galileo and the Leaning Tower of Pisa* (1935), who agreed with Wohwill.

2. This is most likely. It is quite clear that the Holy Office meant business and, furthermore, despite the respect shown to him, that Galileo was well aware of their ruthlessness. The curious source for this particular story was a portrait by Murillo (1617-82) of Galileo in prison, painted about 1640. The suggestion has been made that Galileo did utter the phrase, but on a less perilous occasion than before the Holy Office immediately after his formal abjuration.

3. Milton's comment on the visit in *Areopagitica* (1644) is: "There it was that I found and visited the famous Galileo, grown old, a prisoner of the Inquisition for thinking in Astronomy otherwise than the Franciscan and Dominican licensers of thought." Galileo's comment on his blindness is found in a letter of 1638: "Your dear friend and servant Galileo has for a month been hopelessly blind. You may imagine the affliction this causes me when you stop to consider that the sky, that world, and that universe which, by my remarkable observations and clear demonstrations, I had opened a hundred or a thousand times wider than anything seen by the learned of all the past centuries, is now diminished and restricted for me to a space no greater than that occupied by my own body" (Geymonat:163).

4. Galileo's father, Vincenzio Galileo (1520-91), was not just a performer, but a mathematician and musical theorist. In particular, he was interested in whether musical intervals could be determined on purely mathematical grounds, a belief he argued against in his *Dialogue on Ancient and Modern Music* (Ronan:59-62).

5. Two Latin editions of Archimedes appeared in the 16th century and would have been available to Galileo: the 1543 edition of Niccolo Tartaglia, and Commodino's edition of 1558.

6. The girls were placed in a Florentine convent in 1613 and became nuns. The impression is conveyed that they had very little choice in this decision. His son, Vincenzo, served as his father's amanuensis in later life.

7. In about 1597 Galileo invented his "geometric and military compass," also known as a sector, which readily allowed the division of a line into a number of equal lengths. It had numerous applications in surveying and gunnery. By 1606, when Galileo published his book on the subject, he had made about a hundred, most of which were sold. Details can be found in S. Drake, "Galileo and the First Computing Device," *Scientific American* (April 1976). Other instruments devised by Galileo include a pulsilogium for measuring the human pulse, air thermometer, giovilabio for computing the position of Jupiter's moons, and a hydrostatic balance.

8. There is absolutely no possibility that Galileo invented the telescope, nor does he make such a claim in *SN*. The instrument as a usable object seems to have been developed in the first years of the 17th century by a group of Dutch opticians. It is not clear whether the first workable model was constructed by Hans Lippershey (c.1570-c.1619) of Middleburg or his neighbor Zacharias Janssen (1580-c.1638). Lippershey had made one by 1608, but there are reasons to suppose that Janssen may have been successful in 1604. A third contender, Jacobus Metius (1580-1628), was an instrument maker from Alkmaar.

9. The translation here is inaccurate in that Galileo did not use the term "telescope". Dutch inventors had spoken of a *kijkglas*, from the verb *kijken*, to look, peep, while Galileo spoke indifferently in 1610 of an *organum, instrumentum* or *perspicillum*. The new term, telescope, was coined by Joannes Demisiani and was first recorded in English in 1648.

10. On 11 November 1572 Tycho Brahe was, as usual, observing the heavens: "I noticed that a new and unusual star, surpassing the other stars in brilliancy, was shining almost directly over my head. . . A miracle indeed. . . . For all philosophers agree. . . that in the ethereal region of the celestial world no change, in the way of generation or of corruption takes place" (from "De nova stella", 1573 [Runes:102-3]). It was a supernova, as was the appearance in 1604 of a new star reported by Kepler to be as bright as Jupiter. No supernova has been seen in our galaxy since then. As they were temporary, fading away within a matter of months, it was possible for them to be incorrectly identified as atmospheric events and even as reflections of terrestial events, such as the Massacre of St. Bartholomew's Eve (1572). In contrast, *SN* revealed *permanent* features of the heavens, much less easy for traditionalists to dismiss.

11. How did the scholars of antiquity reconcile their belief in the smoothness and perfection of the moon with the marking on it so clearly visible to the naked eye? According to Plutarch (c.46-c.126) in *The Face in the Moon*, the face was a reflection of the oceans of the earth. It nonetheless clearly worried scholars; a long confused discussion of the problem can be found in Dante *(Paradiso*, canto 2).

12. Galileo's observations here created a whole new field of study – lunar topography. He had included in *SN* a sketch of the moon's surface, giving prominence to the crater later named *Tycho*. Improvements in telescopes permitted Hevelius (1611-87) in *Selenographia* (1647) to produce a more comprehensive work. The first problem was to name the features observed. An earlier worker, Langrenus (1600-75), tried to impose Biblical names; Hevelius tried to apply geographical ones (the Alps, the Mediterranean), but few stuck. G.B. Riccioli (1598-1671) introduced in *Almagestum novum* (1651) the modern custom of naming lunar features after astronomers, philosophers, and other eminent thinkers.

13. Galileo had to face the inevitable challenge to his priority. Simon Marius (1570-1624), for example, claimed to have observed in his provocatively titled *Mundus jovialis anno 1609 detectus* (1614) the satellites in 1609, although his four-year period of silence added little to his credibility. A similar problem occurred in 1606 when Mayr claimed to have pre-empted Galileo in the discovery of the geometrical and military compass. Since he did not succeed in this matter, Mayr exacted a minor revenge. Galileo had proposed to name Jupiter's satelites after the Medici children. Reading outward from Jupiter, he recommended Catharina or Franciscus, Maria or Fernandus, Cosmus Major or Cosmus Minor. Mayr proposed an alternative set, called Sidera Brandenburgica in honor of his patron, and named them Mercurius Jovialis, Venus Jovialis, Jupiter Jovialis and Saturnus Jovialis – as a sort of miniature solar system. He also suggested using the names of four "illicit loves" of Zeus or Jupiter as possibilities: Io, Europe, Ganymede, and Callisto. He produced some Latin doggerel to go along with his suggestion: "Io, Europa, Ganymedes puer, atque Callisto / Lascivo nimium perplacuere Jovi" ("Io, Europe, the boy Ganymede, and likewise Callisto / Aroused to excess the lust of Jove"). These are the names that survived (Ley:262-63).

14. Jupiter actually has fourteen satellites, the last of which is only a few dozen kilometers in diameter and was discovered in 1979. Galileo and other 17th-century astronomers had clearly missed nothing, however, for the fifth Jovian satelite, Amalthea, remained undetected until 1892, when Edward Barnard (1857-1923) discovered it at Lick Observatory on Mount Hamilton, near San Francisco, using the newly-built 36-inch refractor. Galileo's refractor had a diameter of about one inch.

15. The first non-Jovian satellite was discovered by Huyghens in 1655 when he noted what is now known as Titan orbiting Saturn. Four further satellites were discovered by G.D. Cassini (1625-1712) of the Paris Observatory before 1684. This meant that, including the moon, ten satellites were known to 17th-century astronomers. The ability to predict their orbital periods and speeds would serve as crucial experimental tests of later gravitational theories.

16. Another important telescopic observation of Galileo was of the phases of Venus, first announced anagrammatically in a letter dated 11 December 1611. This was an important observation, for Copernicus himself had predicted that if the planets orbited the sun rather than earth, then Venus should present to a viewer on earth the same phases as were presented by the moon.

17. It would perhaps be more accurate to say that Galileo sought for a single frame of ideas within which both terrestial and celestial motion could be fitted but that he got it wrong. He developed a concept of uniform, inertial motion but argued that such motion must be circular. Galileo's innovation was to extend this concept to rectilinear motion. Thus, if all frictional constraints were removed from a ship in a calm sea it would continue to move, apparently rectilinearly; but in fact it would move in a circular path around the earth. Thus all inertial motion is circular. It took Newton to sort this one out.

18. Kepler obviously had his limitations as a telescopic observer. Not only did he see *square* stars, but reported that he saw "ten or more" moons.

19. These were prophetic words indeed. Padua, as part of the Venetian republic, certainly seemed more capable of resisting Vatican pressure than Florence as part of Tuscany, whose resolve could vary considerably with a change of ruler. Thus, when Cosimo II died in 1620, Galileo found himself serving the widowed grand duchess until the ten-year-old Ferdinando came of age. When called to Rome in 1632, Galileo knew he was in trouble and did everything he could to delay his journey. The publication of his *Dialogue Concerning Two Great World Systems*, (1632), in which he openly if circuitously favored the Copernican system, ran against express instructions he had received from the Inquisition. To delay his journey, Galileo solicited help from his doctor, sought the intercession of friends in Rome, and finally, as a last resort, turned to Ferdinand II. The prince replied that he was "unable to arrange that you may not go. Perhaps your prompt obedience. . . may reconcile the minds of those who seem to be stirred up against you" and went on to offer Galileo the use of one of his litters (Geymonat:142).

20. In addition to the usual priority dispute, Galileo claimed the sunspots to be genuine features of the sun's surface, and hence evidence of its mutability; Scheiner, who clearly beat Galileo into print, declared the spots to be small bodies distinct from but orbiting the sun.

21. Here the dispute was with Ludovico delle Colombe and concerned the question whether bodies floated because of their shape or because of their density. In a series of impressive experiments using wax mixed with lead shot, Galileo showed that variations of shape were irrelevant, while the weight of the water displaced by the body was crucial.

22. Galileo replied with his *Il saggiatore (The Assayer)* (1623) to the *Libra astronomica (Astronomical Balance)* (1619) of H. Grassi, another Jesuit astronomer. The issue was on the nature and motion of comets. Galileo mistakenly took them to be atmospheric effects. The orbit of comets whatever it might be was clearly not circular; consequently Galileo could not allow them to be real bodies moving in the heavens.

23. *Discorsi* was not published without difficulty; he very much wished to see it in print before he died. His problem was to find a publisher, since there was a ban on all his works. This meant going to a Protestant country, and for this reason *Discorsi* first appeared in Leyden in 1638.

CHAPTER 8

De motu cordis (1628)

by William Harvey

The heart in all animals has cavities inside it. . . . The largest of all the three chambers is on the right and highest up; the least is on the left; and the medium one lies in between the other two.

–Aristotle, *Historia animalus*

The heart is an exceedingly strong muscle. . . . It contains two separate cavities.

–Hippocrates, *De corde*

The cavities of the heart are four in number.

–Galen, *De usu partium*

The heart has twelve apertures.

–*Nan-ching*, a Chinese medical classic, Han period

Life and Career

The 1628 publication of *De motu cordis* by William Harvey gave the world its first great work of modern physiology. In contrast to *De fabrica*, nothing in the work reveals the author's personality behind the anatomical diagrams; no autobiographical diversions break into its prolonged and anonymous argument. Born in 1578 in Folkstone, Kent, Harvey came from a background of merchants and landowners; his father was sufficiently prosperous to serve in 1610 as mayor of Canterbury. From 1594 to 1597 Harvey was a Cambridge undergraduate, pursuing what was still a medieval course of study. The great days of Cambridge science lay in the future; in the late 16th century, the university library contained only 500 books and manuscripts, and anyone with scientific or medical ambitions found the best training at universities in Italy. Consequently, Harvey traveled in 1599 to Padua, the university of Galileo, in the steps of Copernicus and Vesalius.

At Padua Harvey studied anatomy under Fabricius (1537-1619), the discoverer of the valves in the veins. Although this important finding was not published until 1603 in *De venarum ostiolis*, Fabricius claimed to have known of them since 1574 and would almost certainly have demonstrated his knowledge of them to Harvey. In 1602 Harvey returned to England an M.D., and immediately began to establish a medical career. The 16th century witnessed the first attempt by English physicians to professionalize themselves. In 1518 under the guidance of Thomas Linacre (1460-1524), another Paduan graduate, physicians distanced themselves from barber/surgeons, on the one hand, from herbalist/apothecaries, on the other, by setting up the College of Physicians (later to become the Royal College of Physicians). They gradually acquired the right to license physicians in England, and in the 16th century granted licenses only to those able to pass an examination on the works of Galen. Only M.D.s of Oxford and Cambridge were exempt; accordingly, Harvey submitted himself for examination to the College and was granted permission in 1604 to practice.[1]

In addition to a private London practice Harvey took on a number of other commitments. From 1609 he served as physician at St. Bartholomew's Hospital, and from 1615 regularly gave anatomical lectures at the College of Physicians. The main focus of his life, however, changed in 1618 when he was appointed as physician to the Court. Thereafter he served both James I and Charles I until the latter's downfall in 1646.[2] These appointments were far from honorary and involved Harvey in extensive European travel.[3] He was with Charles at the Battle of Edge Hill in 1642 and was present at the siege of Oxford in 1646. After the downfall and execution of Charles, Harvey more or less retired from the public eye, even declining in 1654 the offer of the presidency of the College of Physicians.

Since Harvey was a public figure, many detailed descriptions of his personality and less-official duties have survived. The best is the affectionate portrait provided by John Aubrey (1626-97) in *Brief Lives*, where we learn that he was "very contemplative," suffered from gout, left £20,000 to his brother, "kept a pretty wench to wayte for him. . . .and made use of for warmeth-sake," and was probably addicted to opium in later-life. Another episode is related of the time when he was called to examine some supposed Lancashire witches in the early 1630s. Belief in witchcraft was widespread; James I had written in *Demonology* (1597) of his deep conviction not only in the existence of witches, but that they had "grown very common amongst us." Harvey himself seemed not impressed by witchcraft. Of one young girl supposedly bewitched, according to Robert Boyle, he thought "her strange Distemper to be chiefly Uterine, and curable only to *Hymeneal* Exercises; he advised her Parents. . . provide her a Husband" (Keynes 1978:212).

On another occasion Harvey visited an admitted witch and pretended to be a "wizard." While she was away on an errand for him he captured her toad, dissected it, and reported "the entrayles, heart and lungs. . . in no way differed from other toads, of which he had dissected many, ergo it was a playne natural toad" (Keynes 1978:214). While the above episode is known only from an 1832 article, and may be apocryphal, Keynes is not prepared to dismiss it and finds it "quite in line with his character."

The characteristic Keynes refers to here is the "direct experiment with his dissecting knife." Above all, Harvey was an experimenter of genius. His predecessors had conducted experiments of greater subtlety and complexity, but like all great experimentalists he adopted a simpler, purer style. For the Harveys and Rutherfords a few wires, some string, and the wisdom to ask precisely the right questions produced experiments of great clarity, capable of illuminating major areas of experience. Nor was Harvey's interest restricted to the cardiovascular system. Years of careful experiment and observation produced a second major work, *Exercitationes de generatione animalium* (1651), and only the destruction of many of his notebooks by parliamentary forces in 1642 prevented Harvey from completing a number of other projects.[4]

Harvey died in 1657, aged seventy-nine and wealthy, from a stroke, although a tradition persists that he actually took an overdose of opium after waking on the morning of his death and finding he could no longer speak. Aubrey strenuously denies the story, while admitting that Harvey had intended to ease death if necessary but that "the palsy did give him an easie passe-port" (Aubrey:291).

Early Views

If Harvey in the 17th century first showed that the heart was a pump forcing blood around the body, what did earlier anatomists and physiologists make of the heart? The same instruments and opportunities available to Harvey – scalpels, probes, ligatures, and experimental animals – had been equally available to anatomists of Alexandria two thousand years before.

We might begin with Herophilus, one of those Alexandrian anatomists, who lived in the third century B.C. Neither the Hippocratics nor Aristotle in extensive writings on the heart had distinguished between arteries and veins.[5] Herophilus, however, pointed out that all veins originated from the right side of the heart and were thin-walled; arteries had thick walls and could be traced back to the heart's left side. A further and more puzzling distinction was noticed at post mortems: veins were invariably found to contain blood, arteries were inexplicably empty – hardly the facts to suggest a circulation between the two.

Instead, a later Alexandrian anatomist, Erasistratus, drew the more natural conclusion that the arterial and venous systems were separate and independent. Each part of the body, every organ, had its own dual supply of arteries and veins; why else this duplication, Erasistratus argued, unless it was to supply to each part the different stuffs carried by the two systems? Veins clearly carried and supplied blood. By contrast, arteries were not really empty but full of air or *pneuma*, carried from the lungs to the left side of the heart by the pulmonary vein and then via the aorta to the body.[6] Erasistratus also noted a third network, the nerves, distributed throughout the body, and consequently concluded that there was some third vital substance transported to all parts of the body.

Given this basic scheme, later scientists could fill in the details as they were able. It was generally agreed that the venous system provided the blood which actually nourished the body. Its source was thought to be the liver: its rich blood supply suggested a tree with its roots in the liver absorbing the nutriments sent from the stomach and distributing them through its venous branches to all parts of the body. A common theory was that the arteries provided some kind of air conditioning. Air could enter into the system from the lungs, while those arteries close to the skin were thought capable of drawing air into themselves directly from the atmosphere, like a suction pump, when they dilated and relaxed at diastole.[7] (The early anatomists thought the heart and arteries were active at diastole.) Present in them was a *vis pulsifica* (innate pulsification) that expanded both the heart and arteries and sucked the *pneuma* into them. Contraction (systole) served the purpose of expelling waste products through the lungs and skin pores. The idea that the heart was a muscle had been consciously considered but rejected by Galen: "The muscles have uniform fibers, but not so the heart. . . the movement of the heart is involuntary and ceaseless. . . while that of the muscles is often suspended." For Galen, it was a cardinal principle that "all parts of the same substance are active in the same way"; it followed that "muscles do not have the same activity as the heart, for they do not have the same nature." They even taste different, he added (Graubard:16-17). It was thus thought, and continued to be believed by early Renaissance anatomists, that the heart and arteries dilated simultaneously and contracted simultaneously.

Galen proposed one further route into the arterial system. It had been often noticed that after death the left ventricle contained traces of blood, which, it was claimed, came through imperceptible pores in the septum dividing the heart. Once postulated, the pores came to acquire a number of uses. The *pneuma* of the arterial system could now be seen as a mixture of air and the more vital and spirituous fraction of the blood. Some of the *pneuma* could even be supposed to pass through to the venous system. Again, the discrepancy between the size of the incoming *vena cava* and the smaller, outgoing pulmonary artery could be understood on the assumption that some blood had been shunted to the left ventricle.

The first point that strikes a modern physiologist or anyone familiar with the cardiovascular system is how the ancients could have been so wrong. Could they not see that there were no pores in the inter-ventricular septum, that the heart and arteries neither contracted nor expanded at the same time, that the arteries contained the same blood as the veins, and that the blood circulates around the body? A number of factors prevented anatomists and dissectors for two millenia from observing a system familiar today even to young schoolchildren. First, it is extremely difficult to gain any understanding of the heart's operation by direct examination. Harvey spoke for generations of frustrated anatomists when he commented on the difficulty of trying to work out the sequence and location of systole and diastole in the heart: "In many animals this takes place in the twinkling of an eye, like a flash of lightning. Systole seemed at one time here, diastole there, then all reversed, varied and confused." So "truly difficult" in fact that, like Fracastorius, he was tempted to conclude that "the motion of the heart was to be understood by God alone" (*De motu cordis*, chap. 1).

Examination of the arterial and venous systems does not advance the matter much further: the links between the two systems are far from obvious. Where they join up in the lungs they are soon lost in the complexity of pulmonary tissue; at other sites in the body the blood vessels can be traced only so far before petering out in a network of ever finer dimensions. Some classical anatomists thought it likely that animal tissue was formed from this final intermingling of nerves, veins, and arteries. The idea that there could be connections, "anastomoses" in technical language, was impossible to support by simple observation. Even Harvey was forced to take the presence of such anastomoses on trust.

Third, even if someone had proposed that the blood were pumped around the body by the heart, he would inevitably have provoked the obvious question, why? It was far from clear to Harvey. As it happened, the circulation of the blood was established long before physiologists were aware of the interaction between blood and air in the lungs. Until then they found themselves in the uncomfortable position of having to argue for the existence of a major system with no apparent use. Until such a use had been plausibly demonstrated it was too easy to reject the very idea of a circulation as unnecessary.

The added difficulty of introducing a new approach, one which admittedly served no purpose, was compounded by an established scheme worked out by Galen, which was both functional and could cope with all the known facts. Galen, a keen experimentalist and experienced dissector – although probably more with animals than man – had established that arteries contain blood, not air, and that there was a connection between the arterial and venous systems. Evidence for this latter claim could be seen when an animal was killed by severing a large artery. He noted "you will find the veins becoming empty along the arteries; now this could never occur if there were not anastomoses between them" (Graubard:39). Such facts failed to suggest to Galen the idea of circulation. Indeed, he dismissed the idea as absurd as it entailed the view that "all the parts of the body would be nourished by one and the same ailment" – and how could the liver and lungs be nourished by the same blood? (Graubard:34). The actual details of the Galenic sytem, which came to be accepted by medieval and Renaissance scholars, can be briefly summarized:

Venous system: Galen based this on the liver, a clearly bloody organ with considerable venous connections and with undifferentiated tissue serving no other obvious purpose commensurate with its size. It is thus the site of blood production, made from food already converted into chyle and transported from the stomach and intestines through the portal vein to the liver. In the liver it combines with other substances to become the "natural spirits," which are then sent on a one-way journey through the venous system to all parts of the body. The natural spirits provide the body with its basic nutrition. As hearts and lungs are part of the body, they too require nutrition through their venous supply.

Arterial system: Galen proposed that some of the blood is carried by the *vena cava* to the right ventricle, where it passes through septal pores into the left ventricle. There it is mixed with air or *pneuma*, drawn into it from the lungs to form "vital spirits." These are distributed in an analogous manner to all parts of the body through the arterial system. Their role is to provide a quality such as animal heat or life, or even to exercise a cooling effect.

Nervous system: Galen theorized that some arterial blood reaches the brain, where it receives a third *pneuma* to form the "animal spirits." These are distributed through the nerves, thought to be hollow like blood vessels, to all parts of the body, where they serve as a basis for sensation.

The system, at least in its macro-physiological aspects, was rather neat. It linked the three main organs of antiquity – brain, heart, liver – with the arteries, veins, and nerves into a loosely connected system that provided a basis for the explanation of such fundamental processes as growth, sensation, heat, and life – all done, be it noted, in purely naturalistic terms. At the same time it took into consideration a number of otherwise puzzling aspects of the finer structure of the body. Why is the liver so richly endowed with blood vessels? Why are arteries and veins structurally different? Why does the blood in them differ? If anything were fundamentally wrong with the system, it was not obvious to experienced anatomists from ancient times to the Renaissance. Had the idea of the circulation of the blood occurred to any of them, it would have been quickly dismissed.

Precursors of Harvey

In the fourteen hundred years separating Galen and Harvey, two significant additions were made to the basic structure of the Galenic system. The first was the discovery, by a number of hands, of the pulmonary or lesser circulation – the passage of blood from the right to the left side of the heart via the lungs. Some of the early discoveries of the subsystem were so well hidden as to have had no real impact on anatomy. The Persian scholar Ibn Nafis,[8] for example, spoke of such a system in the 13th century as, two hundred years later, did the strange Michael Servetus (1511-53).[9] A native of Aragon, Servetus was one of the obstinate figures typical of the Reformation who, convinced of an unpopular truth, insisted on proclaiming it loudly, frequently, and in the most inappropriate circumstances. Servetus's own peculiar compulsion was to deny the Trinity, a dogma he claimed to be unbiblical, but which nevertheless was one of the few beliefs the denial of which could get a man burnt by Catholics and Protestants alike. When in 1553 he published a 700-page *Christianismi restitutio* which argued forcibly for his unitarian position, Servetus was soon imprisoned by the Inquisition of Lyon. Escaping, he traveled with incredible folly to Calvin's Geneva where, with Swiss efficiency, he was burnt at the stake. Servetus had chosen to reveal the details of the pulmonary circulation in *Christianismi restitutio*. As most copies were destroyed by the authorities of various countries, it is unlikely that his views were widely known. Only three copies of the work have survived, and Harvey was likely to have been as ignorant of the Latin of Servetus as he was of the Arabic of Ibn Nafis.

Not until the publication of *De re anatomica* by Realdo Colombo (1510-59), student of and successor to Vesalius at Padua, did the new doctrine receive public attention. Colombo stated clearly that blood did not pass through any septal perforations in the heart, but is "carried to the lung by the *vena arteriosa* (pulmonary artery)... and then, along with the air, is taken to the left ventricle of the heart through the *arteria venalis* (pulmonary vein), a fact which no one thus far has either noticed or recorded in writing" (Graubard:94). Similar views can be found in a number of other late 16th-century writers. One of them, Andrea Cesalpino (1519-1603) in *Quaestionum peripateticorum* (1571) is thought to have used for the first time the term *circulatio* to describe the passage of the blood across the lungs.

The other item of significance was the already mentioned discovery by Fabricius of the valves found in the veins. As an old man, Harvey singled out this particular discovery as influencing him more than anything else. Why, he had asked himself, were valves "so placed that they gave free passage to the Blood Towards the Heart, but oppos'd the passage of the Venal Blood the contrary way?" Nature was unlikely, he added, to adopt such a structure without "Design." So he concluded the "Blood could not well, because of the interposing Valves, be Sent by the Veins to the Limbs; it should be Sent through the Arteries, and Return through the Veins, whose valves did not oppose its course that way" (Keynes 1978:28). As will be seen below, in *De motu cordis*, Harvey quoted another source for the idea of the circulation of the blood.

Even with the lesser circulation and the presence of valves in the veins, it should be emphasized that Colombo and all other anatomists still continued to think in terms of Galenian physiology. Harvey stole no one's thunder in 1628. He may not have worked in a vacuum, and he clearly had learned from the Italian anatomists of Padua and elsewhere; nevertheless, the discovery of the circulation of the blood was Harvey's and Harvey's alone.[10]

The Circulation of the Blood

In *De motu cordis*'s prefactory letter to John Argent, president of the College of Physicians, Harvey claimed: "On several occasions in my anatomical lectures I revealed my new concept of the heart's movement and function of the blood's passages round the body. Having now, however, for more than nine years confirmed it in your presence . . . I have yielded to the repeated desire of all and the passing request of some, and in this small book have published it for all to see." The lectures referred to were the Lumleian, basic anatomical lectures, first given at the college in 1582, and by Harvey first in 1615. His prefactory letter seems to suggest that his revolutionary views had been formulated as early as 1619 and possibly even earlier, and, further, that he had demonstrated them in his classes at the college.

A transcription and examination of Harvey's lecture notes by G. Whitteridge in 1964 failed to confirm this claim.[11] It is always possible, of course, that his lectures differed radically from the preserved notes, but if so it poses the new problem of why Harvey chose to wait so long before publishing his views. A scientist does not normally delay ten years or more before publishing his life's work.[12] Another possibility was that the basic idea had occurred to him ten years or so before publication, and that the intervening years were spent on its development and proof.

Whatever the reason, when *De motu cordis* (*DMC* henceforth) finally appeared, it was a slim volume of only seventy-two pages. Less the dedications, prefaces and introductions, the actual text for the argument runs to little more than fifty pages in seventeen chapters. Harvey begins by dismissing previous accounts of the heart and its functions as "obscure, inconsistent and impossible," yet his own particular argument is also somewhat disorganized. Points are made, dropped, returned to, and developed, only to be dropped again while positions already established are summarized for a third time. Within the repetition and unexpected breaks Harvey employs three main types of argument.[13]

The first consists in pointing out the absurdities, anomalies, and difficulties facing anyone who adopts the traditional Galenic view of the heart and its functions. This method of attack is continuous, and is used in virtually every chapter of the book. In the Introduction a battery of questions, carefully thought out, show that the traditional system could not work: "How may the arteries of the foetus draw air into their cavities through the mother's abdomen and the uterine mass? How many seals, whales, dolphins, other species of cetaceans and all kinds of fish . . . draw in and give off air through the great mass of water . . . If the arteries during systole exhale waste vapours . . . through the pores of the flesh and skin, why not at the same time the spirits said to be contained within them, for spirits are much more volatile than sooty wastes?" Why is so much blood sent to nourish the lungs? Why, if the pulmonary vein connecting the lungs to the heart carries air, is it not made from ringed tubes like the bronchi? The queries are still pouring out in the last chapter: Why should stronger and sturdier animals have a "more thick, powerful and muscular heart" than those with a more slender frame? Why are the arteries nearer the heart "stronger and more ligamentous than those terminal branchings. . . so similar to veins in make-up that it is harder to tell one from another by ocular examination of their tunics." The objections are based on common observations already known to many, but their cumulative impact must have been considerable. Seldom can a theory have been shown to contain so many absurdities and anomalies.

Harvey's second main line of argument describes the results of his anatomical dissections. Little, he knew, could be obtained from the working hearts of apes, pigs, dogs, or any of the other animals normally used by anatomists. Everything was too quick and confused to allow anyone to extract a coherent picture. To avoid this problem Harvey turned to "snakes, frogs, snails, shell-fish, crustaceans and fish,"[14] whose slower and simpler hearts with fewer chambers and vessels allowed the uncommitted anatomist to observe the actual sequence of events in the systole/diastole cycle. From the hearts of eels, serpents, and fish Harvey made the crucial observation: "The heart does not act in diastole but in systole for only when it contracts is it active" (DMC, chap. 2). In support Harvey noted: "At the instant the heart contracts, in systole,. . . the arteries dilate, give a pulsation and are distended. Also when the right ventricle contracts and expels its contents of blood, the pulmonary artery beats and is dilated" (DMC, chap. 3).

This was the first observation in the history of physiology not just to refute the traditional picture, but to suggest a new one, that the heart actually pumps blood. Where did the blood come from? In the lungless fish Harvey noted the blood being squeezed out of the single auricle directly into the single ventricle, from one side of the heart to the other. In the toads and serpents "it is obvious. . . that the blood is transferred from the veins to the arteries by the heartbeat" (*DMC*, chap. 6). The next stage was the mammals, approached cautiously by way of their embryos in which the lungs "are idle as though they did not exist," but in which there was no difficulty in tracing the passage of blood from the right to the left side of the heart.

By direct observation Harvey had thus traced the source of the blood pumped from the left ventricle into the arterial system. In man all that could be seen was the movement of blood into the lungs via the pulmonary artery and the separate arrival of blood in the left ventricle via the pulmonary vein from the lungs; there was an observational gap that Harvey could not bridge. Since the two sides did connect in all the simpler cases, Harvey confidently claimed that they also did in man, even though this involved the passage of the blood through the lungs. Why else, he reasoned, should vessels from the heart to the lungs be so substantial? Why, also, did a complex system of valves at the opening of the pulmonary artery prevent blood flowing back to the heart? It was no more difficult to suppose that blood passes through the spongy lungs than to suppose that "alimentary juices filter through the liver" or that urine passes through the kidneys or that sweat goes through the skin (chap. 7).

Having established the reality of the pulmonary circulation, Harvey developed his third main argument in chapters eight and nine[15] and in so doing produced one of the earliest uses of calculation in physiology.[16] At the beginning of chapter eight Harvey explained how the idea of the circulation of the blood occurred to him. He had begun by pondering how much blood might be lost from a severed artery, on the role of the valves found in the heart, and other "such matters." But then he began to consider "how much blood is transmitted, and how short a time does its passage take. Not deeming it possible for the digested food mass to furnish such an abundance of blood . . . unless it somehow got back to the veins from the arteries and returned to the right ventricle of the heart, I began to think there was a sort of motion in a circle" (chap. 8).

The next chapter gives facts and figures. The left ventricle holds from 1.5 to 4 ounces of blood when full. The heart beats and pumps blood into the circulation 2,000-8,000 times an hour, but because of the valves of the heart no blood can return into the left ventricle from the aorta. Harvey began his calculation. Suppose, he declared, as little as one-eighth the contents of the left ventricle were discharged with each beat, and using the smaller figure of 2,000 heartbeats an hour some fairly astronomical figures would soon be reached. Simple calculations clearly showed that "in a whole hour, or in a day, . . . more blood is continuously transmitted through the heart, than either the food which we receive can furnish, or is possible to be contained in the veins" (chap. 9). If only one dram per beat were expelled and the heart beat 1,000 times per half hour, the liver would have to produce in that period more than ten pounds of blood. Compared to a lactating mother who, according to Harvey, produces *daily* no more than three-four pounds of milk, the figures required by the Galenic system were absurd. He drew the obvious conclusion: "But however, though the blood pass through the heart and lungs, in the least quantity that might be, it is conveyed in far greater abundance into the arteries, and the whole body, than it is possible that it could be supplied by juice of nourishment which we receive, unless there were a regress made by its circulation" (chap. 9).[17]

Harvey had thus established the existence of the pulmonary circulation and had also produced a general argument for the presence of a further circulation from the aorta back to the heart. The details and the direction of the circulation remained to be demonstrated. That blood flowed from arteries to veins Harvey showed by a simple demonstration known to Galen: "If an opening be cut . . . even in a small artery . . . the whole blood content may be drained from the entire body, veins as well as arteries, in almost half an hour's time" (chap. 9). The truth of this argument from phlebotomy was demonstrated with the effect of ligatures. A ligature applied tightly above the elbow caused the arterial pulse to disappear at the wrist, while the artery above the ligature became swollen as if its contents "were trying to break through and flood past the barrier." When the ligature was slackened the veins below it became "varicosed" and "swollen," which showed that blood "passes from arteries to veins, not the reverse."

To establish that blood returned to the right side of the heart through the venous system, Harvey, in chapter thirteen, spoke of the functions of the valves first described by his teacher, the "celebrated anatomist Hieronymous Fabricius of Aquapendente." The siting of the valves would not allow a probe "from the main venous trunks very far into the smaller branches," but presented no obstruction "in the opposite direction, from the branches toward the larger trunks." In other words, the valves permitted blood to flow in one direction "towards the heart, from the periphery inwards."[18] Chapters fourteen-seventeen were devoted either to summaries of positions already established or to a number of miscellaneous points.

Harvey realized that his position did not explain two major problems. The first was the problem of anastomoses. Galen had declared there were pores in the heart's septum through which the blood passed from the right to the left side – unfortunately, according to Galen, too small to be perceptible. In strong and confident language Harvey replied: "But, damn it, no such pores exist, nor can they be demonstrated!" (Introduction). In the same manner the actual links between the arteries and veins could be seen nowhere, and Harvey was forced, like Galen, to assume the existence of imperceptible channels. The point troubled Harvey for the rest of his life, and in a later work on the circulation he confessed: "I have never succeeded in tracing any connection between arteries and veins by a direct anastomosis of their orifices [despite the fact that] by boiling I have rendered the whole parenchyma of those organs so friable that it could be be shaken like dust from the fibres, or picked away with a needle until I could trace the fibres of every sub-division and see every capillary filament easily" (Keele:146). The problem was solved shortly after Harvey's death. In 1661 M. Malpighi (1628-94) identified, under his microscope, capillaries in the lungs of a frog joining the ends of the arteries to the termination of the veins. Harvey seems never to have used a microscope,[19] not even in his embryological work, where it could have been of considerable value. The most he used was a *perspicullum* or lens.

The second major difficulty facing Harvey was to find a use for the elaborate circulation he had detected, since no one, he admitted, could explain why nature needed such a circuit. In *DMC* he suggested that "the blood may go through the lungs, to be cooled by the inspired air and saved from boiling and extinction" but went on to add, "There may be other reasons" (chap. 6). When later he replied to the criticisms of Riolan (see below) he faced the similar objection with the uncompromising response: "the facts manifest to the senses wait upon no views, the works of Nature upon no antiquity" (Keynes 1978:326). The facts were as they were and could not be discounted simply because they could not be fitted into some available explanatory scheme. Once more Harvey's patience and good sense were justified. After his death John Mayow (1640-79), an English physician, published his *Tractatus quinque* (1674), in which the first insight was gained into the complex interaction between the gases in the lungs and the constituents of the blood.

A residual problem about *DMC* begins with the title, which mentions the heart and not the circulation of blood. The first extravagent reference to the heart is in the address to King Charles, where Harvey presents the image of the King as the basis of his kingdom, the sun of the microcosm and the heart of life. It is pursued further in chapter eight in an untypically rhetorical passage: "So the heart is the beginning of life; the Sun of the Microcosm, even as the Sun in his turn might well be designated the heart of the world." Yet when Harvey came to publish his embryological work in 1651, *De generatione animalium* (hereafter, *DGA*), the emphasis had been changed. Detailed study of the sequence in which organs appear in chick embryos had convinced Harvey that blood is "prior to its receptacles," i.e., was formed before the heart and vessels. He consequently began to stress the importance of the blood; no longer was the heart the basis of life, instead it is blood which is "author and preserver of the body, and that element in which the vital principle, the anima or sensitive soul has its dwelling place" (Keele:197).

The historian Christopher Hill has tried to show how Harvey's beliefs fitted into general 17th-century religious and political thought in a way which illuminates other aspects of Harvey's life and work. Hill's unrivaled knowledge of 17th century literature has developed in him a nose for heresy and subversion only slightly less acute than that of Torquemada of the Spanish Inquisition. In Harvey he claims to detect signs of the heresy of mortalism, the view that the soul dies with the body at death and must await until Judgment Day for its resurrection. In the intellectual climate of the day mortalism was no trivial eccentricity; the last Englishman to be burnt as heretic had been Edward Wightman in 1612 for the claim that "the soul is mortal. . as the body is." Thus Harvey's claim in *DGA* that the soul or the anima was intimately connected with the blood could carry the implication that it too died with the blood. But this is mortalism and not something to be declared openly and publicly.

Thus in *DMC*, published in 1628, Harvey had to be careful. A full-hearted dedication to the king might protect his ideas if it could thus be insinuated that the work was written under the patronage of King Charles. Hill suggests Harvey prudently delayed publication of *DMC* and decided to publish abroad; he also confused the issue by emphasizing the role of the heart in the title at the expense of the blood. Hill notes that mortalism flourished in Padua and amongst physicians, two clear links to Harvey.

Such arguments are little more than guilt by association and speculations about Harvey's motives. They do, however, serve to emphasize that there were currents in Harvey's thought related to his background that we can only sense rather than analyze. Hill claims to see in Harvey's physiology an expression of the political conflicts of what was in many respects the most turbulent period of English history. Finding a parallel between Harvey's discoveries and political life, Hill writes, "the heart receives life and being from and ministers to the blood, as a constitutional monarch draws his being from the electorate and ministers to it." Just as Newtonian physics is the "ideological analogue of monarchy limited by law", Harvey's anatomy is "the analogue of monarchy limited by representative assemblies" (Hill: 67).

Reception

After the publication of *DMC* Harvey is said to have told Aubrey "that he fell mightily in his Practize, and that 'twas believed by the vulgar that he was crack-brained; and all the Physitians were against his Opinion... many wrote against him. With much adoe at last, in about 20 or 30 years time, it was received in all the Universities of the world" [Aubrey: 289-90] In actual fact if the published record is examined it appears that many of the leading scholars and physicians of Europe expressed their support for Harvey without undue delay. Keynes has listed about 70 works which refer to Harvey's theories between the date of publication and 1657, the year of his death, and while few were in complete agreement, many were convinced by Harvey's discoveries.

Robert Fludd (1574-1637) was the first to publicly declare himself in *Medicina catholica* (1629). Fludd, a physician and friend of Harvey, was a strange figure on the fringes of 17th-century science. He and Harvey became members of the College of Physicians about the same time and later worked together on *Pharmacopoeia Londoniensis* (1618). Fludd later became a Rosicrucian and Hermeticist, a Renaissance magi rather than an experimental scientist, more concerned with the relationship between the microcosm and macrocosm than the careful observation of hearts and lungs in fishes and frogs. People like Fludd spent their whole lives in a state of constant surprise and excitement at the number of correspondences and analogies they could find between the most unlikely branches of nature. So he was eager to embrace Harvey's new theory and with no delay at all spotted the "impress both of the planetary system and of the zodiac" on the blood: "Thus as the moon follows her unchanging path, completing her journey in a month, she incites the spirit of the blood, and therefore the blood itself, by virtue of its *astra imperceptibilia* to follow in a cyclical movement. Every seaman is acquainted with the influence of the moon on wind and tide. Why should she not exercise a similar influence on the microcosm of man?" (Keynes 1978: 135-6).

Was Harvey himself influenced by the Hermetic tradition? Science historians in recent years have dwelled on the development of experimental science and its relation to the practice of magic during the Renaissance. Many connections have been traced between them although their significance has not yet been fully worked out. So far little has emerged to link Harvey with the work of scientists of Fludd's ilk, although two passages in *DMC,* already quoted, speak somewhat extravagantly of the sun, the heart and the monarch. Are these Hermetic overtones or simply rhetorical conceits so typical of the period? On balance, any judgment must lie with

Harvey the experimentalist as against Harvey the magi. He could hardly have derived his theory of the circulation of the blood from a consideration of the sun's position in the universe or the monarch's in his kingdom; rather, having established the theory, such analogies began to appear both pleasing and in some sense supportive of an undoubtedly controversial position.

The first attack on *DMC* came in 1630 from James Primrose (c.1598-1659), a physician and pupil of Riolan, who dismissed as irrelevant Harvey's investigation of the "pulsatile heart in slugs" and reaffirmed the virtues of Galen without any new evidence. Harvey adopted a posture of public aloofness, declaring, on one hand, "If I am right, sometime, in the end the human race will not disdain the truth"; and, on the other, "no man hath hitherto made any objection to it greatly worth a confrontation" (Keynes 1978:322). At a later date, he was less restrained. "It cannot be helped that dogs bark and vomit their foul stomachs," he remarked in a letter to Riolan in 1649. Nor can one help being sympathetic to Harvey when replying to objections from Caspar Hofmann (1572-1648), professor of medicine at Altdorf, "I. You appear to accuse Nature of folly in that she went astray in . . . the making and distribution of food. . . . II. You appear to disapprove . . . of the universally accepted view of Nature, that she is not lacking in essentials" (Keynes 1978:233). Harvey tried to put some substance into Hofmann's charges but his heart was not really in it and after a couple of pages he excused himself: "My fingers are already painfully tired from holding the pen and other matters call me away."

One critical work elicited a substantial reply from Harvey: *Enchiridium anatomicum* (1648) by Jean Riolan (1580-1657), court physician to Henry IV and Louis XIII. It is not clear why this work aroused Harvey to publish a small book containing two letters of *exercitationes* (1649). It contained nothing new nor was it especially well argued. But it gave Harvey an opportunity to make public further experimental support for his theory. It is salutory to consider one of them for it shows Harvey's unrealized grasp of the nature of the blood's circulation as we understand it today. He correctly explains why the arteries at death are often empty of blood: "When the lungs subside . . ., they are no longer respiring. So the blood cannot pass freely through them. The heart, however, continues for a space of time to force blood out" which inevitably empties the arteries and left side of the heart while filling the veins. Harvey refused to accept that there was any difference between arterial and venous blood, one of them supposedly "more florid," more spirituous than the other.[20] He believed that if both were placed in separate vessels and allowed to clot, the resulting mix was indistinguishable. Harvey was among the first of the moderns in anatomy and physiology; as a chemist he remained in the Middle Ages. "He did not care for Chymistrey, and was wont to speak against them with an undervalue," commented Aubrey [p. 291].

It was in Holland that Harvey received his earliest and most committed support. Descartes, living in Holland, spoke approvingly in 1637 in his

Discourse on Method of a "physician of England" who had established a "perpetual circulation" of the blood; however, he did not view the heart as a muscular pump but rather as an expansion chamber in which the blood is heated, rarefied and thus forced out of the heart. The Harveyan thesis was defended by an English physician, Roger Drake, at Leyden in 1640 and was further supported by publications from Sylvius (1614-72), Walaeus (1604-49) and Du Roy (1598-1679), all emanating from Leyden about the same time. The first public recognition from Germany came from Herman Conring (1606-81), also in 1640, while two years later Giovanni Trullius (1598-1661) first publicly supported Harvey in Italy. Recognition in Spain came only in 1649 in a publication by G. B. Sobramonte (1610-83) printed in Valladolid.

Despite the early enthusiastic response from Fludd, who published abroad, English scholars virtually ignored *DMC* in the 1630s. Apart from Primrose's hostile reaction the only other response came in a thesis defended for the degree of M.D. at Oxford in 1633 by E. Dawson who seems to have supported Harvey. The first published support came only in 1641 from G. Ent (1604-49), and thereafter regular support tended to come from leading scientists of the day. Thus John Wallis in 1641, Kenelm Digby in 1644 and Francis Glisson in 1654, figures connected with the later emergence of the Royal Society, all publicly supported Harvey, though it took rather longer for the circulation of the blood to find its way into the anatomy textbooks. It was not until 1651 in a work of Nathaniel Highmore (1613-85) that the theory escaped from monographs and papers into a textbook. Even then others continued to resist. Alexander Read (1586-1641) in his *Manuall of Anatomy* (1634; sixth edition 1658) and Thomas Winston (1575-1655) in his *Anatomy Lectures at Gresham College* (published in 1659 but given 1613-43 and 1652-55) make no reference to Harvey at all (Keynes 1978: 318-19). David Hume's comment about physicians in his *History of England* may just as well apply to anatomists: "No physician in Europe, who had reached 40 years of age, ever, to the end of his life adopted Harvey's doctrine."

A further indication of the acceptance of *DMC* is evidenced in an analysis of works published during Harvey's lifetime that expressed an attitude to his work (see Keynes 1978: appendix iv). Bearing in mind that it was possible to support Harvey without accepting all details of his system, then the number of works, including letters of support or lack of, written between 1628 and 1657 can be thus aligned:

Years	*Pro*	*Con*
1626-37	8	5
1638-47	20	9
1648-56	20	6 (Riolan supplied 4)

Tabulated in this fashion, we can easily see how widely and speedily *DMC* was accepted throughout Europe.

How may we assess the significance of *DMC* when in reality its impact on medicine, at the level of diagnosis and therapy, was minimal? Harvey had suggested that infections such as V.D. or rabies could spread through the body, explaining "the whole system is vitiated through the relatively harmless touching of but a small part of it" for "the contamination... reaches the heart in the returning blood, and thereafter from the heart pollutes the whole of the body" (chap. 16). Was this a premature and isolated insight? Did he foresee the possibility of therapeutic transfusions? Records of the Royal Society dated 1665 reveal proposals for the transfusion of blood from dogs to dogs and between pigeons, and in 1667 Pepys wrote in his diary that he had heard of a "poor and debauched man" who for one pound had "some of the blood of a sheep let into his body." Pepys met the man a few days later, found him fit but "cracked a little in his head," and still willing to earn a further pound by additional transfusions. In France, a similar work, carried out by Jean Denys, physician to Louis XIV, led to the death in 1668 of a man who received three transfusions of lamb's blood. European surgeons were finally alerted to the dangers of such operations, and transfusions virtually ceased. In 1818 James Blundell (1790-1877), a London physician, revived the operation but made the crucial change of transfusing human rather than animal blood. Although the inspiration for the idea lay clearly with Harvey, Blundell's work, carried out nearly two centuries afterward, belongs in fact to a different tradition (Keynes 1947: 25-49).

If surgeons and physicians were constrained in their approach to blood transfusion, nothing affected their relish for bloodletting and phlebotomy. They had no idea that the body's blood supply was limited, and thus frequently and extensively ordered bloodletting as a regular form of treatment for the body's ills. In the 18th century, for example, seventy-three percent of the patients attending the Bristol Royal Infirmary had some form of venesection imposed, with an average of 12 ounces of blood taken from them. The 19th-century surgeon, Astley Cooper, warmly recommended the letting of 10-20 ounces of blood as the best treatment for inflammation. (Indeed, the title of the leading British medical journal, *Lancet,* founded in 1823, probably comes from the era of bloodletting.) The custom persisted right through the 19th century, and the first edition of *The Principles and Practice of Medicine* (1892) by William Osler (1849-1919)—a standard text for medical students—endorsed Astley Cooper's views.[21]

Nor did Harvey's work lead to any great insights into the recognition of or the treatment of heart disease.[22] This, as an identifiable field of medical study, began only in the 18th century. It arose from the new pathology developed by Morgagni whose work enabled physicians at last to correlate observable symptoms with precise lesions of the cardiovascular system. With the aid of work by A. Auenbrugger (1772-1809) and R. Laennec (1781-1826), physicians began to listen to the chest, heartbeat, and pulse by using simple techniques of percussion and an early stethoscope so that it became possible to talk sensibly of specific heart complaints. Such an advance inevitably presupposed the earlier work of Harvey but hardly rose directly from it.

If Harvey's impact on medicine was of so little immediate consequence, the importance of *DMC* must be sought elsewhere. It was in fact as a physiologist that Harvey had an immediate and lasting influence. Views of antiquity had been questioned before Harvey: Copernicus had persuaded Europe to reject Ptolemaic cosmology while Galileo had been equally successful in overthrowing the physics of Aristotle. Both had not merely destroyed the theories of antiquity; they had replaced them with new and entirely modern theories. It was clearly time that the view of man presented by antiquity should face a comparable challenge and be presented with an alternative account.

Harvey did more than take this step. A new approach to nature had arisen in the 16th century in association with the work of Francis Bacon (1561-1626).[23] Self-consciously experimental, empirical and critical, the new approach remained in the early years of the 17th century a program for thought rather than achievement. Like any other radical campaign only an initial success could endow it with any real credibility. There was also in the early days of modern science the possibility that it might prove limited, capable of dealing with the motions of planets and the rate with which bodies fall to earth but quite incapable of coping with organic forms. Harvey, more than anyone else, achieved the first great success of modern physiology by a deliberate and careful deployment of new experimental techniques, and by a close questioning of nature so powerfully argued by Bacon.

A sign of the impact Harvey made on his contemporaries' minds can be seen from the work of Gaspare Aselli (1581-1625) who while dissecting a dog in 1622 noted a number of white vessels spread throughout the intestines. As they produced a milky fluid he named them "lacteals." In a classic example of a man seeing precisely what he had been well trained to expect Aselli was sure he had found the vessels responsible for carrying chyle, or semi-digested food, from the intestines to the liver where it would be transformed into blood. Aselli, therefore, claimed to trace the lacteals to the liver, not realizing that lacteals are part of the lymphatic system, then unknown to anatomists, and drain into the thoracic duct rather than lead directly to the liver. Jean Pecquet (1622-1674) had demonstrated as much in 1651. But by this time Pecquet could benefit from the modern perspective on human anatomy. Not expecting to find chyle making its way from the intestines to the liver he failed to see it and was free to seek out new structures which before Harvey would have been, if noticed, dismissed as absurd. It is not too difficult to persuade someone that he has seen something that in fact is not really there; Harvey had that rarer talent, an ability to make a generation open its eyes for the first time and see what was truly present.

Publishing History

Frankfurt, 1628: First edition

Published under the title *Exercitatio anatomica de motu cordis et sanguinis in animalibus,* scholars have speculated why Harvey's work was first published outside England. Keynes's claim that the annual bookfair made Frankfurt "more fitted than London to receive and to appreciate a scientific announcement of such importance" (Keynes 1928: 2) presupposes an urgency on Harvey's part which does not account for the long delay in writing the text. Harvey may have taken advantage of better terms offered by continental publishers. His friend Fludd had used de Bry at Oppenheim for some time and wrote of the advantages: "I sent them beyond the seas because our home-born Printers demanded £500 to print the first volume and to find the cuts in copper; but beyond the seas, it was printed at no cost of mine, and that as I could wish. And I had sixteen copies sent over, with 40 pounds in gold as my unexpected gratuity for it" (see Weil: 142-64).

William Fitzer, Harvey's German publisher, was in fact an Englishman, the son-in-law of de Bry who had inherited the business in 1626 and moved it to Frankfurt. As it turned out, Fitzer proved a disastrous publisher. According to Osler, "Good copies of the *De motu cordis* do not exist. The paper is thin and foxed and no book of the same importance has suffered so cruelly from the binder" (Osler: 72). Keynes was no more generous commenting that it was printed "in a small mean type on paper of such quality that . . . it has turned dark and crumbled" (Keynes 1978: 177).

Fitzer's accuracy as a printer matched his munificence. In most copies of the first edition, a list of 126 errata has been added with the apology, "Benevolent reader, the directors of the printer's shop ask your indulgence for the many errors in a book of such small dimensions in view of the author's absence . . . the novelty of the subject to our proof readers, and the strangeness of the handwriting." Even so, this was a substantial underestimate. The editor of the 1766 *Opera omnia* noted another 120 errors. The plates are cribbed directly from *De venarum ostiolis* (1603) by Fabricius.

The size of the edition is not known. Keynes claims to know of many copies, "between 50-60 in Europe and the U.S." (Keynes 1981: 293), but most of these are in major libraries and collections, and single copies seldom appear on the market. When they do, they command high prices. Although Keynes owned eleven copies of *De motu cordis,* including rare editions, his superb collection lacked a first edition. A first-edition copy from the Honeyman collection, described as in "good but not superb" condition, was sold in 1979 for £88,000. Excluding books prized for plates, such as Audubon's *Birds of America,* this is the highest price ever paid for a work of modern science.

Osler, another great collector and bibliophile, was luckier in his search for a first edition of *DMC.* Already by 1906 he had been looking for a copy for ten years, and then, he writes, "Pickering and Chatto sent me one today which they bought for £30. . . . Though a poor copy I took it. Another copy two days later. Then August 23 1906 – yet another copy offered for £7/7/0." Osler also reported that in 1917 he bid at Sotheby's up to £40 for a copy which finally went for £48 (Osler: 72).

Venice, 1635; Leyden, 1639: Reprints

The first two reprints of *DMC* were hostile. Emilio Parigiano (1567-1648), also known as Parisanus, published most of the text in Venice in 1635, adding to each printed paragraph a dismissive comment of his own. Harvey never saw the work, and neither did many others for it remained unnoticed until reported by Keynes in 1928. Both Parisanus's text and comments were reprinted in Leyden in 1639, along with Primrose's hostile tract of 1630.

Later reprints

The first genuine reprint was brought out in Padua in 1643 and again in 1689 (an extremely rare edition, the only two copies known being in the Osler and Keynes collections). In the 17th century, the work was reprinted in Amsterdam (1645); Leyden (1647); Rotterdam (1648; 1654; 1660; 1671); London (1660, with a facsimile being made in 1928 to commemorate the tercentenary); and Bologna (1697; also rare; only one copy is recorded by Keynes). Later reprints were issued in Glascow (1751) and Edinburgh (1824, the last Latin edition as opposed to later commemorative facsimiles).

Collected works

Geneva: *Bibliotheca anatomica* (1685-89, 2 vols.; reissued 1699). This contains not only Harvey's major works but also the writings of others. Editors: Le Clerget and Manget.[24]

Leyden: *Opera omnia* (1737). This first edition of Harvey's collected works was edited by van Kerckhem.

London: *Opera omnia* (1766). The first and only English edition was edited by William Bowyer for the Royal College of Physicians.[25]

——— : *Collected Works* (1847). This first and only edition of its kind in English was prepared for the Sydenham Society with a life of Harvey and translation by Robert Willis. It was reprinted in Annapolis, Md in 1949. Is it time that England's greatest anatomist and physiologist be honored with a modern critical edition?[26]

Translations

Amsterdam, 1650: the first translation into Dutch and into a modern language.

London, 1653: first translation into English, reprinted in 1673, by Zachariah Wood.

——— 1832: a new translation by M. Ryan.

——— 1889: Willis's 1847 translation is revised by A. Bowie in *Bohr's One Shilling Library.*

Canterbury, 1894: privately printed edition of one hundred copies: a facsimile edition, with translation, of the first edition.

London and New York, 1907: first appearance in Everyman's Library (no. 262), using Willis's translation. Reissued in 1923 and 1952.

London, 1928: Two hundred and fifty copies were printed for the Royal College of Physicians of the 1653 translation, edited by Keynes, with a facsimile of the 1628 text.

New York, 1928: a similar work with a 1628 facsimile and a new translation by C. D. Leake.

London, 1957: to mark the tercentenary of Harvey's death, Everyman's Library published a new translation by K. J. Franklin; reissued in 1963.

The first appearance in other European languages can be listed as follows: German (1878); French (1879); Danish (1929), and Russian (1948).

To summarize, *De motu cordis* has been published in at least thirty separate Latin editions, has been translated into six European languages, and into English on at least five separate occasions; extracts have not been mentioned. It has, therefore, remained in print since 1628 and been available in an appropriate form to all classes of readers.

Further Reading

De motu cordis is available in Everyman's Library (1963), translated by K. Franklin. It remains a readable and relatively non-technical text. Substantial excerpts can be found in a number of anthologies and readings including M. Graubard (1964), Rook (1964), M. B. Hall's *Nature and Nature's Laws* (1970), King (1971), and Clendening (1960). Harvey has attracted a number of modern biographers of whom Keynes (1978) and Keele (1965) are unquestionably the best. For Harvey's ideas two studies are indispensable: Pagel (1971) and G. Whitteridge's *William Harvey and the Circulation of the Blood*. For background material Singer (1957) should be consulted together with Crombie (1952), Tibor Dory's *Discoverers of Blood Circulation* (1963) and the still valuable *Lectures in the History of Physiology* (1901) of Sir Michael Foster.

Notes

1. Examinations were oral, conducted in Latin and consisted almost entirely of elementary questions on the works of Galen and Hippocrates. Examples of questions: how many elements are there? what is an element? how many temperaments are there? what are the three faculties? Other questions touched on the typology of fevers, available therapies and some anatomy. One curious feature of an examination, recorded about 1600, involved the examiner in selecting passages from Galen, writing them out, and asking the candidate to find them in a volume of Galen's works (lacking an index); he was then expected to show signs of understanding the passages selected. See Clark: 98-9. The college's support for Galen was not simply a matter of examinations; in 1559 Dr. John Geynes, who had been critical of him, was forced to recant under threat of expulsion. Harvey became a Fellow of the College

in 1607, served as Censor in 1613, 1625, 1626, 1629 and as Treasurer in the period 1628-9. He was offered the Presidency in 1654 but declined on grounds of "infirmity" and age. Founded in 1518 the College attained its present regal status in 1851 when it became the Royal College of Physicians.

2. Relations between Harvey and King Charles (1600-49) appear to have been close and there is evidence that Charles took an interest in Harvey's researches.

3. In 1630 he accompanied the Duke of Lennox to France and Spain while in 1636 he went with the Earl of Arundel on an art collecting expedition to Holland, Germany, Austria and Italy. With Charles I he visited Scotland in 1633, 1640 and 1641.

4. "I have heard him say he wrote a book *De insectis,* which he had been many yeares about, and had made dissections of Frogges, Toades, and a number of other animals, and had made curious observations on them, which papers... were plundered at the beginning of the Rebellion. ... No griefe was so crucifying to him as the losse of these papers, which for love or money he could never retrive or obtaine" (Aubrey: 286).

5. The term *phleps* was normally used by Aristotle to refer indiscriminately to arteries or veins; it survives in such words as phlebitis and phlebotomy.

6. To the obvious objection that people could bleed to death simply by cutting their arteries, Erasistratus replied that when cut, the *pneuma* was free to escape leaving a vacuum behind. This in turn drew blood into the arteries from the adjacent veins.

7. To this strange view Harvey's objection was conclusive: "How may the arteries of the foetus draw air into their cavities through the mother's abdomen. ... How many seals, whales and dolphins... draw in and give off air through the great mass of water" (*DMC*, Introduction).

8. "The benefit of this blood (that is in the right cavity) when it is thinned and attenuated is to go up to the lung, mix with what is in the lung of air, then pass through the arteria venosa (pulmonary vein) to the left cavity of the two cavities of the heart and of that mixture is created the animal spirit": Ibn Nafis (Graubard: 62).

9. Servetus was struck by the size of the pulmonary artery from the right ventricle to the lungs: "it was not made of such sort or of such size, nor does it emit so great a force of pure blood... into the lungs merely for their nourishment." There must be, he concluded, "another purpose" and went on to propose the passage through the lungs. The argument was a common one (Graubard: 86).

10. This view has been challenged. Claims were made in the 17th century that priority be given to Hippocrates, Leonardo da Vinci, Giordano Bruno, Paolo Sarpi, Walter Warner and, more plausibly but with no real foundation, Cesalpino. Cesalpino certainly seemed to use all the right words and spoke of a "continuous motion from the heart into all parts of the body," "anastomoses" and of blood being carried back to the heart through the veins. Examination of Cesalpino's text reveals a more complex picture. For one, it is never really clear whether he was talking of a flow of blood or of heat. Secondly, he seemed to be merely restating a curious doctrine of Aristotle that heat flowed to the body from the heart during the day and back to it during sleep. Cesalpino apparently sought to express this view in the contemporary idiom of veins and arteries (Keele: 120-21).

11. The lectures do contain the following passage: "It is plain from the structure of the heart that the blood is passed continuously through the lungs to the aorta as by the two clacks of a water bellows to raise water. It is shown by application of a ligature that the passage of the blood is from the arteries into the veins. Whence it follows that the movement of the blood is constantly in a circle, and is brought about by the beat of the heart." Whitteridge, however, dates this passage as a later addition made in 1627 or 1628 (Keynes 1978: 93).

12. Unless they happen to be Copernicus, Newton, Descartes, Galileo or Darwin who all, for various reasons, delayed publication of some of their most important work for decades rather than years!

13. As editions of *DMC* are so various and as the chapters are so brief reference will be made to chapters rather than pages.

14. Harvey was reported to have examined about forty different animal species.

15. It is this argument that "is said to form the dividing line between him and his predecessors. His discovery – so it is interpreted – pivots around this point; it is a *sine qua non*" (Pagel: 73).

16. It was by no means the first. Sanctorius (1561-1636), for example, in his *De statica medicina* (1614) reported the results of a prolonged experiment in which he recorded his weight over a period of time, together with accurate measurements of his consumption of food and drink, and his "palpable excreta." His results showed the existence of "insensible perspirations" which accounted for 62% of his weight loss.

17. Some refused to accept the argument. Riolan, for example, objected that Harvey grossly exaggerated the amount of blood expelled from the heart and the rapidity of circulation. Accepting the idea of partial circulation, Riolan seemed to think that the blood would circulate only two or three times a day. As most of his data was obtained from corpses at post mortem ["I have found in a dead man above four oz." (chap. 9)], it was objected that Harvey based his physiology on the quite untypical conditions observed during autopsy.

18. And not simply, as is sometimes thought, "up" rather than "down." Harvey neatly disposed of this red herring by pointing out that valves are present in the veins, "not to prevent blood from falling by its own weight into areas lower down, for there are some in the jugular vein which are directed downwards, and which prevent blood from being carried upwards" (chap. 13).

19. As Copernicus ushered in modern astronomy without the use of the telescope and Harvey founded modern physiology without the use of the microscope, the role of instruments in the development of modern science can be said to be exaggerated. Both Copernicus and Harvey possessed the same instruments that could have been used in ancient Alexandria.

20. Johan Vesling (1598-1649), professor of anatomy at Padua, raised the question in correspondence with Harvey in 1637: "In the arteries the blood is bright red, in the veins dark purple." Were they really, he asked Harvey, the same?

21. Surprisingly, the practice of bloodletting, on an admittedly reduced and more modest scale, was still being recommended in the sixteenth edition of Osler's *Principles and Practice of Medicine* (1947).

22. Nowhere in Harvey's work is there any reference to defective heart valves or any description of diseased coronary arteries.

23. Aubrey described the relationship between Harvey and Bacon thus: "he had been a physitian to the Lord Chancellor Bacon, whom he esteemed much for his witt and style, but would not allow him to be a great Philosopher. Said he to me, 'He writes Philosophy like a Lord Chancellor', speaking in derision" (*Brief Lives,* 288).

24. It contains *De motu cordis* and the letters of 1649, the *De circulatione sanguinis,* and the embryological work, *De generatione animalium.*

25. In addition to the works listed in footnote 24 above, Bowyer's edition also contains a life of Harvey, additional letters, and the *Anatomiae Thomae Parri.* This latter is a report on the autopsy of Thomas Parr carried out in 1635 on instructions from the king. Parr was supposedly the oldest man in England and claimed to be 152 when he died. First married at age eighty, found guilty of adultery at age 112, married a second time when 112, and still working in the fields when 132, Parr interested many. Harvey gave no indication in his report that he disbelieved Parr's unusual longevity.

26. Two further works have been made available since Bowyer's edition: *Praelectiones,* Harvey's lecture notes, edited and translated by Whitteridge (1964) and *De motu locali animalium* (1627), also edited and translated by Whitteridge (1959).

CHAPTER 9

Philosophiae naturalis principia mathematica (1687)

by Isaac Newton

Nec fas est propius mortali attingere Divos.
So near the gods – Man cannot nearer go.

– Edmund Halley

Here lies Sir Isaac Newton, Knight, who by a vigour of mind almost supernatural, first demonstrated, the motions and Figures of the Planets, the Paths of the comets, and the Tides of the Oceans. . . . Let Mortals rejoice that there has existed such and so great an ornament of Nature.

– From the Newton sarcophagus, Westminster Abbey[1]

Qui genus humanum ingenio superavit.
Who surpassed all men in genius.

– From the Newton statue, Trinity College, Cambridge

Does he eat, drink and sleep like other men? . . . I represent him to myself as a celestial genuis, entirely disengaged from matter.

– Marquis de l'Hôpital

A man who, had he flourished in ancient Greece, would have been worshipped as a divinity.

– Dr. Johnson

However little read, either in Newton's own lifetime or since, no book has ever achieved the same secular fame and authority as *Philosophiae naturalis principia mathematica* (hereafter *Principia*). For well over a century not only did it provide the framework within which physics and astronomy operated, but it also laid down the basic patterns for other disciplines.[2] It is still common to dismiss fledgling subjects like psychology or sociology as awaiting their Newton while recognizing figures like Lavoisier and Darwin as the Newton of chemistry and biology, respectively. One hesitates to quote yet again the Pope couplet "Nature and Nature's laws lay hid in night: God said, Let Newton be and all was light," except to affirm that it expressed a widespread and genuine conviction, that there was no irony in the verse. Nor was it simply *Principia* that interested Europe and America so keenly; the figure of Sir Isaac himself was a subject of constant fascination to his contemporaries from St. Petersburg to Pennsylvania. Anecdotes, personal details, and reports of conversations were quickly picked up and spread with remarkable speed. Nearly three centuries later, both *Principia* and its author, judging by the number of new books that appear each year, have lost none of that appeal. The only English figure who has excited a comparable interest and who can rank with Newton in genius is Shakespeare.

Life and Career

Newton was born prematurely and posthumously on Christmas Day 1642 at Woolsthorpe, near the Lincolnshire town of Grantham. His mother remarried when he was only three and left him to be brought up by his grandparents. To one biographer this was "a traumatic event in Newton's life from which he never recovered," and much of his extreme behavior in later life was "generated by this first searing deprivation" (Manuel 1968: 26). At twenty, when Newton listed his earlier sins, the thirteenth was "threatening my father and mother Smith to burne them and the house over them."

In 1661, Newton was admitted to Trinity College, Cambridge, as a subsizar, that is, someone who paid his way by performing simple domestic services. Newton's academic career flourished. He graduated in 1665, became a Fellow in 1667, and two years later succeeded the mathematician Isaac Barrow (1630-77) as Lucasian Professor of Mathematics. However, if Newton was to establish an academic career in Cambridge, he must first be ordained in the Anglican church, an essential state for Fellows of Oxford and Cambridge colleges in the 17th century. But Newton was by then a unitarian of deep conviction and, while he had earlier affirmed his support for the thirty-nine Articles in order to be admitted to Cambridge, it was clear he would go no further. Although some may have suspected Newton's heretical views in later life, at this time they remained his secret. By 1675, Newton was making plans to leave Cambridge when, probably due to some pressure from Barrow, the Lucasian chair was granted a special dispensation by the Crown so that its occupant was freed from the need to be ordained.

This is, perhaps, the first sign of a major conflict in Newton. He was intensely ambitious, but it was position and status he sought rather than wealth and power. At the same time, he was absolutely committed to his religious beliefs which, if known, would not have sent him to the stake, or even prison. They would, however, have ended any chance of rising in society and they certainly would have cost him his chair. Just such a fate befell his successor William Whiston (1667-1752) who was dismissed from the Lucasian chair in 1710 for his uncompromising expression of his unitarian views. Newton's own position was more complex: unwilling to reveal his own unorthodoxy, he could not hide under the cover of Anglican orders.[3] Nor could he ignore all provocation, as the affair of Alban Francis demonstrated. In 1687, with the Catholic James II on the throne, the Cambridge authorities were instructed to award the degree of Master of Arts to Alban Francis, a Benedictine monk, without him first swearing, as statute demanded, his allegiance to the Anglican church. All knew that this was the thin edge of a very Catholic wedge. On an issue as serious as this, Newton was prepared to stand against the king and he was one of those selected to plead the university's case before the king's representative. "Go your way, and sin no more, lest a worse thing come unto you," they were warned. The university stood firm and Newton, who had been prominent in leading the resistance, cannot have been popular at court. As it happened, James's own position was vulnerable and within eighteen months he was forced into exile. Thus, rather than see Cambridge sullied by Catholicism, Newton was clearly prepared to take considerable personal risks.[4]

On a more personal issue some years later, Newton refused to take the slightest risk. In later life, Newton clearly enjoyed the company of young scientists. He enjoyed their admiration and even their flattery, and they also helped him with various chores. In return, they learned much from him and often attained a position as a result of his patronage. There is absolutely no hint in any record whatsoever that anything in these relationships was sexual. According to Manuel (1968: 190) Newton lived a life of total celibacy.[5]

The deepest of these relations was probably that formed with the Swiss, Fatio de Duillier (1664-1753), who arrived in England in 1687 determined to serve and to charm Newton. In 1707, after twenty years close friendship, Fatio became connected with a group of fanatics referred to as the French prophets from the Cévennes. They would stage public demonstrations at which they became possessed and spoke in tongues. They were arrested and pilloried in London. No help came from Newton.

By this time Newton had what he wanted. In 1695 he accepted the post of Warden of the Mint and four years later became its Master. In 1703, he was elected to the presidency of the Royal Society and, in 1705, he was knighted by Queen Anne. As a man of substance, Newton could not afford to be seen defending a religious crank. In 1696, he left Cambridge and moved to London where he could go about the king's business. For the rest of his life Newton looked after the affairs of the Mint and the Royal Society, twice revised *Principia,* and engaged in a number of more secret researches.

Newton's life was long and complex. Only a few incidents can be presented here to throw some light on Newton's strange psyche. The first of these concerns the notorious dispute with Leibniz over the discovery of the calculus. It has recently been investigated once more at some length by A. R. Hall (1980) and those interested in the details should look there.

In brief, Newton and Leibniz (1646-1716) in the latter half of the 17th century had developed the basic mathematical techniques permitting scientists at last to deal with quantities continuously and instantaneously changing: accelerating bodies, for example, or planets in orbit. Dates are crucial in disputes of this kind, and Newton claimed that he had developed his "direct" and "inverse" method of fluxions in his *annus mirabilis* of 1665.[6] Leibniz began to publish his own work on the differential calculus in 1684 while Newton published nothing before *Principia* (1687). Newton had, however, much earlier circulated among his colleagues in England *De analysi* (1669) and other texts, some of which had been seen by Leibniz on his 1673 visit to London. Further, according to Newton, two letters since known as the *Epistola prior* and *Epistola posterior,* and supposedly containing details of Newton's mathematical advances were sent in 1676 to Leibniz through the good offices of Henry Oldenburg, secretary of the Royal Society.[7]

The issue, in other words, had become sufficiently complex to ensure that even men of good will could be in some doubt as to who had the rightful claim, and good will was not in abundant supply between the mathematicians of England and Germany of that time. The Leibnizians claimed priority of publication while the Newtonians insisted that Leibniz had taken, without acknowledgement, ideas from *De analysi* and the 1676 *epistolae.* The aim here, however, is not to pursue the dispute but to use it to illuminate Newton's character. The most significant document produced was the report of a supposedly independent committee set up by the Royal Society in 1712 to examine all the papers of the case and make an unbiased judgment. The committee was, in Westfall's phrase, "a covey of his own partisans." Not surprisingly, the report found Newton to be the "first Inventor" of the "Differential Method," for Newton "carried out its investigation, arranged its evidence and wrote its report" (Westfall: 725). Of this last point there is no doubt: Newton's draft for the report exists in his own hand. At the same time, with no apparent shame, Newton could dismiss Leibniz's own evidence on the ground that "No Man is a Witness in his own Cause."[8]

It might be thought that the death of Leibniz ended the dispute. For Newton, however, it was just one more sign of the justice of his cause and incited him to a further reprise of a long familiar story. In a scholium to Book II, section II of *Principia,* Leibniz is referred to in the first edition as "that most excellent geometer, G. W. Leibniz"; in the third edition the entire reference had been dropped. It was a device Newton adopted against others. Few scientists enjoy others claiming, even in part, to have anticipated their work; even fewer react generously to those who point out errors in their own work. On both of these issues Newton was sensitive to a pathological degree. The amount of time, for example, he devoted to the Leibniz dispute was colossal.[9] Another example can be seen with his *Optics,* first published in 1704. To all who knew Newton the date was highly significant: It was the year after the death of Robert Hooke (1635-1703), someone he came to detest probably more than anyone.

Hooke held a powerful position in the Royal Society both as curator of experiments from 1662 and as its secretary from 1677-82. Precisely how papers and experiments were presented and reported depended to some extent on Hooke. He also had the infuriating habit of indicating that the discoveries of others had in fact been made earlier by himself. In his reports to the Society, there would be a reference to a conversation some ten years before, or a paper he promised to produce later, or an anagram left with someone, or a claim made in a letter – Hooke was never short of evidence of some kind. In 1672 when Newton began to send the Royal Society some of his work on light and color he immediately ran into trouble with Hooke, who was more than a match for Newton. Lacking the full power and prestige of his later years, and finding in Hooke someone as indifferent to the normal rules of debate as himself, all Newton could do was to suffer under Hooke's general interference. Hooke did pick up a number of genuine mistakes in Newton's work which, instead of communicating privately, he deliberately exposed in full sessions of the Royal Society. Where there was something original and admirable in Newton's work, Hooke would invariably claim that he had made a similar suggestion some years before.

Against this kind of treatment Newton could respond in only one way: silence. Most of the material eventually published in *Optics* was over thirty years old. Why the delay? The answer was given in the book's advertisement: "To avoid being engaged in disputes about these matters, I have hitherto delayed the printing." It was a response Newton seriously considered adopting at one time with *Principia* when he feared its publication would lead to the same kind of controversy.

Before dealing with *Principia,* it may be helpful to see Newton's scientific work in relation to his other interests. In his long life, science actually played a much smaller part than is often imagined for it competed with two powerful rivals. The first of these was his ambition for a career and a position in the world. For over thirty years Newton worked at the Mint and while it was in no sense a regular eight hours-a-day job, it was no sinecure either. Through his position at the Mint and as president of the Royal Society, there were sufficient calls on his time to interfere seriously with any prolonged scientific endeavors.

The other main rival to science was Newton's deep interest in religion and history. He read deeply and thought continuously on these subjects, and his writings on them run into millions of words. His views are discussed in the two volumes by Manuel (1963; 1974). Newton seemed incapable of approaching any subject in a casual manner. If he was interested at all, then he needed to master the subject and to this end devoted several hundred, if not thousands, of hours. The result of these private obsessions was that for much of his later life Newton's expenditure on science was rather limited. What did interest him was the reputation and condition of his already published work. He was therefore willing to spend unlimited time preparing new editions of *Principia* or adding fresh *Queries* to later editions of *Optics.* He was also willing to devote as much time as necessary to the protection of his work against those who challenged his priority.

Nor were his less orthodox interests pursued exclusively in later life. Much of the 1670s, for example, was spent working on alchemy. Westfall has noted that Newton's surviving alchemy manuscripts run to over a million words: 20 per cent of them he assigned to the pre-1675 period, 30 per cent to the period up to 1687 with the final 50 per cent coming from the 1690s. Newton also immersed himself in the religious themes that so obsessed him. He had quite early rejected orthodox Christianity with its doctrine of the Trinity. This in turn led him to a deep study of the history of the church in the fourth century, the period when it rejected the views of Arius and accepted the creed of Athanasius. From this point Newton's interest spread to other aspects of church history and biblical interpretation. At the end, with about two million words to show for his labors, Newton was as familiar with the Bible and the history of the church as any professional scholar. Inevitably this mass of writing was kept secret.[10] A few may have suspected his views were unorthodox, but until the 1936 sale of his papers (which will be discussed presently) the full extent of Newton's extra-scientific interests was unknown.

It should also be appreciated that in financial and social terms Newton attained a position of some substance. His years at Cambridge brought him with the income from his chair and the family estates an annual return of about £250. Not the stuff out of which great fortunes could be assembled, but sufficient to provide a handsome living for a bachelor don of the late 17th century. When he was appointed Master of the Mint in 1699 his salary rose to £600 a year. The real perks, however, came from the commission he received on the amount of gold and silver minted. From available data it would seem that this commission averaged nearly £1,500 a year which with Newton's other salary and income put his earnings at over £2,000 a year, a large figure for the time.

On his death, Newton's estate was valued at more than £30,000, of which £18,000 was held in Bank of England stock, £10,000 in stock and annuities of the South Seas Company and about £2,000 in cash. This enormous sum might well have been much higher at one time. According to one source Newton had invested heavily in South Sea stock and may well have lost as much as £20,000 when the bubble burst after 1720.[11]

It was not, however, just stock and cash that Newton left to eight nephews and nieces.[12] There was also his library of 1,896 volumes, and it was quickly sold by his rapacious kin for £300 to John Huggins, warden of the Fleet prison.[13] In addition there was a vast number of manuscripts whose contents and value were unknown. Before the estate could be distributed it was necessary for someone to stand surety against any outstanding debts Newton may have incurred at the Mint. The only figure capable of doing this was John Conduitt, husband of Newton's favorite niece and onetime housekeeper, Catherine Barton.[14] In return for his risk Conduitt received the unrealized assets of the Newton manuscripts. Fear-

context are usually meant Kepler's laws which state that planets move in ellipses, not circles; that their radius vectors sweep out equal areas in equal times; and that the squares of their orbital periods around the sun are proportional to the cubes of their distance from it. Hooke claimed that all three laws could be derived from the single assumption that the force between the sun and the orbiting planets decreased as the square of the distance between them. That is, if one body was twice as near as another, it would be attracted to the sun with a force four times as great; if it was three times as close, the centripetal force[17] (as it was later called) would be nine times as strong, and so on.

As it happened, Halley and Wren were also convinced of the importance of the inverse square law, but although they had tried to derive Kepler's laws from the assumption they had both failed. Hooke's claim was of considerable interest to them but only if he could provide an *actual* derivation. Despite his claim, Hooke declined to produce his proof on the grounds that he wanted others to try first and after their failure, and only then, would he publish the solution.[18] Clearly neither Halley nor Wren believed Hooke's claim. Wren offered what amounted to a forty-shilling book token to whomever could produce a demonstration within two months, which is like offering a couple of theater tickets as the prize for finding the cure for cancer. Wren's book token remained uncashed, for nothing more was heard of the problem until August 1684, when Halley, in Cambridge on some undisclosed business, called on Newton.

What happened between them Newton revealed many years later to the French mathematician Abraham de Moivre (1667-1754): "In 1684 Dr. Halley came to visit him at Cambridge. . . . The Doctor asked him what he thought the Curve would be that would be described by the Planets supposing the force of attraction towards the Sun to be reciprocal to the square of their distance from it. Sr Isaac replied immediately that it would be an Ellipsis, the Doctor struck with joy and amazement asked him how he knew it, why saith he I have calculated it, whereupon Dr Halley asked him for his calculation without any farther delay, Sir Isaac looked among his papers but could not find it, but he promised him to renew it, and then to send it him" (Westfall: 403). The idea that Newton could not immediately lay his hands on such an important document, Westfall dismisses as a charade. It was more likely that Newton wished to check its accuracy. In 1679, a speculation on a related problem in a letter to Hooke had contained an error which Hooke had lost little time in communicating to a public meeting of the Royal Society. It was a humiliation Newton little enjoyed; even thirty years later – and despite his achievements and reputation – he could not bring himself to admit the error and instead attempted to pass it off on de Moivre as "a negligent stroke with the pen."

But, unlike Hooke, Newton did have a proof. Examination showed it to be flawed but not so radically that it could not be put right. In November 1684, he fulfilled his promise to Halley and sent him a nine-page treatise entitled *De motu corporum in gyrum* (On the Motion of Bodies in Orbit).

Halley immediately recognized that it did all that he, Hooke, and Wren, among many others, had so unsuccessfully tried to do. Not only did it derive the three laws of Kepler from the assumption of an inverse square law, but it showed that part of the converse also held, namely, that an elliptical orbit demanded an inverse square law. Shortly afterwards, Halley was in Cambridge again and presumably expected Newton to produce some reworked version of the first draft of *De motu* ready for publication by the Royal Society. It was not until April 1686, that Halley received from Newton a publishable manuscript: It was in fact Book I of *Principia.*

For having once begun to think about the problem posed by Halley he soon saw that its treatment in the first *De motu* was both limited and superficial. For one, it ignored all question of a resisting medium; secondly, he had only dealt with one-body systems whereas he soon came to realize that "the orbit of any one planet depends on the combined motion of all the planets, not to mention the actions of all these on each other." Somewhat gloomily he added, "To consider simultaneously all these causes of motion and to define these motions by exact laws allowing of convenient calculation exceeds, unless I am mistaken, the force of the entire human intellect."

It took Newton some fifteen months of isolation to work out these problems to his satisfaction. Apart from a few letters he wrote during this period and a visit to Woolsthorpe on family business in early 1685, Newton's life at this period is undocumented. It is of this time that the stories of his amanuensis, the unrelated Humphrey Newton, refer in which Newton is supposed to have been so busy that sometimes he would forget to eat. This is one source of the romantic image of the scientist who absent-mindedly would take the wrong turn on a walk and, so engrossed in his abstruse problems, would fail to notice. Or, "When he has sometimes taken a Turn or two . . . has made a sudden stand, turn'd himself about, ran up ye stairs, like another Alchimides [sic], with an eureka, fall to write on his Desk standing, without giving himself the Leasure to draw a Chair to sit down in" (Westfall: 406).

Newton himself described the writing of *Principia* as taking seventeen to eighteen months and that he sent the manuscript to the Royal Society in the spring of 1686. This, however, can only refer to the manuscript of Book I; the other two books, although completed before, arrived in London nearly a year later in March 1687. Shortly afterwards, on 19 May, the Royal Society officially decided to publish the work and instructed Halley so to inform its author.

On 22 May 1686, Halley wrote to Newton informing him that the Society intended to publish the work "at their own charge." After the good news Halley, with some hesitancy, went on to impart what he knew could only infuriate Newton: "There is one more thing that I ought to informe you of, viz, that Mr Hook has some pretensions upon the invention of ye rule of the decrease of Gravity, being reciprocally as the square of the distances from the Center. He sais you had the notion from him. . . ." The price Hooke apparently demanded was "some mention of him in the preface" (Newton 1959: II, 431).

Newton's reply was remarkably moderate and must have come as a relief to Halley. All he wanted with Hooke was "a good understanding," he declared, and he had not referred to him in Book I because that was not the most appropriate place. Suitable acknowledgment would be made elsewhere. He insisted on his priority over Hooke but in a way as if to suggest it were an issue on which there could well be more than one point of view.

Three weeks later, as if almost from another man, Halley received from Newton the kind of letter he must have expected earlier. Whether Newton had needed time to check through his own papers (an unlikely hypothesis), or whether it was the effect of three weeks brooding, is not known. More likely Newton had heard from other sources less censored accounts of Hooke's claims: "I am told by one, who had it lately from another present at your meetings, how that Mr Hooke should make a great stir, pretending that I had all from him, and desiring that they would see that he had justice done him." This was too much for Newton: a private correspondence with Halley was one thing, a public squabble quite another.

Newton began with a general complaint he had made before and would make again: "Philosophy is such an impertinently litigious lady, that a man had as good be engaged in lawsuits, as have to do with her." Hooke, he declared, had "erred in the invention he pretends to." What had actually passed between them Newton recounted with growing irritation (Newton 1959: II, 235-40):

> He imagines he obliged me by telling me his theory, but I thought myself disobliged by being, upon his own mistake, corrected magisterially, and taught a theory, which everybody knew, and I had a truer notion of than himself. Should a man who thinks himself knowing, and loves to show it in correcting and instructing, come to you, when you are busy, and notwithstanding your excuse press discourses upon you, and through his own mistakes correct you, and multiply discourses; and then make this use of it, to boast that he taught you all he spake, and oblige you to acknowledge it, and cry out injury and injustice if you do not; I believe you would think him a man of strange unsociable temper . . . and therefore he may do well in time to consider whether, after this new provocation I be much more bound . . . to make an honourable mention of him in print, especially since this is the third time that he has given me trouble of this kind. . . ."

At this stage Newton seemed unwilling to admit that the letter of 1679 contained any errors at all, even slips of the pen. He did, however, have one more general complaint: "Now is this not fine? Mathematicians that find out, settle and do all the business must content themselves with being nothing but dry calculators and drudges and another that does nothing but pretend and grasp at all things must carry away all the invention as well as those that were to follow him as of those that went before" (Newton 1959: II, 235-40). The point is a fair one. What distinguished Copernicus,

Harvey, Newton, and Darwin from their contemporaries was not that they were the first to hit upon some particular hypothesis, be it heliocentrism, the circulation of the blood, universal gravitation, or evolution, but that they were the ones who applied them with precision, depth, and comprehensiveness, in ways no one before had suspected possible. As Darwin would later find, once he had demonstrated in detail the process of natural selection, there was an avalanche of contenders all claiming to have been first to formulate the principle he described so successfully. Originality and new ideas are seldom scarce in science. Theorists may often get through a dozen or more in a morning's work. To have the courage and the ability to recognize an idea for what it was worth and then to spend several years pursuing it through its complexities and difficulties was a different thing; it was something, Newton insisted, that in this context he and only he had done.

Newton's position at this time was a curious one. His Cambridge chair brought him £100 a year, but hardly the status in society that he longed for. In 1686 he was already forty-four, and although he had some reputation in science at this time it was by no means an exceptional one. Before *Principia,* his best work was in mathematics, but none of it had been published nor was it likely to win more than a reputation for him. *Principia,* he must have realized, was a work, perhaps the only work he would ever produce, capable of winning for him the position that he sought. There was thus a lot at stake on the fate of *Principia.* As well as a deeply ambitious man, Newton was also an intensely proud one. He found it hard to acknowledge that his work was not completely his own or that, as with Leibniz, others could independently have reached his conclusions. Kepler, for example, is only briefly mentioned in *Principia* and his great insight that planets move in elliptical orbits is there perversely attributed to Copernicus (Newton 1966: 395).

Such excessive pride can lead to a quite unprincipled defense of the sole ownership of a particular piece of intellectual property. It can also lead to a brutal reaction. If his work was not recognized unambiguously as his own then, Newton declared, he wanted no part of it. Rather than produce an impure work, Newton was prepared to produce no work at all. It was an extreme reaction, perhaps, but it was also one, as will be seen, that Darwin in a similar position contemplated in 1858. Newton actually threatened to abandon not the entire volume but, as he put it to Halley, "the third I now designe to suppress" while the first two Books, with no claim on them from Hooke, he proposed to publish under the title *De motu corporum libri duo.* Some consideration for Halley must have surfaced at this point for Newton immediately added, "On second thoughts I retain ye former title. Twill help the sale of ye book which I ought not to diminish now tis yours."

The point of this reference was that Halley as clerk of the Royal Society had found himself not only responsible for editing *Principia,* arranging for its publication and even selling the printed volumes, but he had also found himself, somewhat unwillingly, responsible for all of the book's costs. The reason for this curious arrangement was simply that the Society was broke. Unlike the Académie des Sciences, the Royal Society—despite its title—received no funding from the Crown or any official source. Supposedly, the Society was financed by weekly fees of one shilling per member, but they were generally ignored. The result was that in 1676, in not much more than a decade of existence, unpaid membership fees totalled £2,000. To make matters worse, in 1686 the Society had financed the publication of F. Willoughby's *Historia piscium,* a work that sold less well than expected. For some years afterwards Halley's salary of £50 a year was paid not in cash but in unsold copies of Willoughby's work. Much as the Society wanted to publish *Principia,* they simply lacked the funds. Therefore, on 2 June 1686, Halley was instructed to publish the work at his own cost which, the minutes add, "he engaged to do."

The effect of omitting Book III from the completed work would thus be unfortunate for Halley. The first two books were somewhat austere and by themselves would appeal to few; Book III, or *The System of the World,* by no means light reading, did at least claim to "deduce the motions of the planets, the comets, the moon and the sea." Halley tried to reassure Newton that few would take Hooke's comments seriously; that, indeed, "All mankind ought to acknowledge their obligations to you." And what of those "that will call themselves philosophers without Mathematicks," should they be deprived of Newton's work?

Actually, what seemed to convince Newton of the rashness of his ways was not the external pressure of friends like Halley but an intense scrutiny of his own papers. He just happened to be looking at an old letter he had written to Huyghens some fourteen years earlier which, he told Halley, undoubtedly showed that in 1673 he had spoken of gravity decreasing inversely as the square of the distance. Hooke had publicly admitted that he was unaware of this in 1674, and this was sufficient for Newton. However insignificant it might appear to others, it was clear to Newton that only if he could show the falseness of Hooke's claim from his own resources—in this case his own papers—could he convince himself of the legitimacy of his right to publish.

The publication of *Principia* on July 1687 must have come as a great relief to Halley. Newton, for once, had the courtesy to acknowledge fully and publicly the immense debt he owed him, and in the preface spoke of the "most acute and universally learned" Halley who not only "assisted me in correcting the errors of the press and preparing the geometrical figures, but it was through his solicitations that it came to be published. . . ." Hooke, not surprisingly, fared less well. At one point in the manuscript of *Principia* there is a reference to "Clarissimus Hookius," this was first reduced to a mere "Hooke" but on second thoughts Newton decided to omit the reference altogether. The references to Hooke that do remain appear to be straightforward reports of his cometary observations.

This, for Newton, was far from the end of *Principia.* Almost as soon as it appeared he began correcting it and noting improvements he wished to make.[19] To this end he had had a *Principia* specially bound with interleaved blank pages on which corrections and new material could be entered. The post of editor was much sought after; Fatio made an early and promising bid but his religious excesses removed him from serious consideration. Another who made every effort to ingratiate himself with Newton was the Scottish mathematician David Gregory (1661-1708). One of the more pleasant aspects of Newton's personality was his willingness to trust and advance the young scientists around him. Thus, the second edition was assigned to the younger Roger Cotes (1682-1716), a mathematician whom Newton had placed in the newly found Plumian Chair of Astronomy at Cambridge when Cotes was only twenty-four. The task was by no means easy. Newton allowed no one to take liberties with *Principia* and, while it was permissible for him to take his time in answering letters and to deal with other matters, he allowed no such laxness in others. Actually, Cotes was good for Newton, pushing him to decisions, suggesting changes, and even correcting some errors. Newton in his term seemed genuinely grateful. The changes were numerous and wide ranging; they do not lend themselves to easy classification. Full details can be found in Cohen (1978).

Fewer changes were made in the third and final edition published in 1726. It was not until 1723 when he was eighty-one that Newton decided to bring out a new edition. He chose Henry Pemberton (1694-1771) as editor, another loyal young man. Pemberton was a physician by training and largely self taught as a mathematician. He brought with him the most acceptable of all credentials to Newton, a paper published in 1722 criticizing the work of Leibniz. In Cohen's words, "There were no bold and exciting innovations" in the third edition (1978: 280). But Newton took the opportunity to shed what he clearly thought was a quite unnecessary reference to Leibniz. In the scholium following Lemma II, Book II, Newton had described a correspondence between himself and "that most skillful geometer G. G. Leibniz" in which he had passed to Leibniz some of his results, "under transposed letters," on the calculus. Newton added: "That eminent man wrote back that he had fallen on a method of the same kind . . . which hardly differed from mine in anything except language and symbols." A minor change was made to it in the second edition; in the third a totally new entry was substituted lacking all reference to Leibniz. Instead Newton referred to a correspondence with J. Collins in 1672 and a *Treatise* of 1671 in which the same results had been described. For his care, Pemberton was awarded a gift of two hundred guineas from Newton and an acknowledgment in the preface as "a man of the greatest skill."

"Principia": Contents and Argument

Principia begins with a virtually unreadable ode to Newton by Halley and is followed by Newton's brief preface. In the next section Newton lays down the foundations of his system with eight definitions and three laws. The definitions cover the quantity of matter and of motion, the *vis insita* or inertia, impressed, and centripetal forces. The definitions are followed by a scholium in which Newton argues for the need to assume the existence of an absolute space and time in which absolute motion may take place.

The three Newtonian laws state that (1) every body continues in a state of rest, or of uniform motion in a right line, unless it is compelled to change that state by forces impressed upon it; (2) the change of motion is proportional to the motive force impressed, and is made in the direction of the right line in which that force is impressed; and (3) to every action there is always opposed an equal reaction: or, the mutual actions of two bodies upon each other are always equal, and direct to contrary parts.

After the derivation of six corrolaries, Newton had at last reached Book I in which he proposed to deal with "the motion of bodies." It is not entirely general and mathematical. There is no talk here of the moons of Jupiter, comets or tides. Instead the argument deals with lines, angles, and tangents. To a casual and uninformed reader it could well appear as a geometrical treatise and, indeed, such sections as "To describe a conic that shall pass through five given points" (Proposition xxii) are far from untypical of *Principia* and could well come from a text of pure geometry. Newton was trying to lay down the foundations of celestial mechanics. Consequently, many of the fourteen sections are concerned with determining the specific properties of various orbits produced by a central attractive force (viz. sections III-IX). For the remainder of Book I, section X deals with "the oscillating pendulous motion of modies," sections XII-XIII are on attractive forces between spherical and non-spherical bodies, and the book concludes with a section on "the motions of very small bodies when agitated by centripetal forces."

Some of the landmarks of Newton's achievement in Book I can be briefly indicated. Proposition xi shows that if a body revolves in an ellipse "the centripetal force tending to the focus of the ellipse is inversely as the square of the distance." The same holds true if the orbit is an hyperbola (Prop. xii), or the parabola (Prop. xiii). Kepler's third law is derived as Proposition xv. The converse, that bodies whose "forces decrease as the square of the distances from their centres" move in ellipses, was established as Proposition lxv; it also shows that "radii drawn to the foci may describe areas very nearly proportional to the times," or Kepler's second law. Proposition lxvi deals with the famous "three body problem" in which it is proposed to determine the orbits of three bodies all attracting each other in accordance with the inverse square law. An approximate solution is offered.

Book II is entitled "The Motion of Bodies (in Resisting Mediums)" and is divided into nine sections. As an almost independent treatise on hydrodynamics, it cannot but seem an intrusion between the other two books. Sections I-IV and VI deal with the motion of bodies in fluids; section V with the "density and compression of fluids"; and the remaining three sections, VII-IX, on the motions of fluids themselves. However, toward the end, it does contain a discussion on vortices with direct implications for celestial mechanics. In section IX, "The Circular Motion of Fluids," the final scholium concludes that "it is manifest that the planets are not carried around in corporeal vortices."

In the scholium to section VIII, Book II also contains a theoretical derivation of the speed of sound. It has a curious history. In the first edition Newton calculated the velocity of sound to be 968 ft. per sec., a figure in agreement with experiments conducted in the arcades of Trinity College. Unfortunately for Newton more accurate measurements were made by William Derham (1657-1735) in 1705 by noting the discrepancy between the flash of cannons fired 12 miles away and the later arrival of the sound. Derham's figures, which Newton accepted, of 1,142 ft. per sec. were incompatible with those already published in *Principia*. Consequently Newton introduced into the second edition (1713) what Westfall has described (p. 735) as one of the "most embarassing" passages of the whole work in which a number of apparently ad hoc hypotheses were deployed to overcome the failure of theory and experiment to agree. Thus, Newton spoke of "the crassitude of the solid particles of air" and of the distorting effect produced by water vapor in the air. With these he was triumphantly able once more to derive a speed of sound of 1,142 ft. per sec. in agreement with Derham's observations. In just such a way do theory and experiment often support each other in science, whether ancient or modern, whether in the hands of a master or the rawest of students.

The initial failure of Newton was later shown by Laplace to be due to none of his special assumptions but rather to his having ignored the heat generated by the compression of sound waves. When taken into consideration this factor produced the desired results.

Book III is entitled "System of the World" and returns to the problems of celestial mechanics.[21] It sets out to show that, on the basis of results established in Book I, all the phenomena of the heavens can be derived, including comets, tides, the satellites of Jupiter and Saturn, as well as the more obvious motions of the earth, sun, and moon. Much of it is written in a relatively informal style and contains, in fact, the only extended passages in the whole of *Principia* that most readers will find readily intelligible.

It begins with Newton's four "rules of reasoning in philosophy" and is followed by a brief section entitled "Phenomena." There are six in all. The first two show that the satellites of Jupiter and Saturn observe Kepler's third law, as do the planets about the sun (Phen. iv). The next section, "Propositions," is much longer; thirty-three are dealt with. Almost discursively, Newton discusses such problems as "That there is a power of gravity pertaining to all bodies, proportional to the several quantities which they contain" (Prop. vii); or, "That the motions of the planets in the heavens may subsist an exceedingly long time" (Prop. x); or, "That the flux and reflux of the sea arise from the action of the sun and moon" (Prop. xxiv). There is a short section on "The Motion of the Moon's Nodes," and a much longer section concerned with comets, their orbits and how they can be determined from "three given observations." Book III and *Principia* conclude with a "General Scholium" containing Newton's thoughts on God and the cause of gravity. It also contains Newton's famous motto, *Hypotheses non fingo* ("I frame no hypotheses").

It is also in Book III that one appreciates the formal geometry of Book I. Thus, Proposition xvi, to find the eccentricities and aphelions of the planets, one refers back to Book I, Proposition xviii, to a problem involving elliptic and parabolic curves. Again, to find the diameters of planetary orbits (Prop. xv, Book III), the reference is to techniques provided in Propositions xv and lx in Book I.

To deal at all adequately with even a small section of the topics covered in *Principia* would probably require a treatise on mechanics. However, two topics central to the Newtonian revolution should at least be considered: To what extent was the "system of the world" proposed in *Principia* truly a system, and how valid is Newton's fundamental notion of universal gravitation.

It is probably for this latter concept that Newton is so widely known. There can be few who have not heard the story of the apple and the conclusions he is supposed to have drawn from it. It is also through this concept that Newton was able to break away from the formidable mathematics of Books I and II and speak in more familiar tongue about more familiar phenomena. Few of us have perhaps ever wanted to describe a conic passing through five given points, but many have wondered what causes the tides and how distant bodies like the sun can control the movements of the earth, or why bodies fall to the earth as they do. It is the impression that Newton first answered these questions that continues to keep his name alive in the popular imagination.

Gravity is such a familiar concept today that it is difficult to realize that only three hundred years ago it was a strange and unproven theory. Thus, on reading *Principia,* Huygens, the scientist closest to Newton in range and power, noted that it had never occurred to him to extend "the action of gravity to such great distances as those between the sun and planets, or between the moon and the earth. . . ." (Cohen 1980: 80) Others rejected it as an absurd speculation. Leibniz stated it in the simplest terms: "A body is never moved naturally, except by another body which touches it and pushes it; after that it continues until it is prevented by another body which touches it. Any other kind of operation on bodies is either miraculous or imaginary" (Alexander: 66).

What earlier models of the universe so satisfied scholars that they could find a scheme of universal gravitation absurd? The earliest theory, and one powerful enough to survive until the 17th century, was the Aristotelian. Bodies fall to earth because that is their "natural place"; remove anything away from its natural place and if unrestrained it will return there as directly as possible. Against this accepted view Newton presented a new and devastating argument: In their first second of free fall, earthbound bodies dropped just sixteen feet; on the moon they would take longer to fall the same distance. Attraction between bodies was a function, among other things, of their mass. The less massive moon would therefore be less attractive to a falling body by a precisely calculable amount. For the Aristotelian, it would be necessary to assume that a falling body somehow could know and adjust its rate of fall according to the mass of the body it was falling towards.

With the collapse of the Aristotelian system, the new orthodoxy of the mechanical philosophy emerged with Descartes as its most articulate spokesman. "Give me matter and motion," he declared in 1644, "and I will construct the universe." The universe was a plenum in which there was no need for mysterious forces attracting and repelling anything; a body could move only if it was pushed by an adjacent body. The initial cause of motion was God. However, once in motion, a body would continue to move until obstructed, when the motion would be transferred to another body. Descartes proposed a system of vortices to account for the complex celestial motions.

Against the Aristotelian and Cartesian systems, Newton proposed a totally different approach that introduced the idea of an attractive force. One reason why this proved so hard to accept was that scientists had spent the last two centuries trying to expel such "occult qualities" from the domain of science. It was just too easy to explain the operation of some body, magnetic, chemical, gravitational, or whatever, by supposing that it possessed some special power, force, affinity, or innate quality. The emphasis was on primary qualities that could be displayed and measured. Hence, the great appeal of "matter and motion." To anyone familiar with the physics of his day, *Principia* would have looked at first like a backward step.

How then did Newton arrive at his concept of universal gravitation? His first great insight was to see that science could not be built on the austere ration of matter and motion alone. A crucial argument was deployed in the famous Query 31 of *Optics* (1704): "The parts of all homogeneal hard bodies which fully touch one another stick together very strongly. And for explaining how this may be, some have invented hooked atoms, which is begging the question; and others tell us that bodies are glued together by rest, that is, by an occult quality, or rather by nothing; and others, that they stick together by conspiring motions, that is, by relative rest amongst themselves. I had rather infer from their cohesion that their particles attract one another by some force, which in immediate contact is exceeding strong. . . and reaches not far from the particles with any sensible effect" (Thayer:168).

In later life, Newton himself reported to Conduitt the moment that gravity's full import struck him. It was in 1666 in the garden at Woolsthorpe. Having fled Cambridge and the plague "whilst he was musing in a garden it came into his thought that the power of gravity (which brought an apple from the tree to the ground) was not limited to a certain distance from the earth but that this power must extend much farther than was usually thought. Why not as high as the moon. . . if so that must influence her motion & perhaps retain her in her orbit" (Westfall: 154). Further thought suggested to Newton that while the force of gravity would not be "sensibly weakened" over small distance, at the moon, however, its power could differ "much in strength from what it is here." If the moon was held in its orbit by gravity, then it was probably that the other planets of the solar system were carried around "by the like power." Thus, by comparing the distances and periods of the several planets, Newton was led to conclude that "its strength must decrease in the duplicate proportion of the increase of distance." Thus, if Newton's recollections in later life were reasonably accurate, then by 1666 he had already formulated some of the central insights of his gravitational theory. Still he did not label gravity a universal force, or describe it as a function of mass and distance.

Why did not Newton immediately develop his insights into a comprehensive treatise? One obvious reason was that he had other problems to pursue, some mathematical, some physical (like his work on light) and others that were unrelated to orthodox science. For much of the twenty-one years between the *annus mirabilis* and the publication of *Principia,* Newton actually tried to "bend myself from Philosophy to other studies. . ."; the time spent on it was "grutched" (grudged) unless it was "at idle hours something for a diversion" (Newton 1959: II, 300). It is curious to picture Newton at the height of his power concentrating on problems of theology and alchemy, and turning to the question of gravitation for relaxation. His letter, of course, is somewhat exaggerated for it was written to Hooke with whom he was desperately trying to avoid a correspondence. Yet, stimulated by Hooke's letter, he did in fact work on the inverse square law without, as one would imagine, informing Hooke. It was not until the visit from Halley in 1684 that Newton decided to devote all his efforts to a problem he had thought about for nearly twenty years.

The argument for universal gravitation is developed in Book III of *Principia,* Proposition i-ix (Newton 1966: 406-17). Newton begins by setting out in Proposition iv to show that the moon was attracted to the earth in precisely the same way as an apple falling from a tree. Consider the diagram below—which is exaggerated for purposes of illustration—showing the moon moving from *B* to *D* in its orbit about earth.

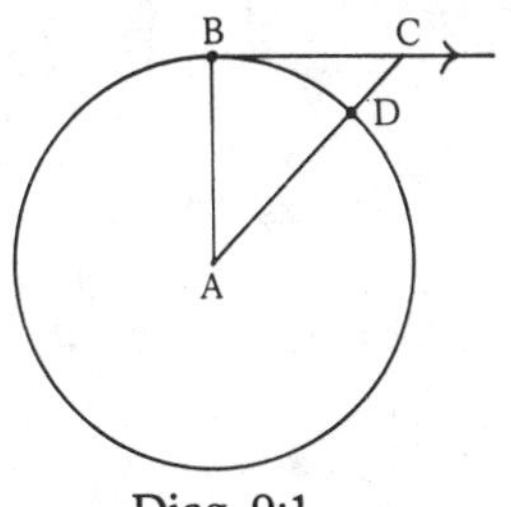

Diag. 9:1

If the moon were moving inertially, it would travel in a straight line from *B* to *C*; attracted by earth (*A*), however, it moves from *B* to *D*. Thus, the distance *CD* is the "falling" of the moon from its inertial path caused by the gravitational attraction of earth. Knowing the dimensions of the lunar orbit and its period, Newton was able to work out, using the method developed in Book I, Proposition xxxvi, that in one minute, the moon would fall to earth a distance of 15 1/12 Paris feet. Taking the distance between the moon and earth to be 60 earth radii, and using the inverse square law, it will follow that the attraction for a body at the surface of earth must be 60^2 times stronger. Working out the arithmetic, Newton concluded that bodies on earth's surface should fall 60^2 x 15 1/12 Paris feet per minute, or, "15 ft 1 inch and 1 line 4/9" per second. Huygens, Newton noted, had worked out with a pendulum the acceleration under gravity and found it to be "15 Paris ft., 1 inch, 1 line 7/9." An impressive testimony to the accuracy of Newton's figures for the dimensions of the solar system.

Proposition v makes the crucial generalization. Newton is now ready to assert that the motions of the satellites of Jupiter, of Saturn, of the Sun are, just like the Moon, "drawn off from rectilinear motion and retained in curvilinear orbits." At this point he appealed to Rule ii of his "rules of reasoning in philosophy": "to the same natural effects we must . . . assign the same causes." It therefore followed that it must be the same cause producing the same effects in the solar system. But, by Law iii, all attraction is mutual and therefore just as Jupiter attracts its satellites, so too will the satellites attract Jupiter. In a concluding scholium to Proposition v Newton added that while the force retaining "celestial bodies in their orbits had hitherto been called centripetal force" he would hereafter call it gravity. It remained for Newton in Proposition vii to show that the "power of gravity pertaining to all bodies" was proportional to their "several quantities of matter."

Gravity, as Newton expressed it in the "General Scholium," "penetrates to the very centres of the sun and planets, without suffering the least diminution of its force. . . . It propagates its virtue on all sides to immense distances, decreasing always as the inverse square of the distances" (Newton 1966: 546). There is clearly something special about anything so pervasive and constant. Three centuries of progress in physics has done nothing to show that the universal aspect of gravitation emphasized by Newton need in any way be limited. Subatomic particles moving in accelerators at speeds almost approaching that of light as well as mighty galaxies and black holes, those creatures in the new menagerie of physics, whatever other traditional orthodoxies they may violate all adhere to the inverse square law which first came to the twenty-three-year-old exile from Cambridge in his garden at Woolsthorpe.

Inevitably, Newton, his followers, and his critics pondered the nature of this peculiar force. Newton's own views varied. In the "General Scholium" which Newton added to the second edition of *Principia,* he adopted a rather aloof attitude, declining even to speculate as to its cause. He did not "frame hypotheses"; whatever could not be "deduced from the phenomena" had no place in experimental philosophy. It was enough, he concluded, "that gravity does really exist, and act according to the laws which we have explained."

In 1693, in a correspondence unpublished until 1756, Newton further issued some dicta about what gravity was not. "That gravity should be innate, inherent, and essential to matter, so that one body may act on another at a distance through a vacuum . . . is to me so great an absurdity. . . .", that no competent thinker could ever accept it. "Gravity," he continued, "must be caused by an agent acting constantly. . . but whether this agent be material or immaterial I have left to the consideration of my readers" (Thayer: 54). If gravity did not act on another body at a distance then how did it act? All Newton offered were speculative thoughts on the possibility of a medium of some kind. In 1675 he had discussed the possibility of there being a pervasive ether distributed throughout space which would allow the propagation of light. At various times Newton returned to this basic idea and proposed a number of mechanisms whereby an extremely subtle, elastic fluid could be used to account for not only the transmission of light but for the transmission of electrical, magnetic, and gravitational forces. But so variable were Newton's views on this issue that in 1702 he seemed to deny outright the existence of a fluid medium throughout space (Newton 1959: IV, 1-3). By the second edition of *Principia,* however, Newton was again toying with the idea of a suitable medium, this time electrical, but abruptly he concluded that "these are things that cannot be explained in a few words," nor did we as yet know enough "of the laws by which this electric and elastic spirit operates" (Newton 1966:547). In the second English edition of *Optics* (1717) Newton once more introduced the notion of the ether, this time so elastic as to be 700,000 times rarer than the air, and proposed a further mechanism to explain the action of gravity (Thayer: 142-44). This was his last published comment on the issue.

If Newton could make little headway in analyzing the nature of gravity, was he any more successful in describing the system of the world? In general, two criteria needed to be satisfied before Newton's theory could be accepted as applicable to the *whole* system, and not just some particular part of it at a particular time. We can only claim to understand the universe if we can predict with reasonable accuracy the position of its major bodies over substantial periods of time. Of course, on this particular issue, only time would tell whether, for example, the Newtonian predictions for the positions of Jupiter and Saturn would be realized as accurately in 1750 as they had been in 1700. In astronomy, errors indetectable in the short run have a nasty habit of growing by imperceptible amounts to become embarrassingly large.

It would also have been essential to show that the universe was a stable system. Popular opinion assumed that the system of planetary motions had been so designed by God that Jupiter would not collide with Mars in the foreseeable future, that the universe would continue – with minor variations – to look the same to generation yet unborn as it had done to Newton.

To the surprise of 18th-century astronomers, anomalies began to appear within the Newtonian system. As early as 1695 Halley had suspected that Newtonian theory could not adequately account for the observed motions of the moon; observed, that is, over a long period of time. Using Arabic and Greek sources, Halley noted that the reported time and the predicted time of past eclipses did not tally: the computed time was preceding the observed time by as much as an hour. Halley proposed that the moon was undergoing a secular acceleration.[23] Work by later astronomers produced a figure for the lunar acceleration of about ten seconds a century. Not large, perhaps, but detectable and difficult to explain within Newtonian dynamics.

Further problems arose over the lunar perigee. Perigee refers to the point in orbit in which the body in question approaches closest to earth. It was known that this point was not fixed but moved around in orbit. Newton had claimed that it would take eighteen years for the lunar perigee to completely move around the earth's orbit; careful observation showed the period to be nine years. In this case, it proved possible to correct Newton. In 1743, A-C. Clairault (1713-65), in his *L'orbite de la lune dans le système de M.Newton,* showed that the addition of some extra terms to the appropriate equations would produce the expected result. Newton had simply failed to take *all* the relevant factors into consideration.

A more recalcitrant problem emerged over a discovered acceleration in the orbital motions of Jupiter and Saturn. Again, it was Halley who made the crucial discovery. Tables published in 1695 showed that Saturn was steadily decelerating and Jupiter accelerating. Halley showed how it was possible to introduce terms into the equations that would yield the observed results. The cause of the anomalies he attributed to the mutual attraction of the superior planets. This was, indeed, a serious difficulty facing Newtonians for the changes in mean motion of the superior planets were cumulative and, if continued, one planet would eventually crash into the sun while the other would escape from the solar system altogether.

For some this was too much. Clairault, for example, questioned the validity of the inverse square law. Perhaps, he argued, it only worked for certain distances; in other cases, we might need a term which varied inversely to the fourth power. Such ad hoc approaches toyed shamelessly with factors and terms, adding them whenever they seemed to suit the occasion. A major mathematician like L. Euler (1707-83) had tried to avoid such ad hoc adjustments, yet he was forced to conclude: "As often as I have tried these forty years to derive the theory and motion of the moon from the principle of gravitation, there always arose so many difficulties that I am compelled to break off my work" (*Theoria motum lunae,* 1772). At this point it must have appeared tempting to conclude that if someone like Euler could make no sense of Newton's planetary system then, indeed, there was something wrong with the basic assumptions of *Principia.*[24]

Such a conclusion would in fact have been premature. A solution of sorts came from the great French school of mathematicians centered around J. L. Lagrange (1736-1813) and P. S. de Laplace (1749-1827). Between 1799 and 1825, Laplace published his five-volume *Mécanique céleste,* mainly a collection of earlier results of his own and others, which for a generation confirmed the validity of the Newtonian system. He showed that the near identity of five Jupiter orbits (59.3 years) with two of Saturn's (58.9 years) meant that they only came into conjunction irregularly. If five orbits of Jupiter had exactly equalled two of Saturn's, then they would have come into conjunction in precisely the same position and whatever effect one planet had on the other would be repeated every 59.3 years until the system collapsed. But, in reality the point of conjunction does not fall at the same point in their orbits but moves around the orbit in a complete cycle. Its effect on one side, to accelerate Jupiter perhaps, would be compensated when conjunction took place on the opposite side of the orbit and Jupiter was retarded. The whole cycle, according to Laplace, took 929½ years.

Further investigation revealed several other periodic inequalities which just happened to balance out over long periods of time. As a result Laplace in 1788 declared, "Thus the system of the world only oscillates around a mean state from which it never departs except by a very small quantity. . . it enjoys a stability that can be destroyed only by foreign causes and we are certain that their action is undetectable from the time of the most ancient observations until our own day."

In truth, of course, Laplace could draw no such general conclusion; all he was permitted to conclude was that all the major anomalies and inequalities known to late 18th-century mathematicians could be shown to behave in this way. As would soon be shown, he was leaping to fairly dogmatic conclusions about the solar system without being aware of some of its major parts. In 1788, he would have just known of the existence of Uranus, but only just for it was barely seven years since William Herschel (1738-1822) had observed the first planet to be discovered in modern times. Shortly afterwards in 1801, the first of the minor planets, Ceres, was discovered by Giuseppe Piazzi (1746-1826), and ominous anomalies had begun to appear in the orbit of the newly discovered Uranus.

Although Uranus was first recognized as a planet in 1781, a search of earlier records revealed that its position had been recorded as a "fixed star" seventeen times before. Unfortunately, however, if its current position was accepted it proved impossible to harmonize the orbit of Uranus with observations made a century earlier, whereas if the early observations were accepted Uranus should not be where it was in 1800. Nor was there any reason to doubt the accuracy of such early observers as Flamsteed (1646-1719), the British Astronomer-Royal. Continued observation of Uranus only made the problem more acute. By the late 1820s, observations of Uranus could no longer be reconciled with those recorded in the 1780s.

A number of solutions were possible, the most obvious being again to deny the accuracy of Newtonian mechanics. This time it was no less a personage than the Astronomer-Royal G. B. Airy (1801-92), who declared in 1846 that "there was a great probability that the laws of force differed slightly from that of the inverse square distance." More perceptive minds weighed another alternative: the apparent anomalies might be due to gravitational forces from some as yet undiscovered planet. The suggestion was not a new one and can be traced back as far as Clairault in 1758. The difficulty was to find it. Planets are small and the heavens are vast. Use a less powerful telescope on a wide field, and the dim light of a planet could well be overlooked; use a powerful telescope on a small field and, unless one knows where to look, it could take decades or centuries to examine the whole orbit. The solution to the problem of where to look came from Newtonian mechanics.

In theory it should be possible to work backward from the observed anomalies in Uranus's orbit to the site and magnitude of the disturbing force. One who was very familiar with the techniques of computing planetary and cometary orbits was Urbain Leverrier (1811-77) who had done some spectacular work on a comet observed in 1844. Was it a new one? Comets with long periods present great difficulties to astronomers as there is no simple way to determine in a particular appearance whether it is a new one or the return of one seen some centuries earlier. As there is little to distinguish one comet from another, the issue was normally settled by determining the orbit and working backward to any earlier appearances. If they coincide, then it is not a new comet. In this way, Leverrier showed that the comet of August 1844 was in fact the same one observed by Tycho in 1581. With this success behind him, he turned to the problem posed by Uranus.

Within two years, Leverrier thought he had found an answer and he wrote to Johann Galle (1812-1910) of the Berlin Observatory to ask him to train the nine-inch refractor on the spot Right Ascension 22° 46′, declination −13° 24′ where he should see an unrecorded three-inch disk. A few days later he received a telegram from Galle informing him that "the planet whose position you have pointed out actually exists."[25] This discovery of Neptune was greeted by some enthusiasts as the ultimate confirmation of Newtonian mechanics.[26] It was certainly a dramatic public display of the power of science in general and of Newtonian mechanics in particular, and it led to much extravagant description. O. M. Mitchell, director of the Cincinnati Observatory, wrote in 1862, "None but the rarest genius would have dared to reach out 1,800,000,000 miles into unknown regions of space, to feel for a planet which had displaced Uranus by an amount... so small that no eye, however keen and piercing, without telescopic aid, could ever have detected it" (Mitchell:165-66).

As if in a rerun of the previous half-century, anomalies began to be noticed in the orbit of yet another planet. This time it was Mercury which presented a confusion.[27] Leverrier tried to repeat his success with Uranus and predicted the existence of a new planet lying somewhere between Mercury and the Sun. So confident was he of its existence that he even produced a name for it: Vulcan. Because viewing conditions so close to the sun are difficult, failure to observe Vulcan could easily be explained away. At least for a time. When he died in 1877, Leverrier was one of the few who still had faith in Vulcan's existence.

Others, like Simon Newcomb (1835-1911), superintendent of the American Nautical Almanac, once more proposed that the inverse square law was inexact. But this time something more radical was required in mechanics. Newtonian physics, after coping successfully with anomalies for over two centuries, finally found itself presented with a problem which would not yield to any of the techniques that had previously worked so well. The solution was in fact found by Einstein through a profound rethinking of the basic concepts of space, time, gravity, and inertia. The price paid for the solution was that mechanics and the structure of the universe could no longer be seen in all their generality as totally Newtonian. Finally in the first two decades of the present century the system developed in *Principia* had to give way to the relativistic mechanics and cosmology of Einstein.

Immediate Reception

Published on 5 July 1687 *Principia* was reviewed anonymously in France in *Journal des sçavans* and *Bibliotheque universelle.* The latter review is attributed to Newton's friend, the philosopher John Locke (1632-1704). Halley reviewed the work for *Philosophical Transactions,* a strange choice considering his role in its preparation. What would we not give for a review from Hooke! Another anonymous review was published in Leipzig in the *Acta eruditorum.* The so-called Locke review was no more than a summary of *Principia* with a translation of Latin headings into French.

Halley's enthusiasm for the book would surprise no one. Newton, he declared, seems to have exhausted the argument, "and left little to be done by those that shall succeed him." So many and so valuable were the truths contained in *Principia,* Halley continued, it was difficult to accept that they are "owing to the Capacity and Industry of any one Man."

The European response was respectful rather than adulatory. The *Journal des sçavans* declared the work "the most perfect treatise on mechanics that one could imagine" yet, true to the Cartesian tradition, the reviewer raised objections against Newton's reliance on the notion of force. Nor had things changed much in 1733 when Voltaire in *Letters on England* noted: "A Frenchman arriving in London finds things very different in science as in everything else. . . . For your Cartesians everything is moved by an impulsion you don't really understand, for Mr Newton it is by gravitation, the cause of which is hardly better known" (Voltaire: 68). Why, according to Voltaire, the French asked, "didn't Newton use the word impulsion . . . rather than the term attraction?" (Voltaire: 79).

The Leipzig review carried a long, fair and appreciative summary of *Principia.* Its author was impressed by the scale of Newton's work yet, like most other European scholars, was unwilling to accept Newton's conclusions without reservations. The full impact of *Principia* on Europe should not be searched for in the journals but rather, according to Westfall, in the correspondence of its scholars. It is there evident, in the letters of such leading figures as Leibniz and Huygens, that their thoughts on mechanics and cosmology after 1687 were dominated by Newton (Westfall: 473).

The response in Britain was more openly enthusiastic. As early as 1690

Locke, in the preface to *An Essay Concerning Human Understanding,* was speaking of "the incomparable Mr Newton." There soon grew up a rush of books that sought to explain either the new philosophy or to extend it to new areas of thought. Through such works as E. Carter's *Newton's Philosophy Explained for the Use of Ladies* (1739) or Tom Telescope's *The Newtonian System of Philosophy Adapted to the Capacities of Young Gentlemen and Ladies* (1761) together with better known and more serious works from W. 'sGravesande, Henry Pemberton, J. T. Desaguliers, John Keil and Colin Maclaurin, 18th century-educated Englishmen became aware of the work and genius of Newton.

Chemists and physicians, not to mention theologians, were quick to see new possibilities opening up in their own disciplines. References to 18th-century chemistry with its demand for a new approach will be found elsewhere in Chapter 11. Physicians were no less excited and pressing in their demands. Particularly significant was the group centered around Archibald Pitcairn (1652-1713) who taught at Edinburgh and Leyden. He called for a "hydraulic theory of sickness" in which disease represented changes in the quantity, velocity or texture of bodily fluids. He was followed by George Cheyne (1671-1743) who actually called for a *Principia medicinae theoreticae mathematica* but, realizing his own limitations, produced the more modest *New Theory of Fevers* (1702). For a brief period in medical circles there arose the wild dream that physicians too would be able to write equations and prove theorems instead of their more prosaic duties of issuing prescriptions. Such was the power of the Newtonian vision that some could see it extending far beyond the domain of mechanics. Desaguliers (1683-1744), once experimental assistant to Newton, could thus produce *The Newtonian System of the World, The Best Model of Government* (1728).

Theologians too, rapidly and eagerly embraced the Newtonian system and deployed favored parts of his mechanics to prove the existence of God. The best example of this were the Boyle Lectures, first given in 1692, and enjoined by their founder Robert Boyle to prove "the Christian religion against notorious infidels." The first lecturer, Richard Bentley (1662-1742), actually approached Newton and received from him four important letters setting out the extent to which *Principia* could be used to support the existence of God. Bentley was just the first of many. He was followed by such other well known Newtonians as Samuel Clarke, the translator into Latin of *Optics,* William Whiston, Newton's successor as Lucasian professor, John Harris and William Derham, all of whom were quite happy to argue for that curious blend of theology and science labelled by Derham in his 1714 lectures as "Physico-Theology." A typical argument in this field would be Derham's recognition of gravity as a "noble contrivance" designed to keep "the several Globes of the Universe from shattering to Pieces. . . by their swift Rotation around their own Axes" (Jacob: Chap. 4).

Outside Europe the spread of *Principia* is less easy to document. The first copy to arrive in the United States was imported in 1708 by James Logan (1674-1751), creator of one of the finest 18th-century libraries in the U.S. A copy of the second edition was donated to Yale in 1714 by Newton himself while a third copy of *Principia* was obtained in 1739 by John Winthrop (1714-79) of Harvard. Newtonian science was probably first taught there by Isaac Greenwood (1702-45), elected in 1727 to the first Hollis Professorship of Mathematics. It was not, however, until 1740 that the study of Newton, with the arrival of the Rev. Thomas Clap, is known to have reached Yale. The first American edition of *Principia* was published in New York in 1848.

There was an initial interest in Newton in Russia during the reign of Peter the Great (1672-1725). He may even have met Newton during a visit to London in 1698. He certainly arranged for Jacob Bruce (1670-1735), of Scotch origin and an engineer in Peter's army, to amass a formidable library of European science. Included were several copies of *Optics,* much secondary literature, and various copies of *Principia.* Bruce was also responsible for the first work in Russian to describe Newtonian science, a translation of Huygens's *Kosmotheoros* (1696) in 1717.

Further east the first hint of interest in Newton in Japan dates to the late 18th century. Shizuki Tadao (1760-1806) spent twenty years translating a Dutch commentary of Johann Lulof on Newtonian mechanics. Completed in 1802, *Rekisho Shinsho* was the first work to present in Japanese the Newtonian laws of motion.

One final reaction to Newton may be briefly noted. It is the curious attempt to apotheosize Newton seen in the comments of scientists, poets and artists alike. No one actually believed Newton *was* divine but nonetheless it rapidly became the custom to speak as if he were not as other men. Some, like Pope, suggested he was a special creation of God; others, like Halley, proclaimed him as close to being a God as man could ever be while a third variation, from Dr. Johnson, noted that in antiquity Newton would indeed have been deified. Such comments were repeated endlessly by figures less well known than those quoted at the beginning of the chapter. Nor were such hyperbole confined to England. The architect E. L. Boulée designed elaborate temples to Newton, whom he called an "être divin," in the 1780s while, even more extreme, F. C. de la Blancherie in 1796 abused the English for their failure to honor Newton adequately. He proposed to introduce a new calendar beginning in 1642, the year of Newton's birth, and to make his home at Woolsthorpe a place of pilgrimage. Saint-Simon followed de la Blancherie in this and proposed in 1803 a new Religion of Newton. Other national heroes, Descartes, Galileo and Kepler, all escaped this particular lunacy and it is perhaps a reflection of the unique reputation of Newton that he and he alone evoked so bizarre a response.

Publishing History

London, 1687: First edition

There were two issues, distinguished only by the title page: the first imprint carries the name of Streater as printer; the second bears the names of Streater and Sam Smith. It is assumed that the second issue was prepared for sale abroad. The size of the edition has been estimated between 250 and 400 copies. Edited by Halley, it was made available to the trade in quires for six shillings (five for cash), and sold to the public for seven shillings unbound and nine "bound in Calves leather and lettred." Twenty copies were given to Newton for his own disposal.

Copies of *Principia* soon became hard to find. There is a reference dated 1708 to a student paying two guineas for a copy, while Hooke's copy, sold at auction after his death in 1703, went for £2 3s. 6d. Copies could be obtained quite cheaply early in this century and Osler mentions the purchase of a Streater issue for eighteen guineas in 1913. By the 1930s a copy sold for £60 to £70. The physicist Andrade, the great collector of modern times, had as many as sixteen copies of various editions of *Principia.* His two first-edition copies were sold for £2,400 and £2,200 at auction in 1965. In more recent times, a first edition went for £12,500 in 1979, while Blackwell, the Oxford bookseller, offered Andrade's Streater issue for £14,000 in the same year.

A number of annotated copies exist. Locke's copy is in the library of Trinity College, Cambridge; copies in Australia and at the University of Texas contain profuse annotations in Halley's hand. Newton's own annotated copy is in the Trinity College library. His specially-bound copy of *Principia,* interleaved with blank pages for corrections and revisions, is in the Cambridge University Library. There is some evidence for another interleaved copy that is missing. The manuscript of 460 leaves, written by Humphrey Newton, with corrections in Isaac Newton's and Halley's own hands is in the library of the Royal Society, London.

Cambridge, 1713: Second edition

Edited by Roger Cotes with many significant changes, including an important preface by the editor. Printed in an edition of 750 copies, it sold for fifteen shillings in quires and a guinea if bound.

Amsterdam, 1714, 1723: Reprints

Two reprints of the second edition were issued to fill the European demand for *Principia.* The 1723 reprint contains four of Newton's mathematical treatises and four of his letters.

London, 1726: Third edition

Edited by H. Pemberton for a 200-guinea payment, this edition contains fewer and less significant changes than the second, but it represents the definitive text as approved by Newton. It was printed in an edition of 1,250 copies in three different sizes. Fifty presentation copies were printed in the largest size (32.6 x 23.4 cm) on Superfine Royal paper; two hundred on large paper (28 x 21.6 cm) and the remaining one thousand copies on paper similar in size to the two earlier editions (24 x 18.5 cm). As a frontispiece is the portrait of Newton painted in 1725 by Vanderbank. It sold for one guinea.

London, 1729: First English edition
Published in two volumes, it was translated by Andrew Motte, of whom virtually nothing is known. It sold for fourteen shillings. This edition served as the basis for all further English translations. It was known that Pemberton planned to publish an English translation but the appearance of Motte's work must have changed his mind.

Geneva, 1739: Jesuit edition
Published in three volumes, this edition is known as the Jesuit edition. The authors of the accompanying commentary, Fathers Le Soeur and Jacquier, were not Jesuits, according to Andrade, but Minims. This edition served as the basis for many future editions and was reissued in corrected form in 1760 in Geneva (in two volumes), in 1780 and 1785 in Prague (two volumes), in 1822 and 1833 in Glasgow (four volumes and two volumes, respectively).

Paris, 1759: First French edition
The first and only complete French translation was done by Madame de Chastellet, Voltaire's mistress, and to which was added a commentary by A-C. Clairault.

London, 1779-85: The Opera omnia
Edited by Samuel Horsley, *Principia* takes up volume two (1779) and three (1782) of this five-volume work.

Glasgow, 1803: Revised English edition
A revision of the Motte translation was done by William Davis and published in three volumes. It contains "System of the World" and a biography of Newton. Originally issued serially – six issues on the first of each month at four shillings each – the edition was published in London in 1819.

New York, 1848: First American edition
The first American edition was prepared from the 1803 Glasgow edition and edited by N.W. Chittenden, who also wrote the fifty-three page biography of Newton which is included. Four further editions appeared before 1850. Between 1803 and 1934 only five English translations of *Principia* appeared. They were all published in New York and they all appeared in the two-year period between 1848 and 1850. Why New York should have had such an exceptional interest in *Principia* at this period remains an open question.

Glasgow, 1871: Reprint
"Finding that all the editions of the *Principia* are now out of print, we have been induced to reprint Newton's last edition without note or com-

ment"–from the statement by the publishers, William Thomson and Hugh Blackburn.

Other notable editions

First foreign editions in the following languages can be noted: *Berlin 1872:* German translation by J.B. Wolfers; reissued 1932; *Petrograd 1915:* Russian translation by A.N. Kruilov; reissued 1936; *Rome 1925:* Italian translation by F.E.U. Forti; a later translation by A. Pala was published in Turin in 1965; *Lund 1927:* Swedish translation by C. Charlier; *Tokyo 1930:* Japanese translation by K. Oka; *Bucharest 1956:* Rumanian translation by V. Marian.

The Motte English translation was revised by Florian Cajori and published in Berkeley, CA in 1934, later reissued in 1946, 1947, 1960, 1962, 1966, and 1973. The two-volume definitive edition was published in Cambridge, England and Cambridge, MA in 1972. The critical text of the third edition with variant readings and a mass of editorial material was prepared by I.B. Cohen and A. Koyré.

Facsimiles

Facsimile editions have been published in London 1953 (first edition); Darmstadt 1963 (1872 German translation), New York 1964 (1848 American edition), Brussels 1965 (first edition, in the Culture et civilisation series), Paris 1966 (1759 French translation) and London 1968 (two-volume Motte translation).

Microprint

Available in New York (1967) from Redex.

Further Reading

Principia itself is available in a Cajori two-volume paperback edition (1966). As a brief and straightforward introduction to the physics of *Principia*, I.B. Cohen's *The Birth of the New Physics* (1961) supplemented with his 1981 *Scientific American* article are strongly recommended. For any anxious to follow in more detail the nature and the development of Newtonian mechanics they are likely to find most illumination in two 19th-century works: *History of Physical Astronomy* (1852) by Robert Grant, and *Analytical View of Sir Isaac Newton's Principia* (1855) by Lord Brougham and E. J. Routh. The relation of Newton's ideas to the intellectual background of the 17th-century is beautifully analyzed in Koyré (1965) while many of the central texts from Newton's various writings have been conveniently collected by Thayer (1953). For the writing of *Principia*, the text, various editions and its reception, Cohen (1978) is indispensable. Two recent biographies, Westfall (1980), an authoritative survey of the life, work and science of Newton, and Manuel (1968), a controversial analysis of Newton's character and career, throw much light on *Principia*. Brewster's *Memoirs* (1855) remains of considerable value. The monograph literature on Newton is by now quite enormous with detailed studies available on virtually every aspect of his life and career. Studies by Hall (1980) and Harrison (1978) on Newton's dispute with Leibniz and on his library respectively contain much of interest while essential primary material has been made available by the Halls in their *Unpublished Scientific Papers of Isaac Newton* (1962). Any work on *Principia* to appear before 1975 will almost certainly have been recorded by Wallis and Wallis (1977). The seven-volume *Correspondence* and eight-volume *Mathematical Papers* are now complete and although both can be dauntingly technical, there remains much in them to illuminate many of the less formal aspects of *Principia*.

Notes

1. The sarcophagus is not to everyone's taste: "a baroque monstrosity" according to Westfall (p.874). It cost £500, money grudgingly extracted from his heirs and shows a reclining Newton surrounded by cherubs displaying such appropriate objects as a prism, a newly minted coin, and a reflecting telescope.

2. Examples can be found in Schofield. Consider, for example, Richard Mead, Newton's own physician, who in his *The Action of the Sun and Moon on Animal Bodies* (1712) sought to find analogues in the human body to the sun and moon's power to cause tidal effects in the natural world; Newton's impact on 18th-century chemistry is considered in more detail in the chapter on Dalton.

3. By any measure, religion was by far the major interest in Newton's life. An analysis of his library shows that 28% of its books were on religion and no more than 25% on science. So, too, with his surviving papers, more of which are devoted to theology than to science.

4. For Newton, the Pope was no less than the anti-Christ. Catholics he saw as worshippers of dead men who also made magical use of the cross. The eleventh horn of Daniel's fourth beast he identified with the Church of Rome. In 1693, Newton seems to have suffered a severe nervous breakdown. He accused his friend of "embroiling" him with various distasteful groups. Samuel Pepys, he declared, had tried to involve him with Catholics (Newton 1959: III, 279).

5. Although Newton accused Locke in 1693 of endeavoring to "embroil me with women" (Newton 1959: III, 280), there are in fact references to women in his life. The best known is the case of Mrs. Vincent of Grantham who remembered some sixty years earlier when, as a Miss Storer, "Sir Isaac. . . entertained a love for her. . . but her portion being not considerable, and he being a fellow of a college, it was incompatible with his fortunes to marry" (More: 19). There is no record of the affair in Newton's papers.

6. Because of an outbreak of plague in 1665, Cambridge University closed and sometime in August Newton returned to Woolsthorpe. In later life he variously remembered his achievements of his *annus mirabilis* of 1665-66: "I found the Method of approximating. . . the Method of Tangents. . . the Theory of Colors. . . and the same year I began to think of gravity extending to ye orb of the Moon. . . I decided that the forces wch keep the Planets in their Orbs must be reciprocally as the squares of their distances from the centers about wch they revolve." Add to these some early work on the method of fluxions and one has a picture of unprecedented and never since matched intellectual power and creativity. Almost as

if in awe of his younger self, Newton later commented: "I was in the prime of my age for invention and minded Mathematicks and Philosophy more than at any time since" (Westfall: 143). He was in fact twenty-three.

7. They also carefully preserved his priority without yielding anything. In the custom of the day, results were given anagrammatically. The simplest of Newton's ran: "6accdae13eff7i319r404qrr4s8t12vx." Its formidable Latin solution, "Data aequatione quotcumque, fluentes quantitates in volvente, fluxiones invenire, et vice versa" ("Given any equation involving fluent quantities, to find the fluxions and vice versa"), clearly did nothing other than protect Newton's priority as did the longer anagram totalling nearly 300 letters (More:190).

8. According to one account Newton was heard "pleasantly tell" a friend: "He had broke Leibniz's Heart with his Reply to him" (Manuel 1968:348); it is quite the most chilling of all the Newton anecdotes.

9. In the Cambridge University Library there are "five-hundred-odd folios of manuscript. . . devoted to self vindication, drafts for the *Commercium*, and attacks on the enemy" (Manuel 1968:327). The reference is to the *Commercium epistolicum* (1713), the official account of the dispute published by the Royal Society.

10. Newton was obsessed by the prophetic writings and returned again and again to the Books of Daniel and Revelation. "If they are never to be understood, to what end did God reveale them?" he asked himself (Manuel 1974:88), and went on to consider at great length the precise meaning of Daniel's dream in which there was a beast with eleven horns. Other manuscripts concern early church history, the dates of Christ's birth and crucifixion, the dimensions and structure of Solomon's Temple and two corrupt scriptural passages (First Epistle of John V:7 and First Epistle to Timothy III:16). Both, affirming unambiguously the Trinity of God, were dismissed by Newton as later interpolations, a result modern scholarship tends to confirm. It should be emphasized that Newton's approach was as scholarly and critical as the time permitted and that nowhere in his theological writings can there be found the gullibility and crassness found in many a modern scientist's attempt to mystify himself with psychic phenomena.

11. There is another interpretation of these figures proposed by De Villamil. The £20,000 loss refers to the profit he could have realized if he had sold out before the Bubble burst. The original £9,000 invested in the Company was, at his death, worth only £5,000 while at one time it had reached a height of £25,000. Thus De Villamil argued (p. 27) while there could have been a paper loss of £20,000, the real loss was only £4,000. Westfall cautiously states that it is not known whether Newton invested and lost such huge sums (pp.861-2).

12. Newton was an only child. His father Isaac had two younger brothers, both of whom left descendants; of them, John Newton, great-grandson of Newton's uncle Robert, inherited the family estate. John was said to be "illiterate and dissolute." He sold the estate in 1732 and died shortly afterwards in 1737, "of a tobacco pipe breaking in his throat in the act of smoking, from a fall in the street occasioned by ebriety" (More:667). There were also descendants from Newton's mother's second marriage to Barnabas Smith. The Bartons, the Smiths and the Pilkingtons, descendants from this connection, inherited Newton's estate on John Newton's death.

13. The fate of Newton's books has been traced in some detail by De Villamil (1931) and Harrison (1978). All 1,896 of them were bought by John Huggins, Warden of the Fleet prison, for £300. Huggins had bought them for his son Charles who had the parish of Chinnor near Oxford where the books were moved. On the death of Charles in 1750, the Parrish went to the Rev. Dr. James Musgrave who also bought the books for £400. Sometime after his death in 1778, the library was moved to the family home at Barnsley Park in Gloucestershire. There it remained until 1920 when the Wykeham-Musgrave family, as they had since become, unaware of the treasures in their possession, sold about one thousand of the books in a sale of 123

lots described simply as a "Library of miscellaneous literature of upwards of 3,000 volumes." (The whole sale realized £170.) Some idea of the value the trade placed on some of the lots can be seen from the fact that Newton's annotated copy of Barrow's Euclid, though sold for only five shillings was soon an offer in a dealer's catalogue for £500. The remaining 858 volumes of the Newton library were traced by De Villamil in 1927 to Barnsley Court by, what he termed, "a series of lucky accidents" (De Villamil: 6). Soon afterwards they were on offer for £30,000. There were no buyers. Maynard Keynes was informed when he made inquiries in 1936 that he could have them for £5,000. Still unsold in 1943, they were finally bought by the Pilgrim trust for £5,500 and deposited in Trinity College, Cambridge. Of the 1,000 or so dispersed in 1920, some still turn up unrecognized in second hand bookshops where they can be had for a few pounds. Harrison reported one such volume was sold in 1975 in a Cambridge bookshop for £4.

14. Catherine Barton, daughter of Newton's half-sister Hannah, was educated by Newton on her father's death in 1693 and later served as his housekeeper in London. In 1717, she married John Conduitt, a man of some wealth, who did much to gather biographical material about Newton and to preseve the bulk of his papers. By all accounts, Catherine Barton was a beautiful and remarkable woman. Her presence in Newton's household inevitably led to gossip. Voltaire, who visited her in the 1720s, could not resist the thought that "fluxions and gravitation would have been of no use without a pretty niece," suggesting that Newton had obtained his post at the Mint by ensuring that Catherine "made a conquest" of the Chancellor of the Exchequer, Charles Montague (1661-1715). This was repeated by Newton's enemies. Actually Montague was an old pupil of Newton's and appointed him to the Mint in 1696, long before he was likely to have ever met Catherine.

15. Thirteen lots, for example, were purchased by a M. Emmanuel Fabius of Paris and have never since been available for examination.

16. It has recently been proposed that Newton's celebrated breakdown of 1693 was the result of some kind of poisoning brought about by his alchemical experiments. It is known that he used to taste many of the compounds he produced. Analysis of his hair by sophisticated modern techniques show in one sample a concentration of mercury forty times the normal level. (P. Spargo and C. Pounds, "Newton's Derangement of the Intellect," *Notes and Records of the Royal Society*, 1979).

17. Huygens had introduced the term *centrifugal force* for the center-fleeing force met with in circular motion. On this analogy, Newton introduced the idea of a center-seeking or *centripetal force*. It is defined in *Principia* as Definition v: "A centripetal force is that by which bodies are drawn or impelled, or any way tend, towards a point as to a centre." By Proposition v, Book III, Newton was finally in a position to note that as "it was no other than a gravitating force we could call it gravity."

18. "Mr Hooke then said, that he had it, but he would conceal it for some time, that others trying and failing might known how to value it, when he should make it public" (More:299).

19. Rouse Ball noted that "of the 494 pages in the first edition 397 are more or less modified in the second edition. The most important alterations are the new preface by Cotes; the propositionon the resistance of fluids II vii 34-40; the lunar theory in Book III; the proposition on the precession of the equinoxes, Book III 39; and the proposition on the theory of comets, Book III 41-2" (Ball:74).

20. The second edition also has a long preface by Cotes which includes some remarks on gravity. Cotes at first wanted Newton to write it although he offered to sign it himself. In the end it was agreed that Cotes should write it. He was instructed not to mention Leibniz. Gravity, Cotes argues, was as much a property of bodies as were extension, mobility, and impenetrability (Newton 1966:xxvi).

21. This should not be confused with *De mundi systemate liber* or its translation *A Treatise of the System of the World* both of which were first published in 1728 under rather mysterious circumstances. It was an earlier draft of what eventually became *Principia*, Book III. Cohen notes that fourteen paragraphs of Book III had been taken almost verbatim from the 1728 work (Cohen 1978:110).

22. The notion of gravity was not new with Newton. Many before had spoken of gravity; Copernicus, Kepler, Gilbert, Bacon, had all considered gravity and attraction. Details can be found in Koyré. Among other things, what they failed to see was that it was universal, everything really did attract everything – without exception. What is more, no one else had appreciated that it was a unique force: whether found on the moon, the sun or in the Grand Canyon, it was the same exact force following the same precise laws. Thus, between any two objects in the universe, of whatever size, of whatever structure and whatever they were made from, however near or far apart, there is an attractive force between them that varies inversely with the square of the distance. In terms of its generality this is a staggering thought, and it was proposed by Newton alone. It is not to be confused with the common view that the sun attracted solar matter, the moon lunar matter and the earth terestial matter.

23. Astronomers use the term *secular* to refer to processes which are continuous rather than periodic.

24. Much of this discussion took place within the framework of international competitions and prizes. Thus in 1751, the St. Petersburg Academy set this problem: To demonstrate whether all the inequalities observed in lunar motion are in accordance with Newton's theory." It was won by Clairault. In 1770, Euler won the prize offered by the Paris Academy for his discussion of the thesis: "It appears established that the secular inequality of the moon's motion cannot be produced by gravity." Thus, throughout the 18th century the Newtonian system was openly debated by the leading mathematicians and astronomers.

25. On his discovery of Neptune, Leverrier suddenly began to argue that, like comets, planets should be named after their discoverers. In support he even began to refer to Uranus as Herschel's Planet. Further confusion resulted from Airy using a third name of Oceanus before the original suggestion of Neptune was finally accepted.

26. Leverrier's work had been largely duplicated by the Campbridge astronomer J. C. Adams (1819-92). He did in fact establish the position of Neptune before Leverrier. Only Airy's crassness prevented Adams' work from being confirmed before Leverrier's.

27. Theory predicted that the line of apsides, an imaginary line joining perihelion and aphelion, should revolve around Mercury's orbit at a rate of 532 seconds per century. The observed rate was 574 seconds. According to relativity theory, there should be an advance of perihelion proportional to $3v^2/c^2$ where v is Mercury's velocity and c the speed of light. This works out to an expected advance of the periheliion of Mercury of 42.9 seconds per century, which is in close agreement with observation.

CHAPTER 10

Systema naturae
(Tenth Edition, 1758)

by Carl Linnaeus

Linnea. . . . A plant of Lapland, lowly, insignificant, disregarded, flowering but for a brief space–from Linnaeus who resembles it.

–Linnaeus, *Critica botanica*

God has suffered him to peep into his secret cabinet. God has suffered him to see more of his created worke than any mortal before. God has endowed him with the greatest insight into natural knowledge, greater than any has ever gained. The Lord has been with him, whithersoever he has gone, and has exterminated all his enemies, and has made of him a great man. . . .

–Linnaeus, *Autobiography*

The tenth edition of Linnaeus's *Systema naturae* marks the beginning of the consistent general application of binomial nomenclature in zoology. The date 1 January 1758 is arbitrarily assigned in the Code as the date of publication of that work and as the starting point of zoological nomenclature. Any other work published in 1758 is to be treated as having been published after that edition.

–Article 3, *International Code of Zoological Nomenclature*

Any mature science possesses a structural background, an intellectual framework, a common vocabulary within which discourse can take place and problems posed. Although such structures are peripheral to the main activity of science, they are a precondition to even the most routine observations and measurements. Astronomers have their stellar catalogs, geologists their eras, epochs, and periods, chemists their periodic tables and handbooks of compounds, and physicists their *International Critical Tables*. The names Friedrich Argelander (1799-1875)[1] and K. Beilstein (1838-1906)[2] may be less well known than those of Copernicus and Dalton, but to the working astronomer and chemist their contributions to science were no less basic. Without such works science would fragment and lack continuity. Observations would become private entries in personal diaries rather than public reports in international journals; discoveries would no longer be communicable let alone topics for discussion and debate.

In the biological sciences there is an equally pressing need to cope with the enormous variety of living forms. A recent work celebrating the centenary of the Natural History Museum in London indicated the magnitude of this problem: "A careful census in 1976 led to an estimate of 22½ million insect specimens as the Museum's holdings, representing about 445,000 named species" (Stearns 1981: 221). But these are only about half the known insect species. The number of animal species already identified is well over a million with 10,000 new species added each year. If the 300,000 known plant species are included, then the biologist's raw material is roughly 1.5 million species with an annual growth rate of about one percent.

Clearly the task of systematically classifying and naming such an abundance of material is a daunting one. Where such preliminary work has not been done the consequences can be drastic. Few families are simpler than the Hominidae with only five living members – the gibbon, orangutan, chimpanzee, gorilla, and man. Yet it is frequently impossible to distinguish one from the other in the early literature. Andrew Battell, who spent eighteen years in 17th-century Angola, called them by vernacular names such as *pongo* and *engecko*. Despite his descriptions, it was not possible to relate them to the primates variously named by later writers as *pygmy, homo sylvestris, orang-outang, satyrus indicus, drill, barris, quoias-morrou, pithecus, ape, boggoe,* and many others. When Thomas Bowdich wrote, in 1819, of a West African biped called *ingena*, "five feet tall and four across the shoulders," it could only be a matter of speculation whether he was describing a new species of Hominidae or adding yet one more synonym to an already confused list.

If such confusion could persist among substantial creatures, even less sense could be made of more elusive and slighter forms. The task of providing the necessary framework within which every animal and plant could be unambiguously placed was undertaken by one of the most creative intellects of the 18th century, Carl Linnaeus.

Life

Linnaeus was born in 1707, the son of a curate in a small southern Swedish town. Although he was expected to follow his father into the church he chose instead to study medicine – a subject more closely related to botany than it is today – first at the University of Lund, and from 1728 at Uppsala. When Linnaeus was born Sweden was still an imperial power, but under the last of the Vasa warrior-kings, Charles XII (1683-1718), the Great Northern War waged against Russia, Poland, Denmark, and Prussia, ended with Sweden bankrupt and expelled from its Baltic domain. The autocracy of the Vasas was replaced by a weak monarchy and a squabbling aristrocracy.

With its loss of empire, economic recovery would have to come from within Sweden itself. This need created a demand to know just what wealth and resources were contained within the national borders; it also created an ideal climate for the practice of botanical research. Linnaeus's two early expeditions to Lapland in 1732 and to the Baltic islands of Oland and Gotland in 1741 were as much economic surveys of the country as they were research trips. The point was expressed by Linnaeus in a 1734 letter to Baron Gyllengrip, the governor of Umea: "If only one could travel like this through all the Swedish provinces. . . how much one could discover that would be of value to our country! How much one could learn from one province of how best to cultivate another. . . . This would be of far greater value to Sweden than all the poetry, Greek and metaphysics taught in our Academies – and it would cost the public far less" (Blunt: 80).

But first Linnaeus spent three years abroad, from 1735-1738, based in Holland. Holland was then the main intellectual center of Europe, and it was the ambition of any student to spend some time there. Students in fact came from all over Europe to attend the lectures of Herman Boerhaave (1668-1738), chemist, physician, botanist, and acknowledged as the greatest scientist of his day. When Linnaeus was invited to accompany a friend as tutor on a trip to Holland, he gladly went.

In addition to scholars, Holland possessed two other attractions. From the Indies came the richest collection of tropical flora then available in Europe. Linnaeus set out to familiarize himself with this material. He soon found work helping J. Burman (1707-1780), director of the Amsterdam Botanic Garden, to prepare an account of the flora of Ceylon, later published as *Thesaurus Zeylanicus* (1737). Holland's second attraction was more personal. In his years at Uppsala he had never completed his doctorate, and to earn a living he urgently needed a medical degree. The simplest place to qualify was at the university at Harderwijk, thirty-five miles from Amsterdam and known throughout Europe for the speed with which it awarded degrees. So Linnaeus sped to Harderwijk on his arrival in Holland[3] and within a week his thesis – already written – on the cause of intermittent fevers was printed, examined, and approved.

Linnaeus was then twenty-eight, lacking reputation, and without any significant publications to his credit. His first approach to Boerhaave was predictably ignored. However, he met in Leyden Johan Gronovius (1690-1760), a physician and botanist, to whom he showed the manuscript of a small work, no more than fourteen pages when first published, entitled *Systema naturae.* Gronovius immediately recognized the modest text to be a work of major importance, and he began at once to arrange for its publication. He also introduced Linnaeus to Boerhaave and other leading Dutch scholars. To publish the manuscript, Gronovius enlisted the support of Issac Lawton (d. 1747), a Scottish physician studying at Leyden. In due course Linnaeus rewarded both men for their disinterested support in the traditional manner of the taxonomist, by assigning their names to the plants Gronovia and Lawsonia.

This was the beginning of one of the most creative and intense periods of his life. Within three years he published nine books. Some of them had clearly been written some years before: *Bibliotheca botanica* (1736, Amsterdam) and *Fundamenta botanica* (1736, Amsterdam); others by their subject matter could only have been prepared after his arrival in Holland. Through Gronovius and Lawson, Linnaeus met George Clifford, a wealthy Englishman and director of the Dutch East India Company. At his country house at Hartekamp Clifford had an important plant and animal collection, and Linnaeus was excited by the range of specimens from Africa, the Far East, and the Americas. For his part, Clifford was impressed by Linnaeus's trick of classifying plants he had never seen or heard of before by a casual glance at their flowers. Clifford wanted Linnaeus to catalog his collection, but first it was necessary to persuade Burman to release him from his work on the flora of Ceylon. The Englishman shrewdly invited Burman to his house and, noting his guest's interest in the second volume of Soan's *Natural History of Jamaica* (1723), offered to swap it for Linnaeus's services. The deal was done and in September 1735 Linnaeus moved to Hartekamp where, living "like a prince," and for a salary of 1,000 florins, he catalogued Clifford's collection. The results of the work was published in *Hortus Cliffortianus* (1737, Amsterdam). That same year, 1737, he published five books totaling nearly 2,000 pages.

By this time Linnaeus was keen to return home. His publishing commitments had been completed and despite tempting and lucrative offers from Clifford he left Holland in 1738. As an indication of his impact on minds already experienced in natural history there is a vivid description of Leyden scientific club meetings provided by Gronovius: "Sometimes we examined minerals, sometimes flowers and plants, insects or fishes. We made such progress that by Linnaeus's Tables we can now refer any fish, plant or mineral to its genus, and thus to its species, though none of us had seen it before. I think these Tables so eminently useful that everyone ought to have them hanging up in this study, like maps. Boerhaave values this work highly and it is his daily recreation" (Blunt: 124).

Although courted by scholars in Leyden, Göttingen, London, and Paris, his reception in Stockholm was more muted. Forsaking, as he put it, "Flora" for "Aesculapius," Linnaeus built up a profitable practice dispensing mercury to a steady supply of patients suffering from venereal disease. His success was soon noted at court and before long he once more had a patron, Count Tessin, and the post of physician to the admiralty. To his patron he expressed his gratitude in terms elaborate even by 18th-century standards: "To my God and my Tessin[4] I offer my love, honour, praise and gratitude, so long as I can whisper, and my children shall praise your Excellency's dust so long as they are on the earth's surface" (Blunt: 132). Such sycophancy was rewarded in 1741 when Linnaeus could at last report that he was "released from the wretched drudgery of a medical practitioner of Stockholm. . . . I have obtained the position I have coveted for so long . . . professor of medicine at Uppsala. . . . If life and health are granted me, you will, I hope, see me accomplish something in Botany" (Blunt: 135).

The remainder of his life was spent at Uppsala teaching, traveling, and producing an enormous number of works. At last Linnaeus seemed to have what he desired so much, and in 1744 he wrote that he had "fame, the work for which he had been born, enough money,. . .a beloved wife, handsome children and an honoured name. . . .What more could a man desire" (Blunt: 151). From his work he obtained an enormous satisfaction, which leads him to boast that none before him had "been a greater botanist or zoologist," "written more books,"[5] or even "been a member of more scientific societies." He claimed his *Systema naturae* was "the greatest achievement in the realm of science." In addition to such achievements Linnaeus went on to claim for himself the virtue of modesty insisting, while bursting with self esteem, that he was "in the highest degree averse from everything that bore the appearance of pride" (Blunt: 177-78).

There was, however, a darker side to Linnaeus's life and personality. The euphoria of 1744 was later replaced by gloom when he in 1758 complained, "I am old and gray and worn out, and my house is full of children; who is to feed them? It was an unhappy hour that I accepted the Professorship; if only I had remained in my lucrative practice all would be well now" (Blunt: 218). In 1772 he was complaining that "he was worn out" and of "corresponding with so many." The starkest expression of his depression was contained in the instructions left in a sealed envelope for his disposal: "lay me in a coffin unshaven, unwashed, unclothed, wrapped only in a sheet. Nail down the coffin forthwith, that none may see my wretchedness. . . .Entertain nobody at my funeral, and accept no condolences" (Blunt: 234-35). Despite his wretchedness Linnaeus was still well aware of his worth, for he also instructed that his epitaph should contain the honorific *princeps botanicorum,* prince of botanists.

After his death in 1778, Frau Linaeus was faced with disposing of his herbaria, library, manuscripts, and the collections of shells, insects, and minerals. By this time the "beloved wife" and "handsome children" were less esteemed by Linnaeus. He left specific instructions that the herbaria were not to go to his son Carl who "never helped me in botany and does not love the subject." An offer of a little more than £1,000 from Joseph Banks (1743-1820) was rejected and, despite his father's mistrust, Carl appears to have made some effort to preserve the integrity of the collection. After Carl's sudden death from a stroke in 1783, the collection reverted to his mother who lost little time in disposing of it. Banks was no longer interested but advised his young friend, the naturalist Sir James Smith (1759-1828), to make an offer. For 1,000 guineas, Smith was able to preempt agents of Catherine the Great and many other collectors; he acquired 19,000 sheets of pressed plants, 3,000 letters, 2,500 books, a number of manuscripts and Linnaeus's own collection of insects, shells, skins, and minerals. The Linnean Society of London was formed shortly afterwards in 1788 with Smith as its first president. On Smith's death in 1828, Linnaeus's library and what remained of his collection was sold by Smith's widow to the society for 3,000 guineas.

Zoological Classification

The opening of *Systema naturae* clearly reveals the unlimited scope of Linnaeus's taxomonic ambitions for the first entry, in suitably large print, under *Imperium naturae* is God whose attributes are listed as eternal, omniscient, and omnipotent. Authorities cited to justify the attributes include Aristotle who described God as the "being of all beings" and Seneca who affirmed him as the "cause of all causes." Eminent though such authorities were, something more than pagan testimony was needed, and Linnaeus quotes the author of Exodus who reported God to be "everything visible, audible and sensible." This entry is followed by brief descriptions of the cosmic elements – the Universe, Stars, Elements, and Earth – before approaching the main concern of *Systema,* the "three grand kingdoms of nature": the animal, mineral, and vegetable.

Although the first edition of *Systema naturae* (hereafter cited *SN* for easy reference) was only a slim fourteen-page volume, it grew through later editions, was modified and changed until it reached by the tenth edition three volumes of more than 1,300 pages. In it Linnaeus described, named, and classified some 4,400 animal species, 7,700 plant species and some 500 minerals. No one had ever attempted anything like this before, and few would attempt it again. For each entry a Latin binomial[6] is provided, in many cases for the first time, a full description is given and, where appropriate, authorities and sources are cited. The idea that one man could take upon himself the task to name and classify every natural object known to man is scarcely credible, and it is difficult to find anything remotely comparable. Stars are, with few exceptions, numbered rather than named; chemical elements are so few that naming them is a simple matter. It is true that there are a vast number of chemical compounds, each needing a name, but this is performed in a purely mechanical manner and, in any case, the compound is more likely to be known by its formula than by an unwieldly name. In contrast, the Linnean binomials are in a real sense often the only specific name an organism possesses. Many algae, insects, and worms, and certainly all bacteria, are either known only in some limited, regional vernacular or carry no designation outside that assigned to them in some technical monograph.

For the sake of comparison, it is tempting to cite the Biblical passage (I Kings IV: 29-34) in which an attempt is made to convey the depth of Solomon's wisdom. After a weak reference to the number of proverbs and songs that he knows, it is recorded that Solomon "spake of trees, from the cedar that is in Lebanon even unto the hyssop that springeth out of the wall: he spake also of beasts, and of fowl, and of creeping things, and of fishes."[7] The "spake of" in this context undoubtedly refers to Solomon's knowledge of the names of all living things. It is from passages such as this one that the ancient and obscure tradition of ritual magic had emerged in which the knowledge of names is frequently associated with the possession of power. In the West it was linked with the name of Solomon and his supposed *Clavicula Salomonis,* a work available in most European languages (see Butler, 1949).

SN begins with Linnaeus's classification of the animal kingdom. The initial problem faced by any systematist is how first to divide the vast totality of animals into meaningful classes. As Aristotle had realized 2,000 years before, it was by no means obvious which characteristics to designate. If the most natural feature, the number of legs, is adopted we have to accept that, as quadrupeds, crocodiles and horses are the same kind of animal; so also, as bipeds, are birds and men. Consequently Aristotle sought for a more distinguishing feature and finally selected the mode of reproduction. He distinguished animals who brought forth their young alive (the vivipara) from the egg layers (the ovipara). In this way he was able to separate crocodiles from horses, birds from men and, no mean feat, fish from whales. For the next two millenia, when men thought of animal classification, it was to this schema that they turned. For much of the time, however, men thought little of such a topic, being content to think of animals in the manner of the medieval *Bestiary* (see White, 1954), namely, as a source of wondrous tales and moral analogies. Pliny (A.D. 23-79), in his *Natural History,* the main source of medieval knowledge of plants and animals, made the barest attempt to order his material. Sometimes he began with the largest element of the collection, the elephant in his account of animals, and worked downward to the smallest. It was apparently sufficient, as Alexander of Neckam (1157-1217) did in *De naturis rerum,* to divide animals into aerial, aquatic, and terrestial.

Nor did the naturalists of the Renaissance advance things much further. Pierre Belon (1517-64) in *Des oyseaux* (1555) did little more than divide birds by habitat: birds of the river, of the plain, and so on. Conrad Gesner (1516-65) in *Historia animalium* (1551-87) was a little more ambitious but broke no new ground. The five volumes dealt separately with major groups like quadrupeds, birds, fish, and in each group there was an initial sorting into natural classes like *Bos* for all cattle, and *Simia* for all apes. Beyond that, animals were arranged alphabetically.

It should be emphasized how misleading it is to compare these works with modern zoological texts. The great aim of the Renaissance scholar was to exhaust his subject, not to illuminate it.[8] Consequently, in Gesner can be found the various names of animals in different languages; their use in commerce, agriculture, and medicine; references to the creatures in literature; and, of course, all the wonders and curiosities recorded by encyclopedists and travelers since Aristotle. To object, as Buffon did two centuries later, "to the vast amount of useless erudition with which they stuff their works," and to ask "how much natural history can one expect to find in this hodgepodge of writing?" (Eliot and Stern: 172-73) was to forget that the period separating Gesner from Buffon was one in which the character and limitations of zoology were worked out. Belon in the 1550s included in a work on the "natural history" of fish such creatures as whales, seals, crustaceans, hippopotami, beavers, and otters. This did not mean that Belon was unable to distinguish between hippos and cod (in the same work he drew a distinction between cartilaginous and bony fish), or that he had some different classification principle by which cod, hippos, and lobsters were all labeled as fish, but rather that principles of classification had yet to emerge in history. What did exist was the custom of grouping organisms into convenient categories which could vary from book to book and should be judged by no other factor than convenience.

The 17th century saw some remarkable advances. Scholars began to examine the anatomy and physiology of organisms rather than simply to record the tales of travelers and repeat the views of antiquity. Relationships between animals at all levels were examined: anatomical, embryological, physiological, and pathological. The period 1550-1700 was one of great progress in anatomy, crowned by Harvey's discovery of the circulation of the blood. It was a period in which the microscope enriched biology. It is also the period in which printing assisted in the dissemination and appreciation of knowledge. The flora and fauna of Europe became clearly and widely known. Further, reports from wide-ranging expeditions replaced medieval fantasies. To give just one example, a single work of the Jesuit Father d'Acosta, *Historia natural y moral de las Indias* (1590, English translation 1604), presented to European scholars the fauna of a whole new continent. Material like this had to be absorbed into the existing knowledge and the mixture reworked into some acceptable whole. It was against this background that 17th-century naturalists began, in both botany and zoology, the ongoing task of classifying all known organic forms.

Table 10:1: Ray's System (1693)

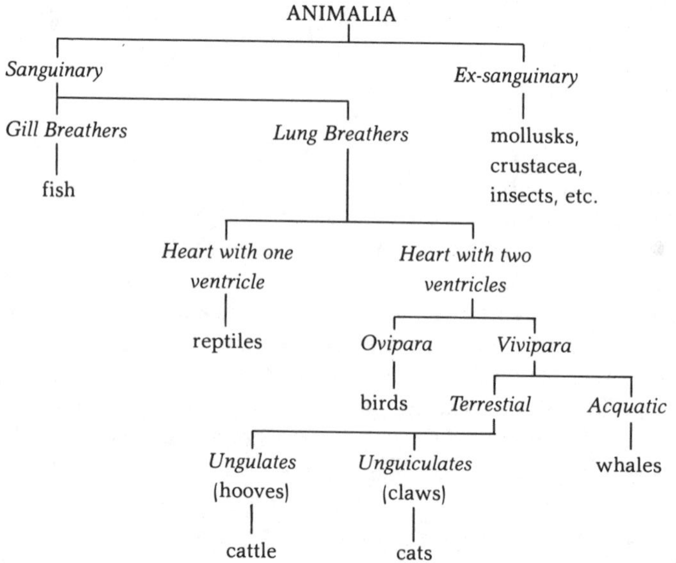

One of the most ambitious systems was proposed by John Ray (1627-1705), the 17th-century's leading British naturalist, in *Synopsis methodica animalium* (1693). Ray's system is somewhat messy in that it relies on a large number of independent characters which are applied in a series of dichotomies. He was inspired by the great advances made by physiologists and comparative anatomists. For a diagram of his scheme, see Table 10:1.

At this point Ray continues with questions about whether the animals have one, two, or four hooves and how many nails are in their claws. Further discriminations are made in terms of horns (permanent for cattle, shed for deer), teeth (two incisors for rabbits, more than two for dogs and cats) and tails (absent for apes, long for monkeys, short for baboons). As can be seen there is little continuity or elegance in the classification, as if Ray were still seeking a system rather than displaying the details of one already established.

The same cannot be said of Linnaeus about a half-century later. He clearly gave the impression that he had a system, and that by the time he published the tenth edition of his *Systema*, that system had been fully worked out. The structure was presented in a diagram suggesting a greater harmony and inevitability than was really justified. Linnaeus recognized six classes, distinguished as follows:

Heart with two auricles and two ventricles blood warm, red	viviparous oviparous	Mammalia Birds
Heart with one auricle and one ventricle blood cold, red	lung breather gill breather	Amphibia Fish
Heart with one auricle and no ventricle blood cold, white	with antennae with tentacula	Insects Worms

Even a casual examination of the classes show that the system fails to hold together, that the picture of six classes emerging neatly from three distinct types of heart is a fictitious one. His class of Worms is clearly no class at all but the traditional catchall for everything remaining after the more obvious life forms have been appropriately located. Containing mollusks, shells (Testacea), worms (Intestina), sponges (Zoophytes), and corals (Lithophyta), it is nothing less than the whole of the modern Invertebrata minus the insects. It is well to remember that, again with the exception of the insects, systematic knowledge of the invertebrates was virtually nonexistent before the 19th century. Indeed it is only with J. B. Lamarck (1744-1829) that the term "invertebrate" was introduced into science. Realizing that the Linnean class of Worms "is a sort of chaos in which the most disparate objects are included. . . I divided the entire known animal world in my first course of lectures at the Museum in 1794 (the year 2 of the Republic) into perfectly distinct groups viz: animals that have vertebrae, animals without vertebrae" (Lamarck 1914: 62-64). The sorting out of the invertebrates, a task begun by Lamarck in *Philosophie zoologique* (1809), required much of the 19th century to complete.

Nor did Linnaeus show much more insight into his class of Amphibia. It was subdivided into reptiles (lizards, frogs, turtles), serpents, and nantes (rays, sturgeon, etc.). It was only with four classes – Mammals, Birds, Fish, and Insects – that any kind of satisfactory groupings emerged. With each of them he saw a significant feature that not only gave unity to the class but permitted the main orders of each to be naturally distinguished: teeth (Mammals), beaks (Birds), fins (Fish), and wings (Insects).

For Mammals, the idea of using teeth as a basis of classification came to him on his Lapland journey when he noticed "the under-jaw of a horse, having six fore-teeth, much worn and blunted, two canine teeth, and at a certain distance from the latter twelve grinders, six on each side. If only I knew how many teeth and of what kind every animal had, how many teats and where they were placed, I should perhaps be able to work out a perfectly natural system for the arrangement of all quadrupeds" (Blunt: 57).[9] In this way Linnaeus was able to construct a plausible sequence of Mammals in the tenth edition of *SN*:

Primates: four "dentes primores," or incisors, in their upper jaw
Bruta: like the elephant, with no incisors in either jaw
Glires: with two incisors in each jaw, like the rat
Ferae: cats and dogs with six incisors
Pecora: like sheep and camels, with no incisors in the upper but many below

The following two orders were not as well defined:

Belluae: horse, tapir, hippopotamus, with "many, obtuse fore-teeth"
Bestiae: pigs, moles, shrews, with an indeterminate number of fore-teeth

While standing rather on its own is:

Cetacea: whales, dolphins, with cartilaginous teeth

The tenth edition of *SN* was the first to classify correctly the Cetacea; in previous editions they had been placed with the fish. It was also in the tenth edition that the term mammal was introduced into zoology; earlier, Linnaeus had used the standard term quadrupeda.

The point of such a classification is not that it agrees with the modern classification of mammals. Judged by this standard, Linnaeus's work fails by a wide margin. Bearing in mind that his orders and those of modern systematists may not be strictly comparable, it is nonetheless interesting to contrast them. As an example, A. S. Romer (1894-1973), a leading vertebrate paleontologist, recognized twenty-six mammalian orders rather than the eight of Linnaeus. Of these, only two, Primates and Cetacea, actually agree in name; of the others, only Linnaeus's Ferae seems to find a modern counterpart in Romer's Carnivora (Romer: II, 416-17).

Of special interest is Linnaeus's treatment of Primates. In the first edition of *Systema,* he defined Anthropomorpha as possessing "incisors four on either side" and divided them into three genera: *Homo, Simia,* and, surprisingly, *Bradypus* (sloth). Man (*Homo*) was defined by his ability to know himself, while the other two genera were distinguished from each other by their digits: *Simia* having five anterior and five posterior, and *Bradypus* having three or two anterior and three posterior. Linnaeus also described – although without much conviction – the satyr which he supposed to be "tailed, hairy, bearded, with a human body, much given to gesticulation, extremely lascivious" and of the same genus as the "tailed men" described by "modern travelers."

Ignoring the rather curious inclusion of *Bradypus*[10] (which by the tenth edition was removed to the Bruta), Linnaeus had already placed man and apes in the same order. Such a grouping attracted a fair amount of criticism. To it Linnaeus replied in a 1747 letter: "Show me a generic character. . . by which to distinguish between Man and Ape. I myself most assuredly know of none. I wish someone would indicate one to me" (Greene: 184). He admitted that apes lacked a capacity for languages, but he denied it to be a "characteristic mark" as it failed to involved "number, figure, proportion or position." Unlike teeth and digits, languages could not be counted, located, or described morphologically; they were, therefore, irrelevant to classification.

By the tenth edition, man's position, though further elaborated, remained basically the same. Corresponding to the 1735 Anthropomorpha, Primates were this time divided into four genera:

Homo: As before, characterized by an ability *nosce te ipsum* (to know thyself).

Simia: Characterized dentally. But it should be noted that Linnaeus's knowledge of apes and monkeys was far from reliable and depended almost entirely on printed sources.

Lemur: Also characterized dentally.

Vespertilio: The inclusion of bats was probably due to the ostensible similarity of their teeth to some monkeys as well as their possession of only two mammae, sited, in primate fashion, on the chest, a combination of properties found in no other mammal outside this order.

The tenth edition enlarged Linnaeus's treatment of *Homo.* He listed just two species: *Homo sapiens* and *Homo troglodytes.*

Under *Homo sapiens* he described the following:

Homo sapiens ferus: Described with admirable economy as *tetrapus, mutus, hirsutus* (dumb, hairy, and quadrupedal). Linnaeus could quote a number of examples from the literature although he was writing long before the discovery of the best attested case of this variety, the Wild Boy of Aveyron, found in 1799.

Geographical varieties: Linnaeus recognized four: *Homo americanus, europaeus, asiaticus,* and *afer.* The features attributed to them can be conveniently represented in the following table:

Table 10:2

	Color	*Temperament*	*Character*	*Clothes*	*Government*
H. americanus	red	choleric	obstinate and free	paint	custom
H. europaeus	white	sanguine	gentle and inventive	clothes	fixed law
H. asiaticus	lurid	melancholic	grave and covetous	clothes	opinion
H. afer	black	phlegmatic	careless and indolent	grease	caprice

Linnaeus also recognized another variety, *Homo monstrus,* and recorded the following examples: *alpini* (small, agile, and timid); *patagonici* (large and lazy); *monorchides* (single-testicled, as the Hottentots); *macrocephali* (conical-headed, as the Chinese); and *microcephali* (flat-headed, as the Canadians).

Under the second species of man, *Homo troglodytes,* Linnaeus included only *H. nocturnus* or *H. sylvestris orang outang.* Using Pliny as his source, he dutifully recorded a creature found in Ethiopia and Java, which he described as white, erect, small, speaking with a hiss, shunning daylight, and foraging at night. Clearly, he doubted its existence.

There are obvious defects in Linnaeus's treatment of man and primates, so obvious as to require no comment. But it is necessary to emphasize nonetheless how progressive and revolutionary his system was. That he realized there was only one basic type of man – *H. troglodytes* can be safely ignored – is no small insight. However, even this insight was dwarfed by his introduction of a major intellectual landmark, the Primate order. The term and the concept are both his. The idea that man, apes, and monkeys sufficiently resemble each other to be classed together is a Linnean innovation and one based on solid anatomical structures. To anthropologists of the day, the Primate order appeared ill-founded; it was rejected, for example, by J. Blumenbach (1752-1840) in *De generis humani variatate* (1776), and in its place he proposed the order Bimana, occupied by man alone, with apes and monkeys together forming a quite different order, Quadrumana. Accepted by Cuvier at the turn of the century, Primates disappear from zoology for almost a hundred years and are not heard of again until the time of Darwin. In *Descent of Man* (1871) Darwin reported: "Recently many of our best naturalists have recurred to the view first propounded by Linnaeus, so remarkable for his sagacity, and have placed man in the same Order with the *Quadrumana* under the title *Primates*" (Darwin 1871: 190).

In general, objections to *SN* tended to be of two kinds. The more radical denied that such a system of nature was at all possible. The idea that there were just six classes into which all animals could be placed was rejected as absurd. Such doubts were strongly argued by the Comte de Buffon (1707-88) in his immense and influential *Histoire naturelle* (1749-67, 15 volumes). The work begins with an "initial discourse" in which Buffon pointed out some of the consequences of Linnaeus's self-imposed restrictions: "Serpents, according to this author, are amphibians, crayfish are insects and all shellfish . . . are worms. . . ." A true picture can only be gained, Buffon argued, by "augmenting the number of divisions . . . since in nature only individuals exist" (Eliot and Stern: 177-79). Attempts to group the individuals into species could only be arbitrary for, Buffon argued, "nature proceeds by unknown gradations . . . she passes from one species to another . . . by imperceptible nuances. As a result, one finds a great number of intermediate species and mixed objects which it is impossible to categorise and which necessarily upset the project of a general system" (Eliot and Stern: 165-66). Such views were in reality modified considerably as Buffon's work on *Histoire naturelle* developed and, in any case, had no effect on someone like Linnaeus with, in Buffon's phrase, "a mania for classification." It also became apparent that extreme individualistic views were untenable, for systematists did succeed in creating systems, admittedly imperfect ones, but systems that should have been altogether impossible if individualism was true.

The second and more damaging criticism raised against *SN* was the kind raised by Lamarck and already mentioned. Zoologists rejected the system of Linnaeus because, in some cases rightly, they thought they could do better. Consequently the first half of the 19th century saw the emergence of a large number of alternative zoological systems. By the time Darwin came to write *Origin of Species* naturalists could choose from twenty or more available systems. Just why there should be such a multiplicity was one question Darwin sought to answer in *Origin.*

Classification, Darwin noted, is not arbitrary "like the grouping of stars in constellations." Yet if people choose to group animals on the basis of resemblance then inevitably there will be disputes over which resemblances are significant and which only superficial. This in turn will lead to a multiplicity of incompatible systems as one zoologist emphasizes the importance of teeth, another swears by embryological development while a third stresses the sensory anatomy. To say instead that systematists should take *all* characters into consideration was merely naive; if followed it would lead to a Buffon style individualism.

There was a further option, Darwin realized, and this was to seek not resemblance but the "propinquity of descent" (Darwin 1968: 397-99). Once it was appreciated that taxonomy showed relationships of descent rather than resemblance, it would no longer be possible to divide the animal kingdom in so many ways. Consider arranging any human family, the Hapsburgs for example, on the basis of resemblance; it will soon be obvious that many arrangements are possible and that no one of them can claim priority. Arrange the same set on the basis of descent and other genealogical relationships and immediately freedom of choice disappears while a unique, correct classification appears at last to be possible.

Linnaeus's achievements in volume one of *SN*, though considerable, are not perhaps as considerable as his reputation might suggest. After all, by the end of the century his zoological system was just one of many, and it had few supporters. Its strength lay in the introduction of Primates into zoology and his demonstration that something like a system was a possibility. The actual construction of such a system was not within Linnaeus's power or that of any 18th century figure. There are, however, greater and more lasting contributions to science in *SN*, but these are best approached by his work on the plant kingdom in volume two.

The Sexual System

In the 18th century Linnaeus was probably best known for the least durable part of his system: the sexual classification of plants. As with zoological classification, early writers of Herbals[11] tended either to ignore classification completely or confine themselves to the most basic categories. Some like William Turner (1508-68), the so-called father of English botany, in his *New Herball* (1568) simply adapted an alphabetical arrangement. At best some fairly simple grouping might be attempted, as with *Historia plantarum Lugduniensis* (1586) by J. Dalechampius (1513-88) in which separate chapters dealt with such classes as climbing plants, poisonous plants, fragrant plants, and foreign plants.[12] Just how plants were sometimes distinguished in practice can be seen by the attempt of the author of *Grete Herball* (1526) to deal with mushrooms: "There be two manners of them, one manner is deadly and sleet them that eateth of them and be called tode stooles, and the other dooth not."

Traditions of this kind are far from dead. There are still many books available today that classify flowers by color. The aim is to lead the reader, swiftly and painlessly, to the plant he wishes to identify. Such books function as indexes rather than catalogues. In the 16th century another tradition emerged. It was then that such scholars as Cesalpino (1516-65) in *De plantis* (1583) began for the first time to try and group all known plants into a fairly small number of classes on the basis of objectively determined anatomical features. Cesalpino began with the traditional division into shrubs, trees, undershrubs, and herbs,[13] but thereafter established seventeen classes, determined in the main by the number of seeds formed by the plant. In this way Cesalpino could distinguish between the Labiatae with four seeds and the Compositae and their many seeds. In the following century a variety of systems similar to that of Cesalpino appeared, but by Linnaeus's time two in particular had acquired a special prominence, those of John Ray and J. P. de Tournefort (1656-1708).

Their work was undoubtedly impressive. Tournefort in his *Institutiones* established nearly 9,000 species grouped into 700 genera and 22 classes.

His system was based on the corolla or petals and established such groups as the apetalous, monopetalous and polypetalous. In contrast, Ray's *Historia plantarum* (1686-1704) detected a staggering 18,000 species in the plant kingdom. He began with the traditional distinction between herbs and trees and thereafter, in broad outline, the system can be represented thus:

Herbae: Imperfectae (flowerless)
Perfectae (flowering): Dicotyledons
Monocotyledons
Arbores: Monocotyledons
Dicotyledons

The distinction between mono- and dicotyledons,[14] introduced by Ray, was to be of fundamental importance, but otherwise Ray's 18,000 species were derived by adopting a large number of criteria; seeds, fruits, leaves, and flowers. The result was a system of great complexity, so great in fact that looking at his massive tomes today it is hard to imagine how anyone could find his way among the thousands of species.

This was not a problem with the Linnean system. So simple was it that when it first appeared in 1735, he could represent it in a one-page table, the so-called *Clavis systematatis sexualis.* It was precisely this capacity for simple representation that one of the period's leading botanical draughtsmen, G. D. Ehret (1708-70), seized upon. Ehret published it on a single sheet as *Linnaei methodus plantarum sexualis.* It was a great success, Ehret reported, "bought by all the botanists in Holland" and much imitated. Today there are only two known copies.

Ray and Tournefort had scarcely been aware of plant sexuality. It was Rudolph Camerarius (1665-1721), a Tübingen botanist, who first began to study plant sexuality. In *De sexu plantarum* (1694), he published the details of some simple but convincing experiments. He demonstrated that pistillate plants (females), when isolated from staminate (male) ones, invariably failed to produce any seed. It is incredible that for centuries man had developed an efficient agriculture, domesticated wild crops, and cultivated new ones, without any idea that plants reproduce sexually. The first published works of western botany came from Theophrastus, Aristotle's successor as head of the Lyceum. In them he discussed the question of plant production and proposed that they developed by "spontaneous growth from seed, from a root, from a piece torn off, from a branch or twig, from the trunk itself" (Rook: 49). It was presumably this last feature of plant propagation that prevented generations of botanists from ever contemplating the possibility of sexual reproduction. If a new olive tree can be produced simply by planting a twig from an old one, it was a little hard to see how sex was involved.

Once plant sexuality was discovered, it quickly assumed a central position in botanical classification. The 18th century embraced the scheme with an eagerness and rhetoric that is difficult to appreciate now. Lin-

Table 10:3
Linnean Sexual Classification of Plants

Class	Sexual Characteristic	Examples	Binomial	Modern Family
A. HERMAPHRODITE FLOWERS (stamen and pistil on same flower)				
Ia Stamen free: Of equal length				
1 Monandria	1 stamen	mare's-tail	*Hippuris vulgaris*	Hippuridaceae/mare's tail family
2 Diandria	2 stamens	common speedwell	*Veronica officianalis*	Scrophulariaceae/ snapdragon family
3 Triandria	3 stamens	autumn crocus	*Crocus nudiflorus*	Iridaceae/iris family
4 Tetrandria	4 stamens	greater plantain	*Plantago major*	Plantaginaceae/plantain
5 Pentandria	5 stamens	cowslip	*Primula veris*	Primulaceae/primrose family
6 Hexandria	6 stamens	bluebell	*Endymion nonscriptus*	Liliaceae/lily family
7 Heptandria	7 stamens	horse chesnut	*Aesculus hippocastanum*	Hippocastanaceae/horse chestnut family
8 Octandria	8 stamens	ling (heather)	*Calluna vulgaris*	Ericaceae/heather family
9 Enneandria	9 stamens	bay tree	*Laurus nobilis*	Lauraceae/laurel family
10 Decandria	10 stamens	meadow saxifrage	*Saxifraga granulata*	Saxifragaceae/saxifrage family
11 Dodecandria	12-19 stamens	wild-ginger	*Asarum europaeum*	Aristolochiaceae/ birthwort family
12 Icosandria	20 or more stamens attached to calyx	raspberry	*Rubus idaeus*	Rosaceae/rose family
13 Polyandria	20 or more stamens not attached to calyx	poppy	*Papaver rhoeas*	Papaveraceae/poppy family
Ib Stamen free: Of unequal length				
14 Didynamia	2 long, 2 short	red dead-nettle	*Lamium purpureum*	Labiatae/dead-nettle family
15 Tetradynamia	4 long, 2 short	watercress	*Nasturtium officianale*	Cruciferae/cabbage family
IIa Stamen united: By filaments				
16 Monadelphia	into a single bundle	gorse	*Ulex europaeus*	Papilionaceae/pea family
17 Diadelphia	into 2 bundles	common milkwort	*Polygala vulgaris*	Polygalaceae/milkwort family
18 Polyadelphia	3 or more bundles	common St. John's wort	*Hypericum perforatum*	Hypericacea/St. John's wort family
IIb Stamen united: by anthers				
19 Syngenesia		corn marigold	*Chrysanthemum segetum*	Compositae/daisy family
IIc Stamen united: attached to pistil				
20 Gynandria		frog orchid	*Coeloglossum viride*	Orchidaceae/orchid family
B. UNISEXUAL FLOWERS				
21 Monoecia	Stamens and pistils in different flowers but on same plant	maize	*Zea mays*	Graminae/grass family
22 Dioecia	Stamens and pistils in different flowers on different plants	hemp	*Cannabis sativa*	Cannabinaceae/hop family

Table 10:3
Linnean Sexual Classification of Plants

Class	Sexual Characteristic	Examples	Binomial	Modern Family
23 Polygamia	Stamens and pistils in same or different flowers on same or different plant	stinging nettle	*Urtica dioica*	Urticaceae/nettle family
C. FLOWERLESS PLANTS				
24 Cryptogamia – Plants with neither conspicuous stamens nor pistils:				
Musci		common clubmoss	*Lycopodium clavatum*	Lycopodiaceae/clubmoss family
Filices		oak fern	*Polypodium dryopteris*	Filicales/fern family
Algae		sea lettuce	*Ulva lactuca*	Chlorophyceae/green algae
Fungi		death cap	*Amanita Phalloides*	Agaricaceae/gill fungi

naeus's first written account of plant sexuality in 1729 clearly introduced the tone and idiom to be followed by later writers. With the coming of spring, he began, "every animal feels the sexual urge. Yes, Love comes even to the plants. Males and females . . . hold their nuptials . . . showing by their sexual organs which are males, which females, which hermaphrodites." The most extravagant language was reserved for a description of petals which contribute "nothing to generation, serving only as a bridal bed which the great Creator has so gloriously, adorned with such precious bed-curtains, and perfumed with so many sweet scents. . . . When the bed has thus been made ready, then is the time for the bridegroom to embrace his beloved bride and surrender himself to her" (Blunt: 34).

The actual details of the system first reported in the *Systema naturae* are remarkably simple. Plants are divided into twenty-four classes.[15] The division is made by counting the number of stamens, the male organ, and source of the flower's pollen. With this insight and the ability to count up to twenty-one, even the dullest student can assign almost any plant to its appropriate Linnean class. A little more than counting stamens is occasionally involved, for in some classes the stamens length and siting are diagnostic factors. Linnaeus's own diagram, the *Clavis,* is somewhat compressed; it can be seen, together with Ehret's plate in Linnaeus (1979: 97 and 101). The following table shows in essence the Linnean sexual system, an example of each class, its binomial and the modern family to which it is assigned.

The twenty-four classes thus classified in terms of their stamens or male sexual organs were further divided into orders on the basis of their pistils or female sexual parts. For the first thirteen classes the process was extremely simple involving nothing more than counting the number of pistils present. Those with a single pistil were termed *monogynia,* those with two *dignyia,* with three *trigynia,* and so on to those with twenty or more pistils which were termed *polygnia.* It is with these first thirteen classes that the Linnean system works with its greatest economy and elegance. All flowers, for example, with six stamens are automatically members of the class Hexandria. To proceed to further divisions it is only necessary to count the number of pistils. If only one is found, as in the lily, then the result is H. monogynia while the meadow saffron, with six stamens and three pistils, automatically becomes H. trigynia. Beyond the first thirteen classes matters become more complicated. Sexuality remains the key, but no one feature consistently dominates. Instead, different strategies are applied throughout the remaining classes and while the system never actually breaks down it loses much of its elegance. The following devices are adopted:

14. Didynamia: divided into Gymnospermia with naked seeds and Angiospermia with enclosed seeds.
15. Tetradynamia: divided into the two orders Siliculosa with long, narrow fruit pods and Siliquosa with short, broad ones.

16-18 and 20-22: these classes were defined independently of their stamens; Linnaeus was thus free to divide these six classes into orders on this very basis. In this way such terms as Monadelphia monandria and Dioecia triandria emerge to describe particular orders.

19. Syngenesia: of which the common daisy is a typical example, required the most complex treatment and inspired Linnaeus's most elaborate rhetoric. In this class the center is normally made up of tiny flowers packed tightly and known as disc florets; they are surrounded by larger ray florets. Given that the florets may be unisexual or hermaphrodite, a number of distinct combinations are possible. Linnaeus recognized the following six orders:
    - i. Polygamia aequalis: all florets hermaphrodite or, in Linnaeus's more flowery language "many marriages with promiscuous intercourse" – goat's beard (*Tragopon pratensis*) – Compositae
    - ii. Polygamia superflua: disc florets hermaphrodite/ray florets female and fertile or "married females with superfluous females" – wormwood (*Artemisia absinthum*) – Compositae
    - iii. Polygamia frustanea: disc florets hermaphrodite/ray florets sterile or "barren concubines" – sunflower (*Helianthus annus*) – Compositae
    - iv. Polygamia necessaria: disc florets male/ray florets female or "fertile concubines" – marigold (*Calendula officinalis*) – Compositae
    - v. Polygamia segregata: each floret separately confined or "husbands live with their wives in the same house but in different beds" – thistle (*Carlina vulgaris*) – Compositae
    - vi. Monogamia: lobelia
23. Polygamia: divided into the three orders: Monoecia, Dioecia, Trioecia, depending whether the male, female and hermaphrodite flowers were found on the same, on two, or on three plants.

24. Cryptogamia: six orders were recognized: filices (ferns), Musci (moss), Hepaticae (liverworts), Lichenes (lichens), Algae (seaweed) and Fungi (mushrooms).

Reception of the Sexual System

The reception of the Linnean sexual system was rapid and extensive. It is true that some found the system itself too distasteful to warrant consideration. The best known such figure was J. Siegsbeck of St. Petersburg who rejected it as "lewd" and "loathsome harlotry" while insisting that so "licentious a method" could not possibly be taught to the young. Taking Linnaeus's exaggerated imagery far too seriously, he even complained that God would "not have allowed such odious vice as that several males should possess one wife in common, or that a true husband should, in certain composite flowers, besides its legitimate partner, have near it illegitimate mistresses" (Barber: 54). Linnaeus responded by naming a particularly unpleasant weed *Siegesbeckia.* Some objected that the choice of terms made it difficult to use them in ordinary discourse objecting to such examples as *Priapus, Venus deflorata* (used of shells), and of *Clitoria* to name a genus of plants of the order Diadelphia decandria.

Another who objected was William Withering (1741-99), the physician better known for his introduction of foxglove or digitalis into medicine but also the author of the popular work *Botanical Arrangement. . . According to the System of the Celebrated Linneaus* (1776). Withering, mindful of his responsibilities, substituted "chive" for "stamen" and "pointal" for pistil and, while following the Linnean classification, dropped "the sexual distinction in the titles to Classes and Orders." After his death his son published a fourth edition in which the full Linnean terminology was used, pointing out that "the Author rather willingly follows than presumptuously attempts to lead the public taste" (Withering: I, ix).

The reactions of Siegsbeck and Withering proved very much the exception. Europe and America greeted the Linnean system with enormous enthusiasm and relief. Although the first translation into English of any Linnean text did not appear until 1759, the system had been incorporated some years earlier into a number of botanical texts. The first was *History of Plants* (1751) by John Hill (1716-75) in which Linneaus was offered "infinite praise" (Henry: II, 92). Hill was one of the more versatile 18th-century botanists: an apothecary in London and a prolific author of operas, plays, and novels, as well as a large number of botanical works. His various activities evoked from the actor Garrick the epigram: "For physic and farces his equal there scarce is, / His farces are physic, his physic a farce is." After he was refused membership in the Royal Society, Hill published in 1751 a work giving undeserved prominence to some of the Society's less memorable publications on mermen, unicorns, and the emergence of maggots from water. In response, the Society began to take a firmer control over its publications.

As a disseminator of the Linnean system, three of Hill's works deserve particular attention. *British Herbal,* originally serialized in fifty-two parts (sixpence each) in 1756, was the first publication to appear in Britain after the 1753 appearance of *SN*; his *Flora Britanica* (1760) was England's first Linnean treatment of the country's flora while his *Vegetable System* (1759-75, 26 vols.) with 1,600 copper plates not only bankrupted Hill, but also provided the fullest account of the sexual system to date.

Hill was quickly followed by many others. William Hudson (1720-93), author of *Flora Anglica* (1762) and a standard work on the subject for over fifty years, accepted both binomials and the sexual system. Academic botanists like Thomas Martyn (1735-1825), professor of botany at Cambridge, adopted the Linnean system in his *Plantae Cambridgiensis* (1763), as did nurserymen like J. Wheeler in his *Botanist's and Gardener's New Dictionary* (1763) and P. Miller (1692-1771), author of *The Gardener's Dictionary* (1731)–one of the leading texts of the century–in which the Linnean system was applied from the eighth edition (1768) onwards. Clearly, British botany had become predominantly Linnean since the 1760s, and this was also true in most other European countries.

In addition to his books, Linnaeus also produced a fair number of pupils, his "apostles," who set out to explore the flora of the world. In 1771, he claimed with some satisfaction, "My pupil Sparrmann has just sailed for the Cape of Good Hope, and another of my pupils, Thurnberg, is to accompany a Dutch embassy to Japan. . . . The younger Gmelin is still in Persia . . ." (Blunt:190). He might have mentioned many more, including Daniel Solander (1736-82) who accompanied Cook to the South Seas and later served as a Keeper of the British Museum, and Peter Kalm (1715-70) who sailed in 1748 to Philadelphia in search of commercial plants. America already had news of the Linnean system several years before Kalm arrived. Cadwallader Colden (1688-1766), for example, a leading political figure in colonial New York with an interest in botany, had earlier attempted to describe the flora on his estate at Coldengham, up the Hudson near Newburgh. "I understood only the rudiments of botany and I found so much difficulty in applying it to the many unknown plants that I met with everywhere that I was quite discouraged." In 1742 he obtained a copy of Linnaeus's *Genera plantarum* (1737) and gained from it "so much light," he wrote to Linnaeus, that he "resolved again to try what could be done" (Kastner:7). The result was a catalog of 141 plants classified sexually and later published through the good offices of Linnaeus as *Plantae Coldengham* (1749), the first such work to originate in America.

Undoubtedly the grandest and strangest reception awarded Linnaeus came from Erasmus Darwin (1731-1802), grandfather of Charles, in *Loves of the Plants* (1789 and subsequent editions before 1824). This work was also translated into French, Portuguese, and Italian, and it was published twice in New York in fewer than ten years, 1798 and 1807. It was obviously saying something people wished to hear. Darwin had previously translated two major works of Linnaeus into English: *The Families of Plants*

(1787) from *Genera plantarum* and *System of Vegetables* (1783) from *Systema vegetabilum*. Both were two-volume works presenting the minutiae of the Linnean sexual system in formidable detail and at great length. In his *Love of Plants*, Darwin had sounded a different note with which he hoped "to enlist Imagination under the banner of Science."

The work is in fact a long poem of some 1,600 lines in four cantos. Whereas Ovid had metamorphosed humans into flora, Darwin proposed to adopt the stylistic conceit that plants in their sexual activity were human. Described as "exquisite" and "delicious" by his contemporaries, Darwin lacked the lightness of touch and wit that might have prevented the poem from palling. The poem surveys the twenty-four Linnean classes from Monandria to Cryptogamia equating wherever possible the stamens and pistils with nymphs, shepherds, virgins, beaux and belles, or any other plausible personification.[17] Thus the plant stargrass or starwort of the family Callitrichaceae or, in Linnean terms, of the order Monandria digynia, is introduced thus (I,45-46):

> Thy love, Callitriche, two Virgins share,
> Smit with thy starry eye and radiant hair.

And so Darwin continued throughout the obscurer corners of Linnaeus with his ponderous verse. Fish, for example are "scaly tenants of the main," delighting his contemporaries but, to a later age, he seems a better botanist than poet.[18]

Some, however, reacted more critically by pointing out anomalies arising from the system or by objecting more radically to its viability. In the main, the system worked reasonably well in the sense that if two independent practitioners applied it to the same sample they would obtain the same results. As long as people could count correctly they would all agree that the autumn crocus had three stamens and consequently belonged to the class Triandria. In some cases, as with the Valerian family, nature was less straightforward. Here the common valerian, also known as all-heal (*Valeriana officinalis*), has three stamens while the red valerian (*Kentranthus ruber*) has only one, and other species, two or four stamens. What, then, was Valeriana? Was it Mono-, Di-, Tri, or Tetrandria? It was a relatively easy matter for Linnaeus to adopt the convention that in such cases the plant should be assigned to the class with the greatest number of species. Linnaeus recognized in *Species plantarum* some sixteen species of which twelve had three stamens and consequently placed all Valeriana in the class Triandria. Such conventions can, however, only dilute the effectiveness of the system. If too many exceptions are made, the system would soon become unwieldly and alternative systems sought. There are few cases of such anomalies, and those sympathetic to the Linnean approach willingly adopted Linnaeus's proposed convention as a reasonable solution. A typical example of the tolerance shown to such anomalies can be seen in J.L. Drummond's *First Steps to Botany* (1823) where he posed the question of what to do if it were found "there were *thirty* different species of valerian, and *twenty five* of these flowers, containing *three* stamens. . . . although the *remaining five* should have only *one* stamen, or *two*, or *four*?" (p.239). Drummond answered himself without argument that it would "be prespposterous to place the genus in any other than the Triandria." Not all, however, felt so well disposed to such flagrant violations of the basic rules of the system.[19]

More substantive was the complaint that some classes were too large and diffuse, combining flora that had little else in common other than the number of their stamens. The best example of this is the Linnean class Pentandria which accounts for 150 of 1,200 pages of *Species plantarum*. A brief glance at any popular work on botany or wild flowers soon shows that a large number of families possess five stamens: parsley (Umbelliferae), bindweed (Convolvulaceae), primrose (Primulaceae), nightshade (Solanaceae) and forget-me-not (Boraginaceae), to name but a few, all qualify for a place in the Linnean Pentandria. With such diversity and abundance the class itself loses much of its credibility and the system becomes more difficult to use. Thus, however useful the Linnean scheme may appear, genuine doubts could still arise as to its soundness and accuracy.

There was, however, a more radical attack emerging from such as Buffon who questioned the very possiblity of *ever* producing a satisfactory classificatory system. Linnaeus's error, he declared, was a "metaphysical" one and consisted of "wishing to judge the whole by a single part."[20] Instead, he claimed it was necessary "to make use of all the features of objects," only in this way could a "natural system" be constructed (Eliot and Stern: 169). It is not possible, of course, to classify a large number of objects in terms of *all* their characteristics for it soon becomes clear that the result will be, not a classification, but a list of individuals. It should be clear that the natural/artificial dichotomy, as discussed in the 18th century, was a far from absolute one but rather marked ends of a continuum. The disadvantage of the purely artificial mode is that objects like the oak, orchid, and octopus can be classified together as beginning with the letter O but have absolutely nothing else in common. The advantage it brings is ease of application and consequently many popular works on wild flowers currently available adopt the artificial characteristic of color to make an initial classification into blue, yellow, green, red, white and purple flowers. At the other end of the continuum, the adoption of a number of characteristics may produce a more natural system, but it will also produce a vastly more complicated one available only to the specialist. Few systematists have sought the extreme positions and most have comprised and applied the guidelines of ease, simplicity, and naturalness to attain a realistic solution.

Whatever its defects the sexual system was readily accepted by workers of all kinds in the late 18th and early 19th century. In contrast later generations of botanists, while agreeing on the value of natural systems, could not agree on which one to adopt. As a result of Linnaeus's influence, a man in the 18th century could travel from St. Petersburg to New York, from Stockholm to Tahiti, and find everywhere a common language for the discussion and classification of plants; after 1830, going from London to Paris or Berlin meant shifting to a different botanical system.

The first system to gain more than local notice was that of A. L. de Jussieu (1748-1836), a member of a famous botanical family connected with the Jardin du Roi in Paris. In his *Genera plantarum* (1789) he distinguished fifteen classes conveniently, albeit suspiciously, divided into 100 orders. The fundamental division was in terms of the cotyledon, the seed leaf, which could either be absent, single, or double. Further divisions depended on whether there were no petals, a single petal, or many of them. A final division indicated whether the petals and the stamen begin below, above, or are level with, the flower's ovary or, in more technical language, whether they are hypogynous, epigynous or perigynous. Following is a brief look at the Jussieu system with examples of the classes provided. The numbers in parenthesis refer to the number of orders found in each class.

Table 10:4
Jussieu Classification

			(sample)		(class)
Acolyledons			fungi algae, moss	(6)	1
Monocotyledons		Stamens hypogynous	grass	(4)	2
		S. perigynous	lily	(8)	3
		S. epigynous	banana	(4)	4
Dicotyledons	*Apetelae*	S. epigynous	birthwort	(1)	5
		S. perigynous	dock	(6)	6
		S. hypogynous	amaranthus	(4)	7
	Monopetalae	Corolla hypogynous	deadnettle	(15)	8
		C. perigynous	heather	(4)	9
		C. epigynous (antlers united)	dandelion	(3)	10
		(antlers distinct)	goosegrass	(3)	11
	Polypetalae	Stamens epigynous	parsley	(2)	12
		S. hypogynous	cabbage	(22)	13
		S. perigynous	pea	(13)	14
	Diclines irregulares	Unisexual flowers	conifer	(5)	15

Just why the above system should prove so attractive to Europe's botanists during the period 1805 to 1830 needs some explanation. It hardly looks more natural than the Linnean system and indeed conveys an impression too tidy and balanced to be anything but artificial. It nonetheless opened possibilities for classifying plants on the basis of increasing knowledge of their form, function, and structure. This knowledge had been accumulating while the Linneans had been exclusively concerned with their own systematics. As an example of the appeal of the Jussieu system, it should be noted that by the 1820s it had reached as far as the U.S. The key figure was John Torrey (1796-1873) who published an *American Flora* (1824) on strictly Linnean lines. With his protégé Asa Gray (1810-88), later to become the leading American botanist of the 19th century, Torrey began the considerable work of rearranging some 20,000 species. Torrey's first publication, using Jussieu's natural system, was in 1826; its further development was pursued with Gray in their *The Flora of North America* (1838-43, 2 vols.).

In Britain the decline of Linnaeus' sexual system can be traced in the publishing history of the period. The first work to break away from Linnaeus and embrace the natural system was probably *A Natural Arrangement of British Plants* (1821, 2 vols.). by S. F. Gray. The work was largely ignored for Gray had fallen foul of Sir James Smith and the botanical establishment at the Linnean Society. The issue was an apparently trivial one. The standard botanical reference work of the period was the thirty-six-volume *English Botany* (1790-1814) with 2,592 plates engraved by James Sowerby (1757-1822). The descriptive text, Linnean in style, was written by Smith. Despite his massive contribution Smith refused to allow his name to appear anywhere in the thirty-six volumes. Not surprisingly the work became widely known by the simple title, *Sowerby's*. Smith, nonetheless, expected his fellow botanists to ignore this popular usage and acknowledge his anonymous contribution by speaking of Smith's or Smith and Sowerby's *English Botany*. Whether from ignorance or as a deliberate snub Gray ignored the convention and spoke simply of *Sowerby's*. Gray subsequently found his own work ignored and even requests to examine specimens in the collection of the Linnean Society were refused. Gray's son and collaborator, J.E.Gray (1800-75), found his application in 1822 for fellowship in the Linnean Society rejected. Wisely, Gray decided there was little future for him in botany and pursued instead a distinguished career as Keeper of Zoology at the British Museum (1840-75). In 1857 the Linnean Society belatedly elected him to a fellowship.[21]

A more tentative approach to the natural system was made by William Hooker (1785-1865) in 1821 in *Flora Scotia* which was Linnean in the text and natural in the appendix. The first acceptable work to adopt the new systematics was probably John Lindley's (1799-1865) *A Synopsis of British Flora, Arranged According to the Natural Orders* (1829; reprinted 1835 and 1841). For a time, authors of botanical works like *London Flora* (1838) by A. Irvine tended to hedge their bets by including the classification of both Linnaeus and Jussieu, but by the 1850s such works are rare. Even *Sowerby's* (third edition, 1863-86, 12 vols.) eventually went "natural" despite the enormous labors involved in re-describing over 2,500 plates.

But it would be wrong to assume that reference to the sexual system was entirely abandoned, or retained only out of historical interest. Even such a straightforward textbook as Balfour's *Manual of Botany* (fifth edition, 1875) devoted two sympathetic pages to a description of the Linnean system. He did go on to point out that while the system was neither natural nor comprehensive, it was still "useful to the student as an index" (Balfour:413).

In the same work, Balfour not only detailed the Linnean system, but also provided extensive accounts of the systems of Jussieu, de Candolle, Endlicher, Lindley, and J. D. Hooker.[22] For this was indeed the problem presented to a mid-19th century botanist: Which of the several "natural" systems should be adopted? As with zoology, so-called "natural" systems tended to proliferate until the time of Charles Darwin when, once more, his great insight that the only natural system was a genealogical one came to be fully appreciated. The first post-Darwinian system to utilize fully his contribution came from H. Engler (1844-1930) and K. Prantl (1849-93) in their 23-volume *Natürlichen pflanzen familien* (1887-1915). Dividing the plant kingdom into fourteen major divisions and 316 families, they began with the Schizophytes (bacteria) and progressed through such groups as the Bryophytes (mosses) to reach the seed plants or Embryophyta. Inevitably, views have since changed as to the precise genealogical relationship between various classes of plants. Nevertheless, for much of the present century it was their view which commanded the widest support and brought back to botany a unity and authority lacking since the time of Linnaeus.

Linnaeus's own view on the need for a natural classification system was straightforwardly pragmatic. To the fundamental objection, pressed in 1737 by the botanist Johann Amman, that "plants which agree in the number of their stamens and pistils, though totally different in every other respect, are placed in the same class" (Blunt:120), Linnaeus agreed that "a natural system is preferable not only to my system but to all that have been invented." The difficulty was that no usable natural system had ever been constructed and until one was available, the artificial sexual system would have to serve.

Nomenclature

While it is true that the sexual system barely outlasted the 18th century, *Systema naturae* and *Species plantarum* also contained Linnaeus's permanent contribution to science. This was the introduction of the binomial system into science, a tool as basic to biology as the microscope. While a botanist of the 16th century could make do with 5,000 plant names by the time of John Ray—a little more than a century later—this number had risen to 18,000. To provide names for such a collection, botanists adopted three strategies.

They could first take the name straight from classical texts. In this way such names as *Aristolochia* for birthwort, *Quercus* for oak, as well as the more familiar *Narcissus* and *Iris* were taken from the pages of Pliny, Vergil and Theophrastus and placed in the pages of herbals. However, there were in many cases no classical names for the plants of more northern climes, and so early botanists frequently used vernacular names. Different writers often followed different policies with the result that a single herb could be known throughout Europe by a dozen different names. To ease this particular problem, writers like William Turner produced such helpful works as *The Names of Herbes in Greke, Latin, Englishe, Duche and Frenche wyth the Commune Names that Herbaries and Apothecaries Use* (1548). The most impressive work in this tradition was *Pinax* (1623) by G. Bauhin (1560-1624), professor of botany at Basel, who attempted to list all the names previously ascribed to the 6,000 plants known to him.

In the absence of classical and vernacular names, a third technique was simply to describe the plants. To this end arose the characteristic terminology of botany with its seemingly never-ending list of terms to describe leaves, flowers, and other essential parts. In this way it became possible to distinguish between *Convolvus folis ovatis* and *Convolvus folis palmatis*, to take one example. The difficulty in this method was the tendency for the descriptive names to grow to unmanageable lengths. They would invariably start in a modest way but, as more varieties appeared and different species became identified throughout the world, the names would grow and grow. Thus *Convolvus folio Altheae*, acceptable to Clusius in 1576, required the addition of *argenteus* by Bauhin in 1623, and by the time of Linnaeus ten names were needed to distinguish it from related forms. One plant listed by Linnaeus in 1737 had collected as many as thirty descriptive terms.[23]

Clearly some better arrangement was called for. Before Linnaeus, valuable progress had been made by J.P. de Tournefort (1654-1708) who in his *Institutiones* (1700) did much to stabilize the use of genera in botany. Generic names had been used earlier, by Bauhin, for example, who was content to use them by themselves. Tournefort provided the genera with characters taken in the main from the plant's flower and fruit.

Against this background Linnaeus began his pithy discussion of the problem of plant nomenclature in *Critica botanica* (1737). He identified a number of problems, the first of which was the multiplicity of species entering botany. Tournefort, he declared, had described ninety-three tulips where Linnaeus recognized only one. The difficulty was to know just what characters constituted a species. If, he complained, we applied the same criteria to man as we applied to plants, we would have not one man but a thousand with "hair white or red or black or grey . . . we have giants, dwarfs, fat men, thin men, tall men, stooping men, gouty men, lame men. . . . But what moderately sane man would call these distinct species?" (Linnaeus 1938; section 259). We should be as restrictive with plants and realize that "size, locality, season, colour, scent, taste, use, sex, monstrosity, hairiness, duration and number do not constitute distinct species." Such a policy could go some way to control the explosive growth in the number of species being proposed by botanists. Whereas Bauhin was satisfied with 5,000 species, a century later Ray was attempting to cope with 18,000, while less profligate was Tournefort with a mere 10,000.

This trend was reversed by Linnaeus who, in equally comprehensive works, coped with fewer than 8,000 species.

Valuable as such an economy was, there still remained two further problems. What could be done about the variety of names assigned to plants by different workers? As Linnaeus phrased it, "For Aconitus, Angelica. . . as I met them in the field of Rivinus were no longer the same plants that I had previously gathered in the field of Tournefort" (Linnaeus 1938: 217). The second problem was how to make the names of manageable length? Linnaeus solved this by the simple step of separating the plant names from their descriptions. It now became acceptable to describe a plant in as much detail as was necessary to distinguish it from all other forms. Thus, three species of plantain *(Plantago)* were described as (a) *Plantago folis lanceolatis, spica subovata nuda, scapo angulato;* (b) *Plantago folis ovata-lanceolatis pubescentibus, spica cylindrica, scapo;* (c) *Plantago folis ovatis glabris.* Since it was also possible to identify the three species as *Plantago lanceolata, Plantago media,* and *Plantago major* respectively, no real harm was done. It is now possible, following W.T. Stearns, to see precisely how Linnaeus moved from the descriptions (a)-(c) to the binomials listed below.

The first sign of a solution appeared in his *Flora Suecica* (1746) – *Swedish Flora* – in which the 1,140 species mentioned in the book are numbered in a continuous sequence. Did Linnaeus at some time intend to number each plant? It is known that students, rather than listing the involved descriptions of plants, would talk instead of *Plantago 1, Plantago 2,* and so on, with the numbers referring back to some textbook or other. But Linnaeus argued against this practice in *Critica botanica:* "The numerical order which the old Botanists stamped on their own brains they assuredly failed to stamp on plants in such a way that anyone can perceive a trace of it" (Linnaeus 1938: 258).

The numbers adopted in *Flora Suecica* had been used in an earlier work, *Olanska och Gothlandska* (1745), an account of his travels to the Baltic islands of Oland and Gotland. Linnaeus refers to the plants around the Othum River which he lists as "Pyrola scapo uniflora, Pyrola racemis, unilateralibus,. . . Pyrola staminibus adscendentibus pistillis declinatis,. . . Pyrola floribus undique racemosis," among others (Linnaeus 1973: 138). In the index of *Olanska,* information about Pyrola, better known as wintergreen, is presented rather differently:

Pyrola 330 irregularis 221
331 Halleri 206, 221
332 secundiflor 221
333 umbellata 221
334 uniflora 115, 221

where the numbers 330-334 in the first column are the numbers assigned to the same plants in *Flora Suecica,* and the numbers in the final column refer simply to the pages in *Olanska.* At this time the binomials, *Pyrola irregularis, Pyrola Halleri,* and so on, are only implicit in the work. They are still tied to an arbitrary numbering system; they are also confined to an index of a not particularly significant book.

The next stage in the development of the binomial system can be seen in Linnaeus's *Pan Suecicus* (1749), an account of the useful agricultural plants found in Sweden. Here the binomials are promoted from the index to the body of the text, but still remain linked to the numbering in *Flora Suecica.* The final stage, allowing the binomials to stand on their own, was taken in the first edition of *Species plantarum* (1753). This work also possessed an important additional feature–comprehensibility. Before 1753, binomials had been applied to the few plants appearing within the confines of a modern monograph. It was not enough to suggest that the procedure adopted in *Flora Suecica* and *Pan Suecicus* be extended to all plant species; in science, as in life, plans, proposals, and good intentions are not hard to find. What was so impressive about Linnaeus was his total approach: He took all 7,700 plant species that he recognized and assigned to each a specific binomial. Such thoroughness is rarely imitated.[24]

Five years later, in the tenth edition of *Systema naturae,* Linnaeus extended his system of binomial nomenclature to the animal and mineral kingdoms. In this way, he authoritatively named about 13,000 different species, an achievement that is beyond the power of superlatives to express. For the most part, the names have been retained, so that if today the European and Arctic foxes are known as *Canis vulpes* and *Canis lagopus,* respectively, it was because of a decision made by Linnaeus two hundred years ago. Few men have had so profound and pervasive an impact on the language of science.[25]

Those 13,000 species named by Linnaeus are but a fraction of the million or more species waiting to be discovered, identified and named. Even if biologists could agree, explicitly or not, to adopt Linnean names, what would happen to the naming of species discovered after 1758? Would the taxonomists of one country assign one set of binomials while those in another chose a different set? What about those names selected by one taxonomist that had been applied to a different species by a rival? These and other problems emerged with increasing regularity in the half-century following Linnaeus's death.

After Linnaeus

The first to appreciate the urgency of the problems and where a solution lay was a little-known figure named Hugh Strickland (1811-53), an ornithologist of comfortable means and an important figure in the newly-established British Association for the Advancement of Science (BAAS). In 1842 Strickland described the difficulty as he saw it: "If an English zoologist. . . visits the museum and converses with the professors of France, he finds that their scientific language is almost as foreign to him as their vernacular. Almost every species he examines is labelled by a title which is unknown to him. . . . If he proceeds thence to Germany or Russia, he is again at a loss" (Mayr: 299). Strickland realized that any taxonomic system needed three properties: names assigned to species must be *stable* in both time and space; the must be *universal*; and there must be a *public manner* by which binomials are established.

I the 1830s, Strickland set about preparing scientific opinion by arguing his views in print and by lobbying at the regular meetings of BAAS. In 1842, he put forward to BAAS proposals that came to be known as the Strickland Code. After a certain amount of rather dubious politicking (which seems inseparable from taxonomy), a subcommittee was set up with Charles Darwin as a member. Although Darwin had reservations, the Code was accepted by BAAS in 1842.

Strickland proposed that the international community agree that where more than one name had been assigned to a plant, the earliest recorded name take precedence and all others be abandoned. This seems like a sensible solution and that botanists and zoologists would give it immediate universal support, but it proved only the prelude to more than a century of bitter controversy. Botanists were the first to seek international agreement under the guidance of Alphonse de Candolle (1800-1893), who organized the First International Botanical Congress in Paris in 1867. He prepared and submitted to the Congress a *Lois de la nomenclature botanique.* Although all agreed in principle with the law of priority,[26] disagreement broke out almost immediately over de Candolle's proposals, dividing the botanical world. To the outsider, the issues may appear trivial, but to those involved no compromise was acceptable. What happens when it is decided that a species has been placed in the wrong genus? Obviously, it will be transferred to its proper genus. But what of its specific epithet, is that transferable too? It must be retained, answered de Candolle; it was a matter of choice, dissented botanists of the Royal Botanic Gardens, Kew. For nearly half a century international botany was split between those who followed the Kew code and the supporters of Paris. Harmony was only attained at Vienna in 1905 when the Paris practice was adopted.

But by this time further dissension had arisen from American quarters. Raising a more subtle objection, American botanists pointed out that when a species is transferred to its correct genus, it occasionally leads to the specific epithet being identical with the generic term and producing what taxonomists call a "tautonym." An example is *Sassafras sassafras.*[27] Vienna banned tautonyms, but many Americans refused to follow, having, in 1892, formulated their own Rochester code which tolerated tautonyms. In revised form, this code was reasserted and published in 1907 as the American code. In fact, the situation was even mroe complex, since American botanists were deeply split. The establishment, mainly students of Asa Gray, tended to be "closet" botanists and largely based at older, eastern centers of learning, such as Yale and Harvard. Such a group felt deeply its responsibilities to international science and consequently set great store in following Europe's lead.

Against them and behind the Rochester code was Nathaniel Lord Britton (1859-1934), a young New York geologist turned botanist, and E. L. Greene (1843-1915) of Berkeley, a botanist who prided himself as a field worker rather than as a museum man. One issue incorporated in the new European code aroused their particular opposition, and it was so basic that it permitted no compromise. The decision was taken – and clearly formulated in Vienna – that the names of all plants described before 1753 should be those published in the first edition of Linnaeus's *Species plantarum.* Priority was the rule. This sensible decision spared botanists the task of searching the earlier literature for any possible prior names.

With this understanding it mattered not how many different names earlier writers had bestowed on maize, to Linnaeus in 1753 it was *Zea mays* and that, sensbily, was the end of the matter – or should have been. Against this, Greene argued that "ante-Linnean generic names ought *all* to be restored to their proper authors" (Depree: 400). Posterity was absolute; it could not be violated at the convenience of "closet" botanists and a distant European establishment. This the Rochester code recognized; it was not in fact until the Fifth International Botanical Congress in Cambridge, England, in 1930, that unity was achieved, and Americans finally agreed to accept the authority of 1753 and the inadmissibility of tautonyms.

The history of zoological nomenclature, though less divisive is a more bitter one. Here the first international rules were drawn up in 1901, and adopted by the Fifth International Zoological Congress in Berlin soon after (although they were actually issued in 1905). The key date for zoology was inevitably 1758 with the publication of the tenth edition of the *Systema naturae*: All species named before this date were to take their Linnean binomial, and for species named later, priority would apply.

A revision of the rules was authorized by the Thirteenth International Congress in Paris. To complete the revision a special colloquium of the world's leading taxonomists was called in 1953, in Copenhagen, shortly before the meeting there of the Fourteenth Congress. The man behind this was an unlikely figure, Francis Hemming (1893-1964), amateur lepidopterist and senior British civil servant, a director of petrol rationing during the war and Undersecretary of the Ministry of Fuel and Power from 1946 until his retirement in 1953. From 1936 to 1958, he also held the powerful post of Secretary to the International Commission on Zoological nomenclature. Out of Copenhagen emerged a 135-page document which, in 1964, became the *International code of zoological nomenclature* (revised edition). Whatever its defects, Hemming did at least succeed in gaining agreement from the international zoological community.

Article 23 of the code states, "The valid name of a taxon is the oldest available name applied to it. . . ." except, and at this point a number of "limitations" are introduced.[28] It was the need for and the nature of these limitations which caused so much dissent at Copenhagen. According to one critic, R. E. Blackwelder, the issue was only carried by Hemming's highhanded manner and his ability to pack meetings when he needed to win a crucial vote. Thus, for the vote on limitations, according to Blackwelder, "over forty persons entered the room, loudly voted for the limitation, and then left without taking part in any other of the deliberations. These forty were all young Germans who had been urged to attend for the purposes of passing one disputed provision. . . . This highhanded action could not have taken place without the acquiescence of Hemming" (Blackwelder: 383). Only by being so "intransigent," Blackwelder declared, could Hemming gain acceptance for his "verbose" and "divisive" code.

There is just one final item of botanical equipment and its production is a happier tale than that of the international codes. It also forms a pleasing link between Linnaeus and Darwin, arguably the two greatest biologists of modern times. For any system of nomenclature to work it is necessary to have some means of recording names already assigned to species. An early attempt to fill this need had been made by E. G. Steudel (1783-1856) with his *Nomenclator botanicus* (1821, revised 1840) which simply listed the Latin names of all known plants. This soon became dated and, noting its inadequacies, Darwin felt it necessary to produce a more comprehensive and modern work. To this end he provided £250 in 1882 for Kew Gardens to provide an entirely new work under the direction of his friend Joseph Hooker. To insure its completion he left instructions to his executors and children "that in the event of his death" they should pay £250 annually "for 4 or 5 years." The first four volumes of the *Index Kewensis*, listing alphabetically all genera and specific epithets used since the time of Linnaeus, were published between 1892-95, listing 380,000 names from the period 1753-1885. Continually updated, by 1959 a further twelve volumes had been added with an additional 421,000 names.

Publishing History

Leyden, 1735: Systema naturae, sive, regna tria naturae systematice proposita per classes, ordines, genera et species. 14 pp.

On its first appearance, this classic work was only fourteen pages long, two of which were blank. Of the remainder, two each were devoted to the animal and mineral kingdoms, three to plants. These three latter pages contain the first, clear, published account of the sexual system.

To give an idea of the development of the Linnean system through various editions, the main changes in the class of Quadrupeds will be indicated. Linnaeus began with thirty-three genera, listed in five orders:

Anthropomorpha: *Homo, Simia, Bradypus.*

Ferae: *Ursus, Leo, Tigris, Felis, Mustela* (weasel), *Didelphis* (opossum); *Lutra* (otter), *Odobaenus* (walrus), *Phoca* (seal), *Hyaena, Canis, Meles* (badger), *Talpa* (mole), *Erinaceous* (hedgehog), *Vespertilio* (bat).

Glires: *Hystrix* (porcupine), *Sciurus* (squirrel), *Castor* (beaver), *Mus, Lepus* (hare), *Sorex* (shrew). Note: The word *glires* comes from Latin, "doormouse."

Jumenta: *Equus, Hippopotamus, Elephans, Sus* (pig). Note: *Jumenta* comes from Latin *jujum*, "yoke."

Pecora: *Camelus, Cervus* (deer), *Capra* (goat), *Ovis* (sheep), *Bos* (cow). Note: *Pecora* comes from Latin, "cattle."

The first edition is a great rarity, only twenty-nine copies are known to exist. Last auctioned in 1959, a copy fetched £2,000. It is consequently none too easy to examine. Facsimiles exist. Published in Stockholm (1872 and 1907) and Berlin (1881), they are also difficult to obtain. It is surprising that such a brief and seminal work has not appeared in any recently published anthology.

Stockholm, 1740: 80 pp.
The five orders of the Quadrupeds remain unchanged, but the genera are reduced to thirty-two. Anthropomorpha are actually increased by the surprising inclusion of *Myrmecophaga* (anteater), while the Ferae lose *Odobaenus* and *Hyaena. Sorex* is transferred from Glires to Jumenta. Pecora is unchanged. Swedish names are used in the zoological and mineralogical parts.

Halle, 1740: 70 pp.
This is basically a reprint of the Leyden 1737 edition, rearranged, to which has been added a German translation in parallel columns.

Paris, 1744: 108 pp.
A reprint of the Stockholm 1740 edition, with French names in place of the Swedish.

Halle, 1747: 87 pp.
Another reprint of the Stockholm 1740 edition, with German names replacing the Swedish.

Stockholm, 1748: 224 pp.
Swedish names of animals and minerals are provided with the inevitable Latin names.

Leipzig, 1748: 224 pp.
A reprint of the Stockholm 1748 edition, with German in place of Swedish names.

Stockholm, 1753, 136 pp.
An account of the vegetable kingdom only. It was reissued in a limited edition of 50 copies in 1919.

Leyden, 1756: 227 pp.
Corrected and amended by Gronovius who made substantial additions to the fish and insects. This edition also contains the first significant change to Quadrupeds since the Stockholm 1740 edition. The genera are expanded to thirty-four, and an extra order is added, Agriae, containing just two genera, anteaters from Anthropomorpha of previous editions and *Manis* (pangolin). Other improvements include the tidying up of *Leo,*

Tigris, and *Felis* from the 1735 Leyden and 1740 Stockholm editions into a single genus, *Felis*; the appearance of *Moschus* (musk ox) in Pecora, *Rhinoceros* in Jumenta, and the return of *Sorex* to Glires. *Didelphis* is moved from Ferae to Glires.

Stockholm, 1758: Vol. 1: Animal Kingdom, 823 pp.; Vol. 2: Vegetable Kingdom, 509 pp.

Although the material for the third volume on the mineral kingdom was available, it was printed later. Volume 1 was reissued in Leipzig (1894) and in a photographic facsimile in London (1954) by the British Museum. The book contains a number of notable landmarks. It is the first work to deal with Mammals, and in that class are included for the first time Primates and Cete (whales, dolphins, etc.). The class of Mammals is divided into eight orders, thirty-nine genera, and 184 species:

Primates: *Homo, Simia, Lemur, Vespertilio.*
Ferae: *Phoca, Canis, Felis, Vivera* (civet cat), *Mustela, Ursus.*
Bestiae: *Sus, Dasypus* (armadillo), *Erinaceous, Talpa, Sorex, Didelphis.*
Bruta: *Elephans, Trichecas* (walrus), *Bradypus, Myrmecophaga, Manis.*
Glires: *Rhinoceros, Hystrix, Lepus, Castor, Mus, Sciurus.*
Belluae: *Equus, Hippopotamus.*
Pecora: *Camelus, Moschus, Cervus, Capra, Ovis, Bos.*
Cete: *Monodon* (narwhal), *Balaena* (whale), *Physeter* (toothed whale), *Delphinus* (dolphin).

Considerable uncertainty is conveyed with the moving of genera from one class to another, the appearance of new ones, and the disappearance of old and established genera. Despite such oddities as the grouping of bats with apes and man, Linnaeus's remarkable insight manages to stand out, and it remains haunting to read these pages and realize that "primates" and "mammals" are appearing for the first time in man's consciousness.

Details of the species and genera grouped in various classes in Volume 1 can be summarized as follows:

	Genera	*Species*
Mammals	39	184
Birds	63	554
Amphibia	16	217
Fish	51	378
Insects	74	2,109
Worms	69	936
TOTAL	312	4,378

Leipzig, 1762

The existence of this edition has been doubted by some scholars. No copy exists in the British Museum.

Stockholm, 1766-68: Vol. 1: 532 pp.; Vol. 2: 736 pp.; Vol. 3: 222 pp.[29]

This was the last edition that Linnaeus prepared, and it was reprinted in Vienna (1767-70), in Leipzig (1788-93), and in Leyden (1789-96), each in a three-volume format.

* * *

How did this formidable work of 1,500 pages without anecdotes or illustrations become known to the general public? A number of strategies were adopted, one of which involved the culling of manageable and popular parts from the main work. They include such books as *Elements of Natural History–Mammalia* (London, 1775), *Caroli Linnaei entomologica* (Leyden, 1789), and *The Shells of Linnaeus* (London, 1855). Beginning with *Index plantarum* (Hafnia, 1781), a variety of indexes to parts of *Systema naturae* was also separately published.

There seemed little point in translating all three volumes of *Systema,* since anyone wishing to use it would almost certainly be able to read Latin. The name and authority of Linnaeus was too great a publishing asset to be ignored; if his own work was too technical for the general public, then his name was assigned to a more popular work. Thus, Ebenezer Sibly produced an immense work, *A General and Universal System of Natural History. . . By the Late Sir. C. Linnaeus* (1794-1810, 14 vols.), originally issued serially in 198 monthly installments. Despite the title, the work had nothing to do with Linnaeus. Whereas *Systema* is unremittingly taxonomic, Sibly's book is a discursive and readable account of the natural world in encyclopedic detail. Sibly tried to give the impression on the title page that he had merely supplemented Linnaeus with material from later writers, but this was false. The Linnean binominals are not used, although his classes, orders, and genera are, on the whole, followed. For the rest, literature of travelers and earlier natural historians, including Buffon, was incorporated, even tales of dragons and unicorns. The demand for this kind of work must have been considerable, for a second equally substantial work appeared within a few years, claiming to be a translation of the "last edition of the celebrated *Systema naturae.*" This was *A General System of Nature. . .* (1802-06, 7 vols.) by William Turton. Similar works can be found in some form or another in most European languages.

Compared with many other classics of science, the tenth edition of *Systema naturae* can still be obtained, if not easily then at less than astronomical prices demanded for other works. At auction in 1979, a good copy went for £750. A greater demand exists for the 1753 *Species plantarum,* which fetched £1,600 in 1977.

Further Reading

There is unfortunately no general history of taxonomy while the best history of the botany of the period remains, despite modern competitors, J. Sachs, *History of Botany, 1500-1850.* The tenth edition of the *Systema Naturae* (1956) is available in a reprint as is the first edition (1964) although, alas, in a limited edition of only 500 copies. The *Species plantarum* (1957-59) is available in another reprint. These works remain untranslated but despite this can still be exciting texts to peruse and can even come to exercise a quite unexpected fascination. Take some unknown flower and attempt to identify it in either the *Systema naturae* or *Species plantarum* before resorting to a popular modern work like the *Oxford Book of Wild Flowers* (1960); Linnaeus does not invariably suffer from the comparison and some insight is gained into just how exciting the Linnean system must have seemed to the readers of his day. Two works which have been translated are the *Critica botanica* (1938), still an exciting work to read, and the more muted *The Sex of Plants* which can be found in *Classics of Science* (1962) edited by W. S. Knickerbocker. A number of current biographies are available: K. Hagberg, *Linnaeus* (1952); N. Gourlie, *The Prince of Botanists* (1953); neither is to be preferred to the more recent work of Blunt (1971) which benefits enormously from the authoritative ten-page appendix contributed by W. T. Stearns. The writings of Stearns on Linnaeus, though somewhat scattered, are in a class of their own. An indispensable source for the vast output of Linnaeus is the British Museum's *Catalogue of the Works of Linnaeus* (1933) by G. B. H. Soulsby. The impact of Linnaeus on English botanical literature can be followed in Henrey (1975) while his wider influence can be traced in F. A. Stafleu's *Linnaeus and the Linneans* (1971).

Notes

1. Argelender, appointed director of the Bonn Observatory in 1837, published the four-volume catalogue *Bonner Dorchmusterung* (1859-62), in which he recorded the position of all stars greater than the 9th magnitude from the north celestial pole to 2°S.–324,189 stars in all. Stars are still referred to by their *Dorchmusterung* number: *BD 31 3932*, for example, refers to star number 3932 in the declination zone 31 north.

2. Beilstein, a Russian, was based in St. Petersburg and published a two-volume work, *Handbuch der Organischen Chemie* (1880-82), describing some 15,000 compounds. Since then, the presses have never stopped running. The fourth edition surveys the literature to 1910; the *Hauptwerk* (1918-37), published in twenty-seven volumes, contains 200,000 entries. It has also spawned an unending sequence of supplements, *Erganzungswerk*, to cover the literature of subsequent decades in matching 27-volume sets. The fourth supplement, covering the years 1950-59, was begun in 1972. Many volumes are in several parts and by now an up-to-date Beilstein must run to some 250 substantial tomes. "Beilstein" covers only organic chemistry; a comparable enterprise is also required for inorganic chemistry.

3. There was apparently a well-known couplet commemorating Harderwijk (Blunt: 97):

> *Harderwijk is en stad van negotie,*
> *Men verkoppt er bokking, blauwbessen en bullen van promotie.*
> Harderwijk is a commercial town
> Where they sell bloaters, bilberries and degrees.

4. Linnaeus's patron Count Carl Tessin (1694-1770), with Gyllenborg, was the leader of the Hats or war party at court, where he served as chancellor. Linnaeus not only catalogued his patron's collection of minerals and snails in his *Museum Tessinianum* (1753), but further immortalized him by dedicating *Systema naturae* to him.

5. Precisely how many works Linnaeus produced is a question that depends on whether books or editions are counted. Are catalogues also to be allowed? A

number of his works were issued under the names of students while in some cases their work was issued as his own. It is a very confusing situation, but it is likely that Linnaeus himself published no fewer than 150 separate works. Some, like *Dissertation on the Sexes of Plants* (1786), are admittedly slim, but many other works extend to several hundred pages. Linnaeus's claim that among major modern scientists he was the most productive author of books is probably true. However, if papers are allowed, paleontologists and mathematicians probably come out on top, since they are remarkably prolific paper-writers. The mathematician Sir Arthur Cayley (1821-95) published 995 papers, while some paleontologists' output exceed a thousand papers.

6. See below, pp. 247-48.

7. It is interesting to note in Marlowe's *Doctor Faustus* that one of the things Faust demands from Mephistopheles is three books: one of spells for raising spirits, one showing the motions and dispositions of all planets, and one "wherein I might see all plants, herbs and trees that grow upon the earth."

8. The title of Gesner's *Historia animalium* indicates its scope by describing itself as a "philosophical, medical, grammatical, philological, poetic and lexicographical work."

9. A view not universally accepted. As late as 1840, Richard Cope in *Cope's Natural History* wrote: "To pretend. . . we have an idea of a quadruped, because we can tell the number or the make of its teeth, or its paps, is as absurd as if we should pretend to distinguish men by the buttons on their clothes" (p. 145). Ignoring such crude distinctions, Cope nonetheless was happy to recognize the whale as a "cetaceous fish" (p. 529 ff).

10. To a modern zoologist, the sloth would be placed with armadillos, anteaters, and aardvarks among Edentata. Linnaeus may have been misled by the fact that sloths are tree dwellers and had been described by *Conquistadores* as possessing a human face.

11. The term herbal usually refers to a class of books that began to appear in Europe in the 15th century and contained simple woodcuts of plants together with brief descriptions of their properties: medicinal, cosmetic, and so on. By the 17th century they had become quite elaborate works with handsome and precise illustrations. A typical example is *Herball* (1597) by John Gerard (1545-1612). An excellent study of this material can be found in A. Arber, *Herbals* (1950).

12. This is reminiscent of Jorge Borge's description of a Chinese encyclopedia which classified animals into the following groups: (a) belonging to the emperor; (b) embalmed; (c) tame; and concluded with a section on those that from a long way off look like flies.

13. This analysis in fact goes back to Theophrastus (c. 370-287 B.C.), pupil of Aristotle and author of *Historia de plantis* and *De causis plantarum*.

14. A cotyledon is a seed-leaf, and whether a plant has one or two is a significant distinction.

15. For those with an eye for cosmic connections, I offer the following: In 1976 Robert Temple produced *The Sirius Mystery* in which he became very excited about the detailed knowledge of the star Sirius that was possessed by the Dogon of West Africa, knowledge only obtained by western astronomers with the aid of a powerful telescope. A comparison missed by Temple is that the Dogon recognize twenty-four vegetable families, exactly the same number of Linnaeus. Details can be found in G. Gieterlen, "Classification des vegetaux chez les Dogon," *Journal de la Société des Africanistes* 22 (1952).

16. The names are all coined from the Greek *andros* (man), *adelphos* (brother), *gyne* (woman), *gamos* (marriage), *oikion* (household).

17. The idiom persists. Two recent books, A. Bristow: *Sex Life of Plants* (1978) and R. Grounds: *The Private Lives of Plants* (1980), remorselessly pursue the sexual analogies one thought had been exhausted by Darwin two hundred years ago. Grounds comments on the bee-orchid that it is "worse than harlots, lower than whores" and has chapters on "Some Sexual Deviations" and "Masturbation."

18. La Mettrie (1709-51), author of the well-known *L'Homme machine* (1748), published in the same year the less popular *L'Homme plante.* A parody of Linnaeus, it actually turned the whole thing inside out and treated men as plants. In this fantasy, man is *Dioecia monandria* and woman *Dioecia monygnia.* Analogies between human and plant anatomy are pursued with as little wit as was shown by Darwin. Linnaeus was not amused and even threatened to stop work on *Species plantarum* unless La Mettrie was punished. Although only forty-two, La Mettrie died suddenly after feasting liberally on pheasant paté and truffles.

19. One such figure was Buffon, who lost no time in pointing out, "But unfortunately for this system there are plants which do not have stamens. There are also plants in which the number of stamens varies" (Eliot and Stern: 169).

20. "For to desire to discern the differences of plants using solely the configurations of their leaves or flowers as a criteria is as if one set out to discern the differences between animals by means of the variations of their skins. . . . It is at very most only a convention, an arbitrary language, a means of mutual understanding. But no real cognizance of things can result from it" (Eliot and Stern: 167).

21. The story of the Grays and their relationship with Smith can be found in the *Dictionary of National Biography* and in A. E. Gunter, *A Century of Zoology at the British Museum* (1975).

22. A. P. DE CANDOLLE (1778-1841), a Swiss who spent the period after 1816 as professor of natural history in Geneva, introduced the term taxonomy into the scientific vocabulary in 1813. His classificatory work, *Prodomus* (1824-39, 7 vols.), is his major publication. JOHN LINDLEY (1799-1865) was one of the few to base his system on physiological criteria: whether the youngest wood was in the center of the stem (endogen) or the circumference (exogen). Details can be found in his *The Vegetable Kingdom* (1846). S. ENDLICHER (1805-49). J. D. HOOKER (1817-1911), son of the already-mentioned William Hooker, friend of Darwin, served as director of Kew Gardens from 1845 (succeeding his father) and prepared with G. Bentham (1800-84) *Genera plantarum* (1862-83) in which 97,000 species of flowering plants are classified.

23. Mention must be made of a fourth strategy adopted by M. Bergeret in a work entitled *Phytonomatotechnie* and quoted by Erasmus Darwin in *Families of Plants* (1789: xi). Bergeret attempted to assign to each plant a fifteen-letter cluster in which each letter "unravels the description of the plant." Thus the name of belladonna, if I have transcribed it correctly, was *Ieqlyabiajisbey* where *i* in the first position means the plant has a five-cleft corolla, *e* in the second that the interstices are not deep, *q* in the third that the stamens are "inserted on the germ," and so on. Not surprisingly, Bergeret's work has left no trace in botanical nomenclature.

24. But by no means impossible. One who did much more than change 7,700 plant names was Carl Kuntze (1843-1907) who, for some strange reason, preferred earlier names and, consequently, in *Revisio generum plantarum* (1891) changed the names of 30,000 species.

25. But *all* his names have survived. In 1758, in the genus *Felis,* Linnaeus recognized seven species: *Felis leo, F. tigris, F. pardus* (leopard), *F. orca* (jaguar), *F. pardalis* (ocelot), *F. catus,* and *F. lynx,* defined by their possession of equal foreteeth, three-pointed molars, retractile claws and a spiky tongue. Later, taxonomists not only added to the seven species (R. F. Ewer, *The Carnivores* [1973], lists thirty-seven), but also subdivided the family into *Panthera* to cover larger cats and *Felis* to refer to

smaller ones. (The distinction was based on whether a part of the hyoid was ossified.) For this reason, the lion has become *Panthera leo* rather than Linnaeus's *Felix leo.*

26. The law of priority: The valid name of a taxon is the oldest *available* name applied to it (Art. 23, *International Code of Zoological Nomenclature,* 1964). But to become *available,* "it must have been published. . . after 1757" (Art. 11A).

27. Here is an example of tautonym: The wren was named by Linnaeus, *Motacilla troglodytes.* In the 19th century, the American wren was assigned to the genus *Troglodytes.* It was later decided to separate wrens from wagtails; wagtails remained as *Motacilla,* but wrens were transferred to *Troglodytes.* Since zoologists permits tautonyms – unlike botanists – the European wren became *Troglodytes troglodytes.*

28. An example of a limitation on Art. 23 is the rule that "a name that has remained unused as a prior synonym in the primary zoological literature for more than fifty years is to be considered a forgotten name (nomen oblitum)."

29. It is also in Volume 3 that Linnaeus published his final classification of the mineral kingdom. The collection and study of minerals was highly popular in the 18th century; as to their classification, Gmelin in his "new and revised" edition of Linnaeus's *Systema naturae* (1771) listed twenty-seven such systems. The basic division was into three classes – stones, minerals, and fossils – eleven orders and fifty-four genera.

Class	*Order*	*No. of genera*
Petrae	Humosae (from vegetable earth)	12
	Calcaire (from animal earth)	
	Argillaceae (viscous marine sediment)	
	Arenatae (more ethereal sediment)	
	Aggregata (mixed)	
Minerae	Salia	23
	Sulphura	
	Metalla	
Fossila	Petrifecata	19
	Concreta	
	Terrae	

Linnaeus's taxonomic activity did not end here. He also produced, for example, in *Genera morborum* (1763) a classification of disease into eleven genera, yielding 325 different conditions. Nothing, it seems, lay beyond his scope.

CHAPTER 11

A New System of Chemical Philosophy (1808-27)

by John Dalton

Dalton may truly be said to be the founder of modern chemistry. As with the indestructibility of matter, so with the indestructibility of energy. What Dalton did for the first great principle, Joule accomplished for the second: and he is therefore the founder of modern physics. And thus the great twin brethren of Manchester did work for the world the like of which hath not been seen, and the importance of which cannot be reckoned.

–H.E. Roscoe,
John Dalton and the Rise of Modern Chemistry (1895)

Atoms are blocks of wood, painted in various colors, invented by Dr. Dalton.

–19th-century student's definition

Life and Career

In broad terms, what are the features which distinguish the practice of science from all other activities? A difficult and controversial question, perhaps: nonetheless, any answer would need to include some fairly obvious factors. Two in particular, the experimental approach and what might be called the "mathematization" of nature, were known to, and further developed by, the scientists of the 16th and 17th centuries and are associated with the names of Galileo, Descartes, and Newton. But astrologers, too, use mathematical skills, and experiment is not confined to the laboratory but can also be found, for example, in the domestic kitchen. There is, consequently, a third central factor, and it concerns the content rather than the methodology of science. It is the assumption that the preferred mode of explanation in the basic sciences – chemistry, physics, physiology and their many subdivisions – is in terms of identifiable, measurable, and controllable atoms.

While the first two factors were identified in the Renaissance universities of northern Italy, developed in the already ancient centers of learning of Paris and Cambridge, and pursued under the patronage of kings and emperors in London and Prague, the third factor – the atomic hypothesis – emerged in the first years of the 19th century from the back streets of the provincial town of Manchester.[1] Its creator, John Dalton, was an equally unlikely figure. He came, not from one of the ancient centers of learning (indeed, as a Quaker he was legally barred from them), but from the small Cumberland village of Eaglesfield in northern England.[2] Born in 1766, the son of a cotton weaver, he was largely self-taught. Unlike France, England had few positions, outside medicine, open to talented yet unorthodox figures like Dalton. He consequently spent virtually the whole of his life, from the age of twelve in 1778, until a stroke forced his retirement in 1837, as a schoolmaster, teaching young children elementary science, arithmetic, French, and Latin.[3] He taught initially in Cumberland and then, from 1793, in Manchester at a dissenting academy founded in 1766. He resigned in 1800 and worked as a private tutor instructing the young of Manchester in Latin, French, Euclid, or anything else they required. In France he would have been ennobled and appointed to a lucrative position with the Académie des Sciences. The point was emphasized by the distinguished French chemist, P. J. Pelletier (1788-1842), discoverer in 1817 of chlorophyll. Pelletier visited Manchester in 1826 in search of the distinguished Dalton. He had, according to an earlier biographer of Dalton, expected to find Dalton lecturing to (Holmyard: 222-23)

> a large and appreciative audience of advanced students. What was the surprise of the Frenchman to find, on his arrival in Cottonopolis, that the whereabouts of Dalton could only be found after diligent search; and that, when at last he discovered the Manchester philosopher, he found him in a small room of a house in a back street, engaged looking over the shoulders of a small boy. . . . "Est-ce que j'ai l'honneur de m'addresser à M. Dalton?" for he could hardly believe his eyes that this was the chemist of European fame, teaching a boy his first four rules. "Yes," said the matter-of-fact Quaker. "Wilt thou sit down whilst I put this lad right about his arithmetic?"[4]

It should not be thought that, because Dalton came from an obscure rural area, his background was illiterate and unscientific. Indeed, examination reveals a considerable interest in natural philosophy in such northern towns and villages. An indication of this interest and of one way in which it was satisfied was the number of itinerant lecturers who roamed around England speaking to packed audiences on such popular subjects as trigonometry and logarithms (a Mr. Hamer in Bury, 1745), mechanics, hydrostatics, pneumatics, and chemistry (Adam Walker in Manchester, 1766); the laws of motion and optics (Caleb Rotheram in Manchester, 1740s); and chemistry (John Waltire in Manchester, 1790s). Some of these lectures could be pretty elaborate affairs. In addition to astronomical instruments, telescopes, microscopes, magnets, and electrical machines, Adam Walker also carried with him models of "Cranes, Pumps, Water-Mills, Pile-Drivers, Engines, the Centrifugal machine, and a working Fire-

Engine for draining Mines. . . a Bucket-Engine. . . a Wind-Mill. . . and a Machine that will raise Water to any Height" (Musson and Robinson: 105). The arrival of such a figure in a smallish community must have been an exciting event, a mixture of circus, trade fair and the laboratory.[5] Not surprisingly, they were not cheap and there are 18th-century reports of lecturers charging as much as ten shillings a course.[6]

Books and journals were also available. The contents of the library of the Kendal school, where Dalton first taught, included *Principia* and other works by Newton, Buffon's *Natural History,* the works of Boyle, and *Historia coelestis* by Flamsteed. Kendal also possessed a telescope, a microscope, and an air pump. In Manchester, Dalton reported: "There is a large library furnished with the best books in every art, science and language, which is open to all, gratis" (Musson and Robinson: 113).

Dalton himself, however, seems to have received his initial interest in science neither from books nor itinerant lecturers but from local residents. Two in particular he later stressed as being especially significant. The first, blind Jack Gough, was sufficiently well-known to gain a mention in the *Excursion* of Wordsworth, who lived in nearby Grasmere.[7] According to Dalton, Blind Jack was "a perfect master of the Latin and Greek and French tongues. . . . Under his tuition, I have since acquired a good knowledge of them. . . . He understands well all the different branches of mathematics. . . . He knows by touch, taste and smell almost every plant within twenty miles of this place. . . . He is proficient in astronomy, chemistry, medicine etc. . . . He and I have been for a long time very intimate; as our pursuits are common – viz, mathematical and philosophical" (Henry: 9). Above all, under Gough's influence, Dalton's interest in meteorology was aroused and from 1787 he began to keep daily records, a task he continued until the day he died.[8] Another figure mentioned by Dalton was Elihu Robinson, a wealthy Cumberland Quaker who maintained contact with a wider scientific field. He had, for example, corresponded with Benjamin Franklin. Robinson apparently made available to those he considered suitable the latest books and periodicals. He certainly encouraged Dalton in his scientific interests, although perhaps not as much as Dalton might have hoped.

About 1790, there are clear signs that Dalton was unhappy and frustrated with his position as schoolmaster. He wrote to various relatives and to Robinson asking for their advice whether to abandon teaching for a career in medicine. He may well have been hoping for more than advice, but all that he received from Robinson was the unwelcome praise of how well his talents were suited to "that noble labour of teaching youth." Neither was his family sympathetic to Dalton's ambitions. Without the backing of family or patron he resigned himself to remaining a teacher and pursuing his private researches as and when he was free.

Dalton's first serious interest was meteorology and in 1793 he published his first scientific work, *Meteorological Observations and Essays.* This was not, however, his only book outside the field of chemistry; in 1801 he published his little known *Elements of English Grammar.* His main interest really lay in the field of meteorology and the study of the atmosphere, work which was to yield a rich dividend.

Physically, Dalton was not attractive. Humphry Davy (1778-1829) actually described him as "repulsive," lacking in all "gracefulness" and with a voice which was both "harsh and brawling" (Henry: 217).[9] He was also color-blind; indeed, Daltonism or Dalton's Disease refers to the condition, especially the red-green type. He became aware of the defect in 1794 when he first noticed that geraniums, which looked sky-blue by day, changed to red at night. His brother shared this condition. Dalton was, in fact, the first scientist to note and study this phenomenon. He discovered other sufferers and, in 1794, he wrote the first important study of the complaint. His own explanation was that the various liquids or humors in the eye somehow absorbed the red light rays and made things look blue instead. After his death his eyes were examined and it was found that the humors were quite clear. The physicians responsible for the autopsy actually went a little further; they removed the posterior part of one eye and looked through it, with its humors intact, at differently colored objects. Dalton's eye made no "appreciable difference."[10]

Dalton's work did not go unrecognized nor was he unrewarded by his fellow countrymen. He was certainly held in esteem in Manchester as president of the Literary and Philosophical Society from 1819 until his death in 1844. Outside Manchester, he was elected to the Royal Society in 1817, and in 1826 he became the first recipient of its Royal Medal. Even the government recognized his merit by awarding him in 1833 an annual pension of £150 (later increased to £300 in 1836). Monuments to Dalton can still be seen in Manchester. In 1833, £2,000 was collected by public subscription and used to commission a statue that now stands in the entrance to the Town Hall; inside the hall a visitor may still see a fresco by Ford Maddox Brown showing Dalton collecting marsh gas. According to Lyon Playfair, at Dalton's funeral Manchester awarded him the honors due to a king. As many as 40,000 were reported to have filed past his body in the Town Hall, while the cortege of mourners on the journey to the cemetery stretched for nearly a mile. Only the death of Darwin provoked a comparable public response. In our own time such treatment is usually restricted to presidents and film and rock and roll stars.

The Chemical Background: Atoms, Elements and Affinities

The atomic hypothesis is an ancient one, older in fact than Christianity. It can be traced back to the fifth century B.C. when it was propounded by Leucippus and his pupil Democritus of Abdera. The fullest account of ancient atomism is to be found in the *De rerum natura* of Lucretius (95-55 B.C.). At one level the atomism of Lucretius was a straight attempt to explain much of common experience by the action of unseen bodies. Sails are driven, clothes dry, smells travel, statues wear away, all by the motion of bodies which "elude the most attentive scrutiny of our eyes." It is the nature of these invisible bodies that they be everlasting, otherwise the process of erosion would have continued so far as to reduce them to nothing. Equally obvious for Lucretius, there must be a limit to the extent to which things can be broken down, for if worn too small nothing could ever again be generated from them. So, Lucretius concluded, atoms are "absolutely solid and unalloyed," "everlasting," "the least parts of which are themselves partless."

If atomism is in reality so old, why was it so long before it became incorporated into modern chemistry? The answer is that if atomism is old, so too are the objections to it. It has never, until recent times, proved to be a particularly attractive theory. To begin with, it had to face the crucial objection that it was a purely speculative theory, nothing more than the most specious of metaphysics. It is too easy, generations of critics have argued, to introduce entities which are by definition impossible to see. Thus, in the 19th century, several leading chemists of the day rejected atomism. F. Kekulé (1829-96), for example, the discoverer of the benzene ring, stated the position adopted by many chemists: "The questions of whether atoms exist or not has but little significance from a chemical point of view; its discussion belongs rather to metaphysics." Plato was an atomist of sorts; chemists, on the whole, were not.

A second major difficulty implicit in any form of atomism is the question of indivisibility. By definition, atoms are the irreducible units of matter which can be divided no further. But this idea of indivisibility was to prove incomprehensible to many. Descartes, for example, was quite happy to accept the idea of "tiny particles," arguing in words reminiscent of Lucretius: "Who can doubt the existence of a multitude of bodies so small as to be undetectable by sensation? One has only to consider the question of what it is that is added to a thing that is gradually growing. . . . A tree grows day by day. . . . Now who has ever detected by his senses which particles are added to a growing tree in a single day?" (Crosland 1971: 69). Yet Descartes went on to insist that however small the particles may be they would always remain divisible "in thought." If atoms are extended bodies then surely, Descartes and others insisted, we can conceive of an atom divided into two. But once having conceived of it divided, it is difficult to see why, other than for reasons of practical difficulty, it should not be divisible in reality (Crosland 1971: 71).

A third basic objection, raised first by Aristotle but heard throughout the centuries and forcibly argued by William Harvey in the 17th century, was atomism's failure to throw light on the biological processes of growth and development. A house could perhaps be understood in terms of bricks and timber but, for Harvey, it was ridiculous to suppose that "Generation were nothing in the world, but a mere separation or Collection, or Order of things." Atomism was thus dismissed as representative of a more general error: "assigning only a material cause, deducing the causes of Natural things from a . . . causal concurrence of the Elements or from the several disposition or contriving of Atomes." For Harvey, as for many others before and after, the essential point about biological processes was that they were controlled and directed towards some end. It was not necessary to conclude as Harvey did, that such processes were under the control of the "highest artifice, Providence and Wisdom" to see that vital processes did present many problems for any naive materialism, whether it be atomistic or not.

And so, for much of history, atomism was never a viable theory.[11] For too many it could be dismissed as purely speculative or fundamentally incoherent. In any case, even for those who were atomists, there were few connections between their atomic theory and attempts made to reduce matter to its identifiable elements. Thus, rather than speculate about the ultimate structure and units of matter, chemists were more concerned with identifying the elements of common experience that, mixed together, produced the numerous substances found in nature. The earliest answer was phrased in familiar terms and, as has already been seen, saw everything as made from earth, air, fire and water. The four-element theory had a good run and when, during the Renaissance, new theories emerged, they did so as modifications of the four-element theory rather than alternatives to it. The exception was the three-element theory, a genuinely chemical theory, linked with the alchemists of medieval Islam and supported in Europe by the dominating figure of Paracelsus (1493-1541). According to this theory, matter was composed of salt, sulphur, and mercury. The basic chemistry of the *tria prima* was relatively simple. As described by Paracelsus, "Take first, for example, wood. Now let it burn, that which burns is sulphur; that which vaporises is mercury, that which turns to ashes is salt" (Crosland 1971: 142).

The following century was a time of considerable confusion over this issue. Chemists of any standing invariably had their own systems ranging from one to about a dozen elements. Many produced their own terminology and if two chemists did seem to agree about salt and mercury, it was more than likely that the terms were being used in completely different ways. To give some idea of the variety of views, a few examples will prove instructive:

G. Cardano (1501-76): better known as a mathematician; earth, water, and air; fire for him was simply a form of motion.

Telsius (1509-88): passive matter plus heat and cold which condensed and rarefied the passive matter.

J.B. van Helmont (1577-1644): the man responsible for the introduction of the term gas (from the Greek *chaos*) saw water as the essential element. In a famous experiment van Helmont proved his point by developing a five-pound willow sapling into a tree of 169 pounds, fed only on distilled water.

Annibal Barlet (circa 1657): the seven elements, fire, sal ammoniac, water, mercury, sulphur, air, and salt.
J. R. Glauber (1604-1668): salt and fire.
J. J. Becher (1635-82): proposed just air, water, and earth, but distinguished three types of earth: *terra pinguis, terra mercurialis,* and *terra lapida,* or the fiery, metallic, and stony, aspects of matter. It was clearly an advantage to introduce new terms, they were less likely to be found missing in common materials. It was Becher's *terra pinguis* that later became the phlogiston of G. Stahl (1660-1734), the notion which so confused chemists of the following century.
N. Lemery (1645-1715): author of *Cours de chymie* (1675, and some fifty subsequent editions), arguably the first modern textbook of chemistry, took five principles or elements: water, spirit, oil, salt, and earth.

In a variety of guises the five-element theory was to find the most support in the 17th century. It was basically the Paracelsian system plus water and earth.

In addition to such elementary theories, the 17th century also witnessed a revival of atomism, or as it was more commonly known at the time, corpuscular philosophy. The sources, varieties, and interactions of the mechanical corpuscular philosophy throughout Europe are extraordinarily complex. It should, however, be stressed that it appealed more to the physicist and philosopher than to the chemist. A physicist like Descartes, in need of a medium for the transmission of forces through space or the propagation of light, looked eagerly at models of the universe with a plenum of particles. Descartes might insist they could not be indivisible; nonetheless they played a vital role in his dynamics and optics. Also, the discovery of the atmosphere through such workers as E. Torricelli (1608-47) was another physical focus that fitted neatly into a space occupied everywhere by corpuscles.

When chemists did turn to atomism it tended to be, as with Robert Boyle, at a very general and abstract level. For Boyle, there was "one universal matter, common to all bodies, an extended, divisible and impenetrable substance." How, then, did diversity arise? Boyle's answer was that, "We must, therefore, admit three essential properties of each entire part of matter, viz magnitude, shape and either motion or rest." Inevitably, the assignment of magnitude and shape to submicroscopic particles had to be a speculative, if not arbitrary, business and of little interest to working chemists.

It was Newton's influence that, although far from straightforward, was to prove the most decisive. For one thing, Newton, in Query 31 of *Optics,* placed his enormous authority behind some form of atomism. "God in the beginning," he declared with typical Newtonian assurance, "form'd Matter in solid, massy, hard, impenetrable, moveable Particles." There were, however, two other aspects to Newton's work that were of more immediate chemical consequence. First, there was his success in providing a deductive mathematical model for mechanics. Using Newtonian principles, scientists could accurately calculate the motions of heavenly bodies and the paths pursued by terrestrial projectiles. Chemists aspired to an equally spectacular success in their work. Consequently, the 18th century was much preoccupied by the need to develop a deductive mathematical or Newtonian chemistry.

It was thought that Newton's mechanics supplied a model for accomplishing this. By concentrating on and quantifying the force holding between bodies, Newton had revolutionized physics; others tried to do the same for chemistry. Hence developed the distinctively 18th-century attempt to base a new mathematical chemistry on the concept of affinity or the attraction various bodies held for each other.[12] It began with the simple attempt to order substances by the amount of their attraction to another substance. Thus, in 1718 E-F. Geoffroy (1672-1731) published a table of affinities from which were derivable such propositions as lead has a greater affinity for silver than copper. With time, statements of affinity became more precise. Guyton de Morveau (1737-1816), Lavoisier's collaborator, placed small metal discs of equal diameter on a surface of mercury and measured the force needed to remove them. As it took 446 grains to lift the gold disc from the surface while 204 grains sufficed to remove the zinc, Guyton could conclude that "the affinity of mercury with gold is to the affinity of mercury with zinc as 446 is to 204" (Thackray 1970: 213-14). This was just one of the many attempts made to quantify the notion of chemical affinity,

A whole new terminology, peculiar to 18th-century chemistry, had to be developed to describe the newly observed reactions. It received much attention from P. J. Macquer (1718-84) in his *Dictionnaire de chymie* (1766), one of the first such works ever published. In the long entry on affinity, distinctions are made between double, compound, reciprocal, and intermediate affinities. Traces of this approach can also be seen in literature, as in the title of Goethe's novel *Die Wahlvanderschaften* or *Elective Affinities* (1809).[13] Inevitably such a preoccupation with the forces *between* substances and the manner in which they combined and separated tended to draw attention away from the nature of the substances themselves. Atomism, as such, was not something much discussed in the 18th century.

This is not to say that the subject was totally ignored. There were some like William Higgins (1763-1825), a Dublin chemist who, in a work published in 1789, seemed to anticipate many of Dalton's ideas. He did not, however, attach much importance to atomic weights, Dalton's crucial innovation. Having once written the book, he did not bother to push his views until 1814 when he sought to challenge Dalton's priority. The works of the leading chemists A. Lavoisier (1743-94) and Joseph Priestley (1733-1804), for example, showed a total indifference to the atomic hypothesis. The 1771 English translation of Macquer's *Dictionary* has an eleven-page entry on phlogiston but none at all on atoms or corpuscles.

It has been suggested that any earlier interest in atomism, as witnessed by Higgins, for example, would have been premature. It was first necessary to exclude the confusing presence of phlogiston from chemical theory and to establish that what was significant in a chemical experiment were precisely those substances that could be weighed throughout their various transformations. In other words, what was conserved in a chemical reaction was mass. It was also necessary to develop an authoritative list of the known elements. With the work of Lavoisier and his school, these requirements were largely met. His revolutionary work, *Traité élémentaire de chymie* (1789), listed thirty elements. One of the great achievements of 18th-century chemists had been to show the multiplicity of elements as opposed to the rather speculative entities listed earlier. Thus, for Newton and his contemporaries, there was only the air, the atmosphere, an homogeneous elastic fluid.

The simple three- or four- or five-element view was shattered forever in 1754 when Joseph Black (1728-99) discovered "fixed air," more familiarly known as carbon dioxide. In 1766, Henry Cavendish (1731-1810) distinguished between Black's "fixed air" and his own newly-discovered "inflammable air" or hydrogen. This was followed in the early 1770s by the discovery of oxygen ("fire air") by Karl Scheele (1742-1819) and of "dephlogisticated air" by Joseph Priestly. The final common gas, nitrogen, was isolated by Daniel Rutherford (1749-1819) in 1772 and named by him "mephitic air."[14] Since it failed to support life, it was renamed "azote" by Lavoisier who also demonstrated that ordinary air was a mixture of oxygen, nitrogen, and carbon dioxide.

Dalton's heritage was thus markedly different from a Priestly or a Lavoisier, born a generation later. Dalton belonged to the first generation not constrained by the confusions imposed by the phlogiston theory. He was also aware of and, for the most part, had the means to identify and distinguish between about thirty distinct elements. Further, Dalton knew that many of the supposed "elements" of antiquity – like water and air – were, in fact, composed of more fundamental units.

Origins

Just where, precisely, lay the origins of Darwin's atomism? France was the center of European chemistry, and with Lavoisier and J. Gay-Lussac (1778-1850) paving the way, French chemists were in all ways better equipped than Dalton to make crucial discoveries. They were better trained, better chemists, better experimentalists, more highly motivated, and they had better support and equipment than a middle-aged, provincial schoolteacher. Yet it was Dalton, in the obscurity of Manchester, who furnished the central insight of modern chemistry. It is as if the much sought-after quark were found, not by the billion dollar accelerators of Brookhaven, Geneva, or Chicago, but by some radio ham in Madras using nothing but an induction coil. Was Dalton just lucky, or were there deeper reasons for his success?

The person best able to indicate what led to the atomic hypothesis in 1808 is Dalton himself. He duly obliged with at least three different accounts of his discovery. Dalton's innovation consisted in showing how the relative weight of atoms themselves could be determined. Chemists had long known how to determine the weights of the various elements combined in a compound substance. By 1799, such skills had been formulated in the law of constant or definite proportion associated with J. L. Proust (1754-1826) who made the point that compounds are the same wherever they are found. Copper carbonate, for example, is the same whether extracted from the oceans, dug out of the earth, or synthesized in a laboratory. Wherever found, copper carbonate will always contain exactly the same proportions of copper, oxygen, and carbon–5.3:4:1 to be precise. Despite some opposition, Proust's law was rapidly accepted as holding for all compounds.[15]

It was still a big step to advance from Proust's law to the appropriate atomic weights. How, for example, it is possible to pass from the statement that 103 units of copper carbonate always breaks down into 53 units of copper, 40 of oxygen and 10 of carbon, to a clear conclusion about the relative weights of copper, oxygen, and carbon atoms. This was the magnitude of Dalton's insight.

It appears, however, that he approached it by a route other than Proust's. Ignoring the accounts he is supposed to have related to various friends and colleagues, the most reliable evidence of Dalton's views is contained in notes he made for a lecture delivered at the Royal Institution in 1810. They were first published by H. E. Roscoe (1833-1915) and A. Harden (1865-1940), both distinguished Manchester chemists, in their *A New View of Dalton's Atomic Theory* which made available many unpublished Dalton manuscripts found in the possession of the Manchester Literary and Philosophical Society. As some of these were destroyed in 1940, they are now known only through Roscoe and Harden's account.

As with much else in his work, Dalton began with Newton. He quoted *Principia* II: 23 to the effect that "an elastic fluid is constituted of small particles or atoms of matter, which repel each other by a force increasing in proportion as their distance diminishes." The idea of repulsive atoms was persuasive only as long as it was assumed that the elastic fluid of the atmosphere was completely homogeneous. Since Newton, however, it had become clear that the atmosphere contained not one but at least three different elastic fluids: oxygen, nitrogen, and carbon dioxide. Why then if they repelled each other did not the three gases arrange themselves in layers according to their specific gravities? Yet, wherever a sample of air was collected, it seemed always to contain the same amounts of the three gases. In 1801, Dalton declared that he had found a possible solution. Suppose, "the atoms of one kind did *not* repel the atoms of another kind, but only those of their own kind." In this way, "the diffusion of any one gas through another, whatever might be their specific gravities" could be permitted (Crosland 1971:200).

Inevitably, the question had to be faced why atoms repelled only other atoms of the same kind? At this time Dalton's view of atoms was considerably different from today's. In the manner of the period, he pictured each atom as a minute, hard particle surrounded by an "atmosphere of heat" or caloric. Thus, when Dalton drew an atom as he thought it really was, as opposed to the conventional symbols used for the various elements, it would be shown surrounded by closely packed lines radiating outwards to represent the heat content of the atom. This suggested to him a repulsive mechanism. Suppose atoms, including their heat halo, to be of different sizes, then Dalton concluded: "On the supposition that the repulsive power is heat no equilibrium can be established by particles of unequal size pressing against each other. This idea occurred to me in 1805."[16] Stimulated by the idea Dalton began to think about the sizes, weights, and number of atoms present in a given volume.

Shortly before, another problem had also led Dalton specifically to consider the question of atomic weights. He had been working on the solubility of gases in water and other liquids, a topic related to his discussion of how gases in the atmosphere can diffuse through each other. What is more, he was thinking about both of these problems as a physicist and deliberately excluding cases where chemical affinities were present. "Why," he asked, "does not water admit its bulk of every gas alike?" He found that air-free water would, for example, admit greater quantities of carbon dioxide and nitrous oxide than nitrogen, hydrogen, and carbon monoxide. Such results suggested to Dalton that the bulkier, heavier compounds, like carbon dioxide and hydrogen sulphate, were more absorbable than lighter gases like hydrogen. "I am nearly persuaded that the circumstance depends upon the weight and number of the ultimate particles of the several gases: those whose particles are lightest and single being least absorbable, and the others more according as they increase in weight and complexity." He was aware of the import of his insight for he added, "An inquiry into the relative weights of the ultimate particles of the bodies is a subject, as far as I know, entirely new: I have lately been prosecuting this inquiry with remarkable success" (Crosland 1971: 202). Dalton went on to list in a table the "relative weights of the ultimate particles of gaseous and other bodies" – the first such ever published. The actual paper, "On the Absorption of Gases by Water," was read in 1803, although it was not published until 1805; the atomic weights given in it are listed in Table 11.2.

These were the questions that directed Dalton's attention to the problem of atomic weights. There still remained the problem of how to determine them. Clearly there was no way individual atoms could be weighed, measured, or scrutinized. What he did have was Proust's law of constant proportions and a knowledge of the proportions, either from the literature or from his own work, in which the elements combined to produce certain common compounds like water, ammonia, carbon dioxide, nitrous oxide and so on. There is still a big gap between knowing that nitrous oxide is composed of 62 parts nitrogen to 38 parts oxygen (Dalton's 1803 figures), to making judgments about the weight of individual nitrogen and oxygen atoms. At this point, the gap could only be bridged by making some bold (if not arbitrary) and unfounded assumptions. Once made they could be applied to a wide class of chemical elements and compounds, and on the basis of such applications the assumption can be discarded as nonsense, modified in the light of experience, or found to fit reasonably well. The essential point is to make the assumption first.

Dalton's Atomism

The details of the system were fully revealed by Dalton in the first part of his *New System of Chemical Philosophy,* published in 1808. Most readers first reaction to it is one of surprise for in a work of 216 pages Dalton's discussion of his atomic hypothesis does not begin until page 211. The first 210 pages contained a long section (140 pages) on heat in which space is devoted to relatively trivial issues such as elementary thermometry.

Dalton began by emphasizing the importance of the conservation of mass for the study of chemical reactions: "We might as well attempt to introduce a new planet into the solar system, or to annihilate one already in existence, as to create or destroy a particle of hydrogen. All the changes we can produce, consist in separating particles that are in a state of cohesion or combination, and joining those that were previously at a distance." In the past chemists had succeeded in obtaining "relative weights in the mass"; they had made no attempt to go further and establish the "relative weights of the particles or atoms." This Dalton did; it took him just two pages.

It is science at its most prescriptive. No evidence is advanced; no arguments produced. Instead fairly arbitrary rules and terminology are laid down and followed.[17] First, the terminology. Combinations of the elements A and B can be such that

1 atom of A + 1 atom of B = 1 atom of C, a binary combination
1 atom of A + 2 atoms of B = 1 atom of C, a ternary combination
2 atoms of A + 1 atom of B = 1 atom of C, a ternary combination
1 atom of A + 3 atoms of B = 1 atom of C, a quaternary combination

And so on. With the definitions, the scene is set for the introduction of a few simple yet fruitful assumptions (Hurd and Kipling: II:31):

> 1st. When only one combination of two bodies can be obtained it must be presumed to be a binary one.
> 2nd. When two combinations are observed, they must be presumed to be a binary and a ternary.
> 3rd. When three combinations are obtained, we may expect one to be a binary, and the other two ternary.

It was time for Dalton to draw some dividends from his investment in definitions and assumptions. Oxygen and hydrogen, O and H in chemical symbols, were known to combine in only one way, water, and the relative weights of the combining masses were 7:1. It was immaterial whether the amount of water taken was a gram, a microgram or the whole Pacific ocean, any sample of water would still be a binary compound of 7 parts by weight of oxygen to 1 part by weight of hydrogen. Thus 8 grams of water from anywhere would always divide into 7 grams of oxygen and 1 gram of hydrogen while 16 grams would split into the same ratio of 14 grams of oxygen and 2 grams of hydrogen. The ratio, Dalton, declared, was invariable. The consequences of Dalton's elementary arithmetic were profound. For, if the sample of water was divided and divided until eventually the chemist was left with one atom of oxygen and one atom of hydrogen he would find them to be in the same invariable ratio by weight as all other samples. Thus he could conclude that an atom of oxygen was 7 times heavier than a hydrogen atom. In the same way, Dalton was able to establish for another binary compound, ammonia, that the relative weights of the atoms of hydrogen and nitrogen were 1:5.

This must have appeared somewhat arbitrary not only to readers of *New System* but to Dalton himself. For if his initial assumption that water was a binary compound was wrong, as in fact it was, all his weights would be correspondingly inaccurate as well. He did have one way, however, which seemed to offer some control over and support his system. Thus, given the values H = 1, O = 7, and N = 5, it should be possible to work out what the relative weights of a binary nitrogen/oxygen compound would be. That is, given H:O as 1:7 and H:N as 1:5, how are O and N related? The answer is 7:5, and this is exactly the ratio Dalton claimed he found in the binary compound he termed nitrous gas (NO).

Further support for his initial assumption came from the whole family of nitric oxides presented by Dalton in a somewhat modernized form:

Table 11:1
Dalton's Family of Nitric Oxides

Name	*Compound*	*Formula*	*Relative Weights*	*Modern Name Formulae and Weights*	
Water	binary	HO	I : 7	Water (H_2O)	1 : 16
Ammonia	"	HN	I : 5	Ammonia (NH_3)	1 : 14
Nitrous gas	"	NO	5 : 7	Nitric oxide	14 : 16
Nitric acid	ternary	NO_2	5 : 14	Nitrogen dioxide	14 : 32
Nitrous oxide	"	N_2O	10 : 7	Nitrous oxide	28 : 16

The neatness and exactness of the ratios presented by Dalton for the nitrogen oxides indicate that Dalton had somewhat coaxed them to fit his initial assumptions. Looking at them demonstrates why some of his contemporaries complained that Dalton tended to get the results he wanted. Such a charge, while no doubt true, is largely irrelevant. It was not for Dalton to test his atomic theory against exact analyses to any number of decimal places. His task was to illustrate the plausibility of his hypothesis, and then let analytical chemists test, improve, and modify it where necessary. The second part of *New System* attempted this task. Part 1 had closed with an indication of how his hypothesis could fit the troublesome nitrogen oxides. Part 2 was consequently concerned with showing how the hypothesis ccould be employed over a large range of elements.

Dalton's commitment to the importance of atomic weights never wavered through the remaining thirty-odd years of his life, despite the suggestive volumetric work produced by the Parisian chemist J. Gay-Lussac. One of the curious features of Dalton's system was that he had obviously been misled about the nature of water and the corresponding ratio of hydrogen to oxygen. In 1808 he had no way of knowing that water is in fact H_2O and not simply HO. This correction would provide a more accurate ratio of H:O as 1:14 rather than Dalton's 1:7. While Dalton had weighed the amount of H and O in water, Gay-Lussac had measured the volume of the two gases found in steam. He found a simple ratio: 100 volumes of O combined with 200 volumes of H to produce 200 volumes of steam. Comparable results were obtained for other compounds.

Thus, there are now two premises regarding water vapor: (a) H and O combine by weight in the proportion 1:7; (b) H and O combine by volume in the proportion 2:1. The conclusion should have been obvious: there are twice as many hydrogen atoms in water vapor as there are oxygen atoms, and the relative weights of hydrogen and oxygen is 1:14 rather than the 1:7 favored by Dalton for the rest of his life. Others understood this: A. Avogadro (1776-1856), for example, declared in 1811 that "the mass of the molecule of oxygen will be about fifteen times that of the molecule of hydrogen" (Crosland 1971: 226). Berzelius was another who made the inference.

Dalton, however, resisted. In his last chemical publications in the 1840s, he still attributed to oxygen a relative atomic weight of 7. One reason for his disagreement with Gay-Lussac's ratio was his refusal to accept the experimental work it was based upon. In *New System* he declared, "In no case, perhaps, is there a nearer approach to mathematical exactness, than in that of one measure of oxygen to two of hydrogen; but here, the most exact experiments I have ever made, gave 1.97 hydrogen to 1 oxygen" (Dalton 1810: 559). Modern chemists have found this judgment odd; E. J. Holmyard, for example, has commented: "It is rather piquant to listen to Dalton chiding Gay-Lussac for experimental inaccurary when . . . he used to get a different result for the "atomic weight" of carbon almost every time he determined it, while Gay-Lussac's manipulative precision has scarcely ever been surpassed" (Holmyard: 253).

Dalton did have other reasons. The conclusion only follows from the premises (a) and (b) above if it is also allowed that (c) equal volumes of gases contain equal numbers of atoms. But (c), Dalton declared, he had considered and rejected. For one thing Dalton was convinced that atoms of different elements were of different sizes and would not therefore be likely to occupy equal volumes. There was, however, a deeper and more puzzling reason. Suppose, Dalton proposed, "equal measures of azotic and oxygenous gases were mixed and could be instantly united chemically, they would form nearly two measures of nitrous gas (nitric oxide)." The number of atoms formed, however, would be at most only "one-half of that before the union." This, Dalton realized, contradicted the assumption that equal volumes contain equal numbers of atoms.

The point is not at first obvious. Consider the following illustration, using arbitrary numbers and volumes:

(a) Take 100 cc. of nitrogen combined with 100 cc. of oxygen to form 200 cc. of nitric oxide.

(b) Let the two equal volumes of nitrogen and oxygen each contain the same arbitrary number of 100 atoms.

(c) If 100 cc. contains 100 atoms, by Gay-Lussac's law of equal volumes, 200 cc. should contain 200 atoms. But the union of 100 cc. of nitrogen and 100 cc. of oxygen may well result in 200 cc. of nitric oxide but, Dalton argued, it could not possibly yield 200 atoms of nitric oxide. Nitric oxide (NO) for Dalton was a binary compound and consequently 200 atoms of NO could result only from 200 atoms of N and 200 atoms of O. The initial 100 nitrogen atoms and 100 oxygen atoms could, for Dalton, combine to make only 100 atoms of nitric oxide and not the 200 atoms required by Gay-Lussac.

The contradiction is puzzling to those who, like Dalton and his contemporaries, failed to make a distinction between atoms and molecules. This insight came from Avogadro in 1811, and it was generally ignored. Equal volumes contained not equal numbers of atoms but of molecules. Suppose, instead of the above, that the basic units of oxygen and nitrogen were not individual, isolated atoms but the molecules O_2 and N_2. Run the simple arithmetic of (a) - (c) through once more and this time no contradiction emerges. For now 100 molecules of N_2 combined with 100 molecules of O_2 equals 200 molecules of NO.[18]

Why could not Dalton see and accept what is now obvious to every schoolboy? One reason: Dalton had retained a model of the atmosphere in which like atoms repel each other; the image of diatomic stable structures of the form H_2, O_2 or N_2 seemed not only distinctly odd but likely to be more destructive than helpful. Such molecules could only exist if there were attractive forces between atoms, but then what happened to the "spring of the air"? Again, why stop at O_2? Why were not the molecules O_4 or O_6 equally as common? These were legitimate questions for someone in Dalton's position to ask; nor, given the chemical background, were they easy to answer. Such in fact was the scientific climate that Avogadro's hypothesis, published in 1811, was ignored for half a century. In 1858, S. Cannizzaro (1826-1910) finally recognized that it provided the solution to many prolonged disputes over atomic and molecular weights. Announced in dramatic fashion at the First International Chemical Conference at Karlsruhe in 1860, Avogadro's hypothesis led to the periodic table and has dominated chemistry ever since.[19]

Reception

Despite the disturbed times of the Napoleonic era, Dalton's ideas became widely known throughout Europe. C-L. Berthollet (1748-1822) in Paris, for example, had a copy of the first part of *New System* by August 1808, just two months after publication. He no doubt showed it to Gay-Lussac who, in a memoir read in December 1808, referred to Dalton's views. Berzelius in Sweden had trouble getting a copy. He first heard of Dalton's work in a paper by Wollaston published in 1808. He wrote to Berthollet and Davy asking for help in obtaining a copy of *New System* but despite the usual promises it was 1812 before he received one from the author. "Never has a present given me such pleasure as this did," was his initial response (Russell: 261).

David Knight, in his study of 19th-century matter theory, divides Dalton's critics into three main groups beginning with those who still persisted in their Newtonian vision of a quantified chemistry (Knight: Chap. 2). One such visionary was Laplace who, in 1797, looked forward to a time when the science of chemistry would be "brought to the same degree of perfection, which the discovery of universal gravitation has procured to astronomy" (Knight: 22). By 1813, Laplace's views had been somewhat modified. He was then visited by Humphry Davy who discussed his view that chemistry "would ultimately be referred to mathematical laws, similar to those which he had so profoundly and successfully established with respect to the mechanical properties of matter." Laplace greeted Davy's idea "in a tone bordering on contempt" (Levene: 54).

One feature of the new atomism that alarmed those who sought a fully quantified chemistry was the growing number of substances being recognized as irreducible elements. The fewer the number of elements, the more amenable chemistry would be to mathematical analysis. Thus, in his *Elements of Chemical Philosophy* (1812), Davy was willing enough to accept the law of multiple proportions without committing himself to the atomic conclusion that Dalton derived from it. Davy preferred to speak of "chemical equivalents" rather than "atomic weights." Davy, in fact, sought a reduction in the number of elements. There is only one "principle of inflammability," he declared; could there not also be just one "metallizing principle"? Davy's vision was not Newtonian; under the influence of the poet Coleridge and German idealism he had become attracted by the idea of the unity of matter. In *Elements of Chemical Philosophy,* he wrote (Davy 1840: IV, 364):

> There is, however, no impossibility in the supposition that the same ponderable matter in different electrical states, or in different arrangements, may constitute substances chemically different. . . . Thus steam, ice, and water, are the same ponderable matter. . . . Even if it should be ultimately found that oxygen and hydrogen are the same matter in different states of electricity, or that two or three elements in different proportions constitute all bodies, the great doctrines of chemistry, the theory of definite proportions, and the specific attraction of bodies must remain immutable.

Compare this with Dalton's unambiguous pronouncement of the same period: "There are a considerable number of what may be called *elementary* principles, which can never be metamorphosed, one into another, by any power we can control" (Roscoe and Harden: 112). Davy's vision of the unity of matter was by no means uncommon and is to be encountered throughout the century, inspiring some of the best science of the period. Michael Faraday, who began his career as Davy's assistant, was one who shared a similar vision.[20]

A third general objection came from those cautious minds who adopted the approach of Osiander when presented with the heliocentrism of Copernicus (see pp. 000). Throughout the century there was never a shortage of chemists, physicists, or philosophers prepared to deny the ultimate reality of atoms while, at the same time, affirming their intention to use Dalton's assumptions. Some of them forebore the use of the term *atom* and preferred more neutral terms like *equivalents* or *combining proportions.* Thus, William Whewell (1794-1866), an influential figure in British science, declared, "So far as the assumptions of such atoms. . . serves to express those laws of chemical composition which we have referred to, it is a clear and useful generalisation. But if the atomic theory be put forwards . . . as asserting that chemical elements are really composed of atoms. . . we cannot avoid remarking that for such a conclusion, chemical research has not afforded nor can afford, any satisfactory evidence whatever" (Crosland 1971: 206). This attitude was not monopolized by philosophers and historians of science. It was held, for example, by F. Kekulé and by W. Ostwald (1853-1932), Nobel laureate in chemistry (1909). It persisted well into the 20th century with the important positivist school known as the Vienna Circle finding atoms somewhat dubious entities and consequently seeking a variety of ways in which they could be eliminated from the language of science.

So much for the reactions of the natural philosophers. How, by contrast, did the working chemists see *New System*? As chemistry, Dalton's contemporaries found his work less than impressive. Thus, Davy declared him to be a "coarse experimenter" who always "found the results he required, trusting to his head rather than his hands" (Henry: 217). Henry placed him "as an experimenter. . . far below his great contemporaries, Wollaston, Proust, Davy, Gay-Lussac, and Berzelius" (Henry: 231). Elsewhere, this had become Dalton's "vast inferiority" as an experimental chemist (Henry: 143). Berzelius also shared the common view, noting, "in the chemical section he allows himself deviations from the truth that are indeed surprising, one sees how he seeks everywhere to shape nature to his hypothesis." Berzelius's response was simply to provide the accuracy missing from Dalton's work and, "After ten years' work I was able to publish in 1818 tables containing the composition of 2000 substances, calculated from the results of my own atomic weight determinations" (Hartley: 78).

It is not necessary to be a chemist to gain from *New System* the impression of a rather tentative, variable, and inconclusive approach. Each part of *New System,* as can be seen from Table 11.2 below, contained a different list of elements providing different atomic weights, often with an attached question mark expressing Dalton's own doubts about the accuracy of his work. The variability in the atomic weights is due to a number of facts. In many cases he was working with misleadingly impure samples, while his uncertainty for many compounds, whether they were binary or ternary, led Dalton to give alternative figures for such elements as nitrogen, zinc, and mercury. The published results, compared with those published by Berzelius, look decidedly messy.

Having made the point, it should also be said it bears little significance. Indeed, in recent times, eminent physicists have competed with each other as disastrous experimentalists. Wolfgang Pauli (1900-58), discoverer of the Exclusion Principle and the 1945 Nobel laureate, was reported to have been so bad that the special Pauli effect was named after him. According to Frisch, "When Pauli appeared anywhere near a laboratory a dreadful thing was likely to happen. Bits of apparatus fell to pieces or exploded" (Frisch: 48).

It was Dalton's insight that was so valuable, not his dexterity with a balance. The irrelevance of such criticism is further emphasized by the practice of leading chemists who reworked Dalton's material and expressed their results in terms of a different base. Thus, Thomson took oxygen rather than hydrogen as 1 on the grounds that oxygen forms more compounds than hydrogen. It did mean using fractional weights; hydrogen, for example, came out as 0.132. In contrast, Berzelius took a base of oxygen, for reasons similar to Thomson, but made it equal to 100. Even those who did accept Dalton's base, H = 1, were nevertheless compelled to rework the atomic weights out for themselves.[21]

One feature of *New System* which did not survive was its symbolism. Dalton was by no means the first to use conventional signs to represent elements. The chemists of antiquity had, for example, represented their four elements – water, earth, fire, and the air – by the respective figures: 🜄 🜃 🜂 🜁. Other symbols were used through the centuries to stand for the seven metals known to antiquity. Thus Geoffroy, in his 1718 affinity table, has as symbols for gold and iron the planetary signs of the sun and Mars ☉, ♂ used 1,500 years earlier in Alexandria.

Table 11:2
A Comparison of Dalton's Atomic Weights
with Modern Weights

	1803	*1808*	*1810*	*1827*	*Modern*
Hydrogen	1.	1	1	1	1.008
Azote (nitrogen)	4.2	5	5	5 or 10?	14.007
Carbon	4.3	5	5	5.4	12.011
Oxygen	5.5	7	7	7	15.999
Phosphorus	7.2	9	9	9	30.974
Sulphur	14.4	13	13	13 or 14	32.064
Magnesia		20	17	17 *S or C*	24.312
Lime		23	24	24 *S or C*	
Soda		28	28	28 *C*	
Potash		42	42		
Stratites		46	46	46	
Barytes		68	68	68 *C*	
Iron		38	50	25	55.847
Zinc		56	56	29	65.37
Copper		56	56	56 or 28	63.54
Lead		95	95	90	207.19
Silver		100	100	90	107.87
Platina		100	100?	73	195.09
Gold		140	140?	60	196.97
Mercury		167	167	167 or 84	200.59
Bismuth			68	62	208.98
Antimony			40	40	121.75
Arsenic			42?	21	74.922
Cobalt			55?	37	58.933
Manganese			40	25	54.938
Uranium			60?	50 or 100?	238.03
Tungsten			56?	84 or 42?	183.85
Titanium			40?	49?	47.9
Cerium			45?	22?	140.12
Alumine			15		26.982
Silex			45		
Yttria			53		88.9
Glucine (beryllium)			30		9.0122
Zircon			45		91.22
Nickel			25 or 50?	26	58.71
Tin			56	52	118.67
Calcium				17	40.08
Sodium				21	22.99
Molybdenum				21 or 42?	95.94
Rhodium				56	102.91
Tellurium				29 or 58	127.6
Barium				61	137.34
Chromium				32	51.996
Potassium				35	39.102
Strontium				39	87.62
Iridium				42	192.2
Palladium				50	106.4
Columbium (Tantalum)				107 or 121	180.95
Total	6	20	36	42	

The first attempt to develop a systematic symbolism was made by J. H. Hassenfratz (1755-1827) and P. A. Adet (1763-1832) in *Méthode de nomenclature chimique* (1787), a work produced by Lavoisier and his colleagues to provide chemistry with a stable, expressive, and uniform terminology. Hassenfratz and Adet sought to introduce some uniformity by using the same symbolic form for the various classes of chemicals: alkalis and earths were represented by triangles, metals by circles, acid radicals by squares, and so on. Distinctions were made within the classes by the inclusion of letters, usually the initial letter of the element's name. They would thus distinguish between the metals arsenic and silver as: ⒶⓈ, Ⓐ.

The big disadvantage of all the above systems is that none was easy to incorporate in a printed text. They all needed special type which printers were reluctant to invest in as long as there was no general agreement among chemists as to which symbols would be used. Dalton made precisely the same error with his little circles enclosing numerous patterns and letters. He first used them in 1803 but they underwent a number of changes before appearing in their final form in 1808. Hydrogen, for example, was earlier shown as ○ and Ⓗ before becoming the ⊙ in *New System.* By the time Volume 2, part 1 appeared, Dalton had 37 such different circles. The advantage for him was that they represented exactly what he took to be the underlying atomic structure. Atoms, for Dalton, really were round; consequently the symbols ⊙ for hydrogen, ○ for oxygen, and ● for carbon could be used to represent the compounds water (⊙), carbon monoxide (●○), and carbon dioxide (○●○).

Dalton's symbols soon had to face competition from the even simpler, more convenient, and ultimately successful system of J. J. Berzelius. In 1813-14 he proposed the following: "Let us express by the initial letters of the name of each substance a determinate quantity of that substance.... When two bodies have the same initial letter, I add the second letter, and should that also be the same, I add to the initial the first consonant that differs" (Crosland 1962: 270). Thus Berzelius referred to oxygen as O, hydrogen as H, nitrogen as Az (from Azote) and about fifty by now very familiar designations. He could not, however, resist the temptation to corrupt an otherwise simple system by the addition of various extra signs. Where an atom appeared in a compound more than once Berzelius sometimes used a numerical superscript, as in S^2O or SO^3. With the more common oxygen, he began to use dots to indicate the number of oxygen atoms present. Thus SO_2 and SO_3 would be written S and S. Where an oxide was "built up of two atoms of a radical and one of oxygen," rather than doubling the letter, as with HHO for water, Berzelius proposed instead a barred letter, as with H for water. Such supposed refinements tended to detract from the system's prize asset, namely, the ease with which it could be read and printed.

The main disadvantage, however, lay elsewhere. It was all very well for Berzelius to talk of taking the "initial letters" of a name, but names of elements varied from country to country. What could be done with sodium/natrum, tantalum/columbium, kalium/potassium, beryllium/glucinum, or wolfranium/tungsten? Some French chemists, for example, quite naturally took O for gold (*or*) and were thus forced to refer to oxygen as Ox. Compromises were made and over the years agreements were reached with the familiar consequence to the English speaker that sodium is referred to as Na, and potassium as K. Such solutions were once deemed outrageous, but by the 1830s this was considered a trivial price to pay for the overwhelming advantages offered by the system.

Dalton, however, wanted no part of it.[22] In a letter written in 1837, he denounced Berzelius's symbols as "horrifying." A chemistry student, he added, "might as soon learn Hebrew as make himself acquainted with them." His objection was that "they appear like a chaos of atoms. Why not put them together in some sort of order? Is not the *allocation* a subject of investigation as well as the weight? If one order is found more consistent than another, why not adopt it until a better is found? Nothing has surprised me more than that such a system of symbols should have ever obtained a footing anywhere." Dalton was being somewhat ambitious at this point; he was certainly premature in his demand for a formula that was not just descriptive but also structural. Such a demand would eventually be satisfied, but it would require a whole range of chemical concepts and insights unknown in Dalton's day. In the meantime, Dalton notwithstanding, it was the symbolism of Berzelius that proved to be the most convenient and adaptable to the needs of chemists.

Publishing History

Manchester, 1808: Volume 1, Part 1 (pp. 1-220)

It is dedicated to professors of the universities of Edinburgh and Glascow and to members of the Manchester Literary and Philosophical Society. The preface provides an account of its genesis:

> In 1802, [the author] was gradually led to these primary laws, which seem to obtain in regard to heat, and to chemical combination, and which it is the object of the present work to exhibit and elucidate. A brief outline . . . [was] first publicly given the ensuing winter in a course of lectures on Natural Philosophy, at the Royal Institution in London. . . . The author has since been occasionally urged by several of his friends to lose no time in communicating the results of his enquiries to the public, alleging that the interests of science, and his own reputation, might suffer by delay. In the spring of 1807 he was induced to offer the exposition of the principles herein contained in the course of lectures which were twice read in Edinburgh, and once in Glascow.

As they were well received, Dalton continued, "Upon the author's return to Manchester he began to prepare for the press. Several experiments required to be repeated; other new ones to be made. . . . These considerations, together with the daily avocation of his profession, have delayed the work nearly a year; and, judging from the past, it may require another year before it can be completed." As a matter of fact, it took another nineteen years.

Volume 1, part 1 of *New System* is divided into three chapters:

Chap. 1: "On Heat and Caloric" (pp. 1-140) covers such topics as specific heat, temperature of the atmosphere, and instruments for measuring temperature.

Chap. 2: "On the Constitution of Bodies" (pp. 141-211) deals with the constitution of pure elastic fluids, mixed elastic fluids, liquids, and solids. It contains a brief reference to the atomic hypothesis (p. 143).

Chap. 3: "On Chemical Synthesis" (pp. 211-16) develops in a few pages the nature, rules and symbolism of his atomic theory. It also contains a table of twenty known elements, their atomic weights and their symbols. Some compounds are included.

This was not actually the first publication of Dalton's views. Thomas Thomson (1773-1852), a chemistry lecturer at Edinburgh, had learned from Dalton details of his theory during a visit to Manchester in 1804. Thomson accordingly discussed them sympathetically in the third edition of *System of Chemistry* (1808).

Manchester, 1810: Volume 1, Part 2 (pp. 221-560)

It is dedicated to Humphry Davy and W. Henry (1774-1836). Dalton ascribed the delay, in the preface, to "the great range of experiments which I have found necessary to make." The work is a strictly chemical treatise in which the theory of the previous part is applied throughout the chemical domain. This part is divided into two chapters:

Chap. 4: "On Elementary Principles" (pp. 221-70) deals with oxygen, hydrogen, carbon, sulphur, phosphorus, and the various metals.

Chap. 5: "Compounds of Two Elements" (pp. 269-549) deals with the binary compounds of all elements, excluding the metals dealt with in Chap. 4. This is followed by sections on the fixed alkalis (potash, hydrates of potash, soda, among others) and the earths (lime, magnesia). A revised table of the elements is given which grew from twenty in number to thirty-six. Atomic weights are given for all thirty-six, together with those of an additional twenty-four compounds.

Manchester, 1827: Volume 2, Part 1 (pp. 1-357)

It is dedicated to Dalton's Manchester colleagues, John Sharpe and Peter Stewart. According to the preface, half of the volume was ready and printed by the end of 1817 and the rest, apart from the appendices, by September 1821. He also noted his intention, which came to naught, to add a second part to Volume 2 in order to deal with more complex compounds such as acids and salts. The bulk of the text (pp. 1-268) completes Chap. 5 of the second part of Volume 1.

Chap. 5: "Metallic Oxides" deals with earthy, alkaline, and metallic "sulphurets" and "phosphurets"; carbonates, and metallic alloys. A much revised table of elements contains thirty-seven entries, with many from Volume 1, part 2 missing and replaced by new elements.

London, 1842: Second edition, Volume 1, Part 1

Dalton noted, "The first edition of the work having been out of print for some years, the author has been induced. . . to publish a second edition without making any alteration to it."

London, 1953: Facsimile edition

This facsimile edition of the entire work was printed in 1,000 copies.

Translations

There was little demand for *New System* in other languages. Although the whole work has never been translated, both parts of Volume 1 were translated into German by F. Woolf and published in Berlin (1812-13).

Extracts

Extracts were issued in Leipzig in 1889 (30 pp.) and 1894 (221 pp.); in Edinburgh (1899: Alembic Club reprints, nos. 2 and 4); in New York (1902) and in Moscow (1940); but the question can be asked: why was *New System* so ignored? Why should *System of Chemistry* by Thomson, first published in 1802, attain seven editions by 1831, be translated into French not once but twice (1809; 1818), and also appear in two American editions (1803; 1818), while Dalton's work was largely ignored? The answer is that Dalton was not in fact a good chemist. His work in *New System* was tentative, inaccurate, and soon superseded. It also took so long to complete that by the time the final volume appeared, the first was hopelessly out-of-date. It was a publisher's nightmare: 900 pages of technical chemistry that no one wanted to read. This does not mean that there was no interest in Dalton and his atomism, as contained in five pages of his Chap. 3 of Volume 1, Part 1. This is the part that people want to read, and it has been published many times. Some of the more notable anthologies that contain these pages are: Hurd and Kipling, Vol. 2 (1964); M. B. Hall (1970); D. D. Runes (1962); Leicester and Klickstein (1952); and L. K. Nash (1950).

Further Reading

The complete *New System* is available in a 1953 facsimile reprint but most reader's are likely to be satisfied by the extracts from it included in such anthologies as Hurd and Kipling, Leicester and Klickstein and Marie Boas Hall's *Nature and Nature's Laws* (1970). Greenaway (1966) has provided a readable and reliable modern biography although the *Memoirs* (1854) of Henry is still a work of considerable interest and could well deserve a reprint. For background there are the relevant volumes of Partington's *History of Chemistry* (1961-64, 4 vols.) and the more specific study of 18th-century chemistry by Arnold Thackray (1970). Dalton's own contribution to science was critically examined by a number of leading scholars at a conference held in Manchester in 1966 to mark the bicentenary of his birth; the results were published in a volume edited by D. S. L. Cardwell (1968). The struggle to establish atomism can be followed in Knight (1967) while the opposition to atomism in mid-19th-century Britain is the subject of a special monograph edited by W. Brock, *The Atomic Debates* (1967). The final victory of atomism is described in Mary Jo Nye's *Molecular Reality* (1972).

Notes

1. The contribution of Manchester to modern science is remarkable. Not only was it the birthplace of modern atomic theory, but it was also there that James Joule (1818-89), a pupil of Dalton, laid the foundation in the 1840s of the new science of thermodynamics by demonstrating the mechanical equivalent of heat. It was also in Manchester, between 1910 and 1913, that Ernest Rutherford (1871-1937) and Niels Bohr (1885-1962) made the first advances in the new discipline of nuclear physics. These are just the peaks of a thriving and persisting scientific tradition, equalled by no other provincial town in Europe. The reasons for this prominence are dealt with at length by R.H. Kargon in *Science in Victorian Manchester* (1977). One major factor was the Manchester Literary and Philosophical Society, founded in 1781 and, before the establishment in 1851 of Owen's College, the precursor of Manchester University, a focus for the discussion, dissemination and teaching of science. Dalton was president from 1819 until his death in 1844 and read to it 116 papers. Most British towns in the 19th century possessed similar societies yet none equalled in reputation and achievement the Manchester "Lit and Phil."

2. It was not until 1828, with the repeal of the Test Act, that Dissenters were free to hold public office. They were not admitted to the ancient universities until 1856. It was largely to escape this discrimination that the University of London was founded in the 1820s. Dalton was not the only major British scientist to emerge from such a background. Michael Faraday (1791-1867), a greater figure than Dalton, was also a Dissenter, self-taught, and of even more modest origin.

3. In 1785, Dalton and his brother acquired the Kendal School at which they had taught since 1781. They advertised it as a school where youths would be "carefully instructed in English, Latin, Greek and French; also Writing, Arithmetic, Merchants Accompts and the Mathematics."

4. Financially, Dalton seems not to have done badly. At one time he was able to consider leaving £2,000 in his will to Oxford to endow a chair of chemistry. His earnings came from a variety of sources: chemical analyses, for which he charged "often merely a few shillings: but never exceeded a sovereign" (Henry: 206); teaching, for which he received eighty guineas per annum while on the staff of the Manchester Academy; and tutoring, for which he charged two shillings and sixpence per hour, or, if two attended, only one shilling and sixpence each. He received eighty guineas from the Royal Institution in 1810 for lecturing, while a course in Manchester brought him a profit of £58 2s. As for expenses, there is Dalton's report of a trip to London in 1805 in which he spent £200 on equipment for demonstrations at his lectures.

5. The lecture circuit also attracted the less orthodox: charlatans peddling nostrums, conjurors, and those who stressed the wondrous as opposed to the practical and useful aspects of science. One such was Gustavus Katterfelto who described himself as "the greatest philosopher. . . since Sir Issac Newton," and who specialized in the "solar microscope," an instrument that permitted the projection and magnification of maggots and other insects. He also performed experiments "optical, electrical, physical, chymical, pneumatic, hydraulic or hydrostatic," as

well as demonstrating the common ploys of card sharps. Katterfelto flourished in the late 18th century, but the tradition persisted and in Hull, in 1845, limelight replaced the solar microscope to project cheese mites "as large as Cats! with Bristles as large as Hedgehog's!" Also shown on the screen was "the crystallisation of salts, by which a single drop of solution is made to shoot into beautiful crystals and to cover the space of 120 sq ft in a few minutes." More details on the scientific fringe world can be found in E. Dawes, *The Great Illusionists* (1979).

6. Dalton's own lecture courses were fairly regular. Between 1805 and 1835 he is known to have lectured twelve times in Manchester and once in Edinburgh, Glasgow, Birmingham, and Leeds. The lectures were first given at Kendal in 1787 and 1791, for which he charged ten shillings and five shillings respectively.

7. Methinks I see him now, his eyeballs roll'd
Beneath his ample brow – in Darkness pained
Whilst the voice
Discoursed of natural or moral truth
With eloquence and such authentic power,
That in his presence humbler knowledge stood
Abashed, and tender pity overawed
– Wordsworth, *The Excursion*

8. He died about 6:30 A.M. on 27 July 1844. It seems that he had arisen at his usual time and in tremulous writing, "entered in his meteorological journal the temperature, maximum and minimum, of the previous night." He collapsed immediately afterward (Henry:199).

9. Humphry Davy's judgment may not be all that sound. As the man who blackballed Michael Faraday in 1821 when Faraday applied for membership in the Royal Society, it is certainly suspect. Davy was an intolerable snob who might have been tempted to patronize Dalton.

10. "The aqueous. . . was found to be perfectly pellucid and free from colour. The vitreous humour. . . were also perfectly colourless. The crystalline lens was slightly amber coloured, as usual in persons of advanced age" – from the post mortem report on Dalton (Henry: 202). Ironically, Dalton's brain on post mortem presented a "remarkable prominence on the frontal portion of the orbitar plates," the phrenological site of the "organ of colour" (Henry: 201).

11. Many other more specific objections have been repeated down the centuries: How could anything soft be composed of hard, indivisible atoms? How are changes of states, such as melting, to be explained? What of the cohesion of atoms in solid bodies? It was no good supposing that atoms could be held together by some sort of glue, or even that they should have little hooks on them for, as Newton pointed out, glue and hooks would themselves be made of atoms and thus require even more basic glue to hold themselves together. Those who were atomists usually also believed in the general emptiness of space within which atoms could move; there were many, Cartesians for example, who objected to the void. Inevitably, this led to a rejection of atomism.

12. Examples of 18th-century chemical vision: "The laws of affinity. . . are the same with that general law by which celestial bodies act upon one another." That was written by Buffon who also held that the attractive force between chemical substances was the same $1/r^2$ as held between the earth and the moon. Guyton de Morveau sought a chemistry as completely predictive as "the paths of the stars." While even Lavoisier looked forward to a time when improved data would allow the mathematician "to calculate any phenomena of chemical combination in the same way. . . as he calculates the movement of the heavenly bodies." Details can be found in Thackray 1970 (Chap. 7) and Thackray 1968.

13. Elective affinity was the jargon used to describe a situation where compound AB in contact with substance C yielded the new compound AC or BC. It was almost

as if the substance A *elected* to join C rather than B. Why and how substances had such propensities was a question of some depth. Readers of Goethe's novel will spot the analogy.

14. The term nitrogen was coined by J. A. C. Chaptel (1756-1832) in 1790 on the ground that it was part of niter. The new term proved acceptable everywhere except in France where nitrogen is still known as azote and its chemical symbol is accordingly Az.

15. Lest it be thought that Proust's law is too obvious to merit attention, mention should be made that the French chemist Berthollet thought the law false and developed a theory of *indefinite* proportions. He considered, for example, that the proportion of oxygen in a metallic oxide can "vary progressively from the term where the combination becomes possible up to that where it attains the last degree" (Partington: III, 645).

16. The idea will be clearer if Dalton's own diagram of the mechanism he had in mind is examined:

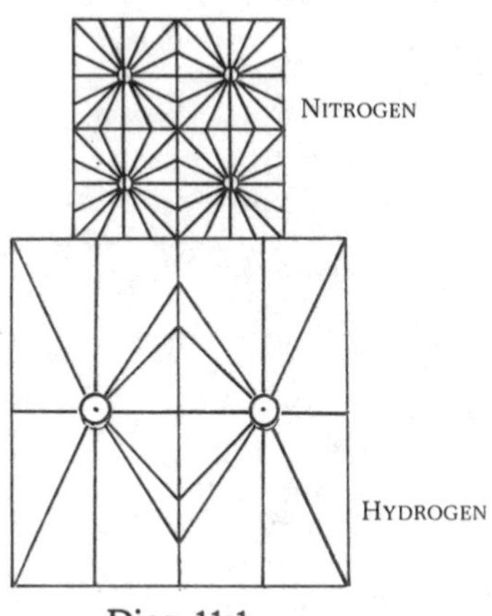

Diag. 11:1

Here it is obvious that the lines of force emerging from nitrogen atoms meet and repel each other, as do hydrogen atoms. But when different atoms meet, being of different size, the lines of force do not mesh and repel.

17. Dalton actually had one very general argument in support of the view that binary and ternary compounds would be preferred to more complex forms. Since like atoms repel each other, then the more atoms of the same element present in a compound the greater would be the repulsive forces present; ultimately they would be so great as to make such a structure impossible. Berzelius developed a comparable argument but this time expressed in terms of atomic sizes: if you imagine a compound ABn as formed from a central atom of A and surrounded by n atoms of B, then n could not be bigger than 12 for, given the estimated atomic diameters, no more than 12 could actually fit.

18. There was ample room for confusion with the new terminology. Avogadro, for example, spoke of a "molecule élémentaire" for an atom of an element; "molecule" by itself could mean either atom or molecule in the modern sense. The distinction as it is now understood seems not to have been in use before 1833 when it was introduced by Gaudin.

19. D. Mendeleyev (1834-1907), the first to formulate, in 1866, the periodic table, commented on Cannizzaro's work: "I regard him as my immediate predecessor because it was the atomic weights which he found, which gave me the necessary reference material for my work" (Hartley: 193).

20. In more recent times, if there is no longer the same desire to establish the unity of matter, there is a reluctance to embrace unnecessarily its profligacy. Thus, in the immediate postwar world, when it seemed that there were just twelve subatomic particles from which all matter was composed – proton, neutron, electron, neutrino, and their antiparticles; the photon; and the three pions: neutral, negative, and positive – this seemed about right; rich enough to provide the necessary variety, yet not so abundant as to cause embarrassment. Then things began to happen, and more and more elementary particles began to be discovered. Matt Roos, a physicist from Finland, published in 1963 the first comprehensive table of new particles in *Review of Modern Physics.* It ran to five pages and covered seventeen particles and twenty-four resonances. The 1976 list occupied the whole 245-page issue of the journal and described literally hundreds of supposedly elementary particles. Inevitably, in such a situation some would seek for deeper and more fundamental particles and hence the quark model arose. Originally there were three quarks out of which all other particles could be constructed. The number has already grown to six, and if history is any precedent, it could multiply further. Details of the modern counterpart of the Davy-Dalton dispute can be found in J. S. Trefil, *From Atoms to Quarks* (1980). See also the significantly titled "The Structure of Quarks and Leptons" by Haim Harari in *Scientific American* (April 1983).

21. Other bases were also in use. Wollaston, for example, used a base of 10 for oxygen; the modern standard of $O = 16$ was introduced into chemistry by the Belgian chemist J. S. Stas (1813-91) in 1850.

22. He maintained his opposition literally to the end. On the evening before his death, he had an "animated" discussion with a "distinguished chemist" on the subject of "chemical notation and symbols; he contending that his own plan of representing the probable position of the atoms of a compound was preferable to the system then and now universally adopted, which merely gives the initials of the elements and their number. . . . He appeared more excited than usual and unable to let the subject drop" (Henry: 200).

CHAPTER 12

Principles of Geology
(1830-33)

by Charles Lyell

To CHARLES LYELL, Esq., F.R.S., this second edition is dedicated with grateful pleasure, as an acknowledgement that the chief part of whatever scientific merit this journal and other words may possess, has been derived from studying the well known admirable PRINCIPLES OF GEOLOGY.

–Charles Darwin, dedication, *Journal of A Voyage Around the World*

I cannot but believe that Lyell, for others, as for myself, was the chief agent in smoothing the road for Darwin. For consistent uniformitarianism postulates evolution as much in the organic as in the inorganic world.

–T. H. Huxley, *Life and Letters*

Life and Career

Lyell was the son of a Scottish landowner. The 5,000-acre family estate of Kinnordy, at Kirriemuir in Forfarshire, was bought by his grandfather, Charles Lyell (1734-96), for £38,000 with funds somehow obtained during twenty years of service with the Royal Navy.[1] Lyell, an eldest son, was born there in 1797. Soon after, the family moved to England where they settled in 1798 on an eighty-acre estate at Lyndhurst in Hampshire, on the fringes of the New Forest. His father, also named Charles Lyell (1769-1849), was a man of some learning. He was well known as a Dante scholar and as a competent botanist who discovered several species of moss including the eponymous *Orthotricium lyelli*. From his father Lyell gained some early fieldwork experience: for example, there is a description of Lyell accompanying his father and William Hooker (1785-1865), later director of the the Royal Botanic Gardens, Kew, on a plant collecting expedition in the Scottish highlands (Wilson 1972:31).

Lyell's interest was diverted to geology while a student at Oxford. One influence was Robert Bakewell's *Introduction to Geology* (1813), a popular work, which he claimed "gave him an idea of the existence of such a science of geology" (Lyell 1881:I,132). A second influence was William Buckland (1784-1856), then reader in mineralogy at Oxford. Buckland, a noted eccentric, is perhaps best remembered for his habit of feeding guests such rare delicacies as mice on toast, crocodile stew, or roast ostrich. He was not to everyone's taste. The normally mild Charles Darwin dismissed him as a "vulgar and coarse man" (Darwin 1974:60). He was nonetheless a stimulating lecturer and teacher as well as an energetic field worker who spent much of his time touring England, either alone or with several students in search of traces of the Deluge. The results were published in his best known work, *Reliquiae deluvianae* (1823).

While at Oxford, Lyell made the first of many geological expeditions. With his parents and sisters, he spent three months in 1818 traveling through France, Switzerland, and Italy, visiting a number of important sites. Later expeditions took him to the Continent many times, with the 1828-29 trip to France and southern Italy being of particular significance. He also visited Scandinavia (1834-35), and made two lengthy visits to the United States (1841-42 and 1845-46) as well as many field trips to geological sites in Britain. These were strenuous trips, often covering fifty miles a day on horseback and hikes of up to forty miles a day were not uncommon. Lyell clearly enjoyed this aspect of geology and happily took any opportunity to get into the field. Age did little to dim his interest. In 1872, at the age of seventy-five, he visited the caves of Aurignac, his last trip to Europe. Nor for that matter did marriage. In 1832, he married Mary Horner[2] in Bonn, and they set off on their honeymoon down the Rhine, visiting sites and museums in Germany and Switzerland.

After graduating from Oxford in 1819 with a second class degree, Lyell began to read for the bar. His father felt it important that his son should be able to support himself, particularly since harvests and markets were so bad that tenants were frequently unable to pay their rent. Consequently, Lyell agreed to pursue a law career and after being called to the bar in 1822 actually began to practice.[3] An entry from an 1827 notebook indicates how successful he was: "13 barristers of whom one makes £30. Eight only £5 each or less. Expense £8 or more each?" (Wilson 1972:72). Lyell was probably not even covering his costs. By this time, however, he began to earn more money from journalism than the law. To such journals as *Quarterly Review,* he began to submit reviews, accounts of geological research, and articles on more general topics. The Lyell estates once more began to prosper and, before long, Lyell was receiving an annual allowance of £400 from his father.[4] With this and his income from writing, it was agreed in 1827 that he should leave the bar and pursue a career as author and geologist.

Lyell only held one job, and that briefly, as professor of geology at King's College, London, from 1831-33;[5] the rest of his long life was devoted to fieldwork and writing. During his lifetime, *Principles of Geology* went through twelve editions, *Elements of Geology* and its various transmutations eight editions, and *Antiquity of Man* another four editions. These were all substantial volumes and in most cases the different editions involved considerable rewriting. Much of Lyell's life after 1830 must have been spent in seeing one of other of these works through the press.[6] Despite this, he seems to have led a fairly busy social life. Darwin reported that Lyell told him that he would not dine out more "than three times a week on account of the loss of time" (Darwin 1974: 58-59). In 1875, two years after the death of his wife and by then showered with honors, Charles Lyell died at the age of seventy-eight.[7]

Geological Background

The early 19th century was a good time to be a geologist. For one thing, geology as a subject was becoming organized and professionalized. The Geological Society of London had been founded in 1807, and it immediately set to work to produce a geological map of England and began publishing its *Transactions.* Lyell's connection with the Society was long and close: elected a Fellow in 1819, he served as secretary from 1823-25 and as president, first in 1835 and a number of times thereafter.

It was also the period in which the basic methodology was being worked out. Lyell, in fact, belonged to the first generation of geologists who could read (or, often as not, misread) the records of the rocks. In the 18th century it was known that fossils or "figured stones," as they were earlier called, were nothing but organic remains. That they were also the key with which the history of the earth could be deciphered, was a major intellectual advance taken later in the early 19th century by a number of scholars. One of the first was William Smith (1769-1839) who, as a surveyor, had become familiar with the geological strata of much of Britain. On his travels he had noticed that the continuity of certain strata could be traced not only by their distinctive rock formations, but also by their equally distinctive fossil content. That strata could be indentified, traced, dated, and compared was probably the most important geological truth ever formulated; without it, there was only mineralogy; with it, there could be geology.[8] Smith knew the value of his insight, and he produced the first authoritative geological map of England and Wales which was published in 1815, predating by four years the Geological Society's map. With his map, he included a "memoir" describing his new method. It was followed by another monograph on the subject (Smith 1816) and a further work illustrating the most common fossils found in Britain (Smith 1817). Progress was also being made in France about this time. Cuvier (1769-1832) and his colleague Alexandre Brogniart (1770-1847) had worked out the geology of the Paris basin in terms of its fossils before Smith had published his British material (Cuvier and Brogniart 1811).

Lyell also had the advantage of working in an extremely popular discipline. Geology had not yet become so specialized and technical that it was beyond the reach of the layman who, with limited experience and effort, could not only master the new subject but even contribute to it. Almost any journey to Europe was likely to lead to some discovery.

Serious books on geology became best sellers. It is still something of a shock to us, looking back, to see just how extensive this interest in geology was. Consider, for example, the case of Roderick Murchison (1792-1871), a fox-hunting squire with few interests outside the sporting scene. One day in 1823, he happened to be in the same shooting party as Humphry Davy, the leading chemist, inventor of the miner's safety lamp, and president of the Royal Society. Davy must have spoken of science with rare potency for, within a matter of weeks, Murchison had sold his horses and moved to London where he studied chemistry with Davy and attended meetings of the Geological Society. Before long, he was accompanying Lyell and Buckland on field trips. In the 1830s, Murchison identified and described the Paleozoic periods, the Devonian, Silurian, and Permian.

At the other end of the social scale, Hugh Miller (1802-56), a Scottish stonemason and an aspiring poet, was one day working in a quarry cut into the rocks of the Old Red Sandstone (paleozoic rocks of the Devonian period). Breaking open some pieces of blue limesone, he found "in one of them there were what seemed to be the scales of fishes, and the impression of a few minute bivalves, prettily striated; in the centre of another there was actually a piece of decayed wood. Of all Nature's riddles, these seemed to me to be at once the most interesting and most difficult to expound" (Miller: 40). Unlike Murchison, the young stonemason needed to earn his living and was forced to seek the answer to his riddle in his spare time. He became a journalist and produced a number of geological works one of which, *Old Red Sandstone* (1841), a fairly formidable work on the fishes of the Devonian period, went through twenty editions before 1900.

Another factor adding to the excitement of geology was the dramatic discovery of great dinosaurs and other giant reptiles. It is in this period that the term paleontology[9] entered the language. The Oxford English Dictionary gives 1838 as its first recorded use; before that, the word in use was oryctology, from the Greek *oruktos* (dug up), as used in the title of one of the earliest works on the subject, *Outlines of Oryctology* (1822) by James Parkinson.[10] The first Cretaceous fossils had been discovered in the chalk quarries of Maestricht in 1770. It was not, however, until the end of the century that the pieces were examined by Cuvier in Paris who claimed that the four-foot jaws belonged to a 25-foot marine lizard, later named *Mosasaurus* in commemoration of its discovery on the banks of the Meuse. More astonishing was the discovery, in 1784, by an Italian naturalist in the Bavarian limestone, of a skeleton of a lizard with claws like a bat, teeth like a crocodile, and a beak like a bird. Cuvier named it *Pterodactylus*, or wing-finger.

These were isolated discoveries. Shortly afterwards in southern England remains of extinct monsters began to turn up on a more regular basis. Mary Anning, a barely literate girl, found in 1812 at her home in Lyme Regis a 17-foot skeleton of a fish-like lizard, *Icthyosaurus.* She went on to make other major finds, including the first *Plesiosaurus,* another marine lizard, in 1821. The Duke of Buckingham was reported to have paid £200 for it.[11]

Some of the astonishment with which scholars contemplated such finds is manifest in Buckland's description of a pterodactyl found by Mary Anning in 1828. "A monster resembling nothing that has ever been seen or heard-of upon earth, excepting the dragons of romance and heraldry" (Buckland: 218). It was common to link pterodactyls with dragons in those days. Charles Kingsley, for example, commented in *Water Babies* (1863) that "people call them Pterodactyls; but that is only because they are ashamed to call them flying dragons, after denying so long flying dragons could exist." Prince Albert, consort of Queen Victoria, proposed that Crystal Palace, the elaborate building in London constructed for the Great Exhibition of 1851, should be stocked with life-size replicas of *Megalosaurus, Iguanadon,* and other extinct beasts of the Mesozoic. The completion of the task, in 1853, was celebrated by twenty-one of those involved in the project holding a banquet *inside* the model of *Iguanadon.*[12] Although the Crystal Palace burned down in 1936, the mighty figures can still be seen at Sydenham. Their designer, Waterhouse Hawkins, attempted to persuade New York authorities to install a comparable set in Central Park, but this proposal was opposed by the Commissioner of Public Works.[13]

Thus, when Lyell, in the 1820s, decided to become a full-time geologist, the institutions, methods, and audience were already awaiting him. Yet there was a further and deeper reason why geology should prove so attractive to a scholar and layman alike. Geology dealt with questions that concerned the age of the earth, the history of its organic forms, the nature and development of man and, most significant of all to the 19th-century scientist, their relation to God. In general, three different theoretical positions evolved to deal with these problems: the traditional approach of A. Werner (1750-1817), the catastrophism of Cuvier, and the uniformitarianism developed by Lyell in *Principles.*

Traditionalism

"The number of true species in nature is fixed and limited and, as we may reasonably believe, constant and unchangeable from the first creation to the present day" (Greene: 128). So spoke John Ray, the leading British naturalist of the 17th century. When presented with the obvious objection that some fossil remains, particularly shellfish, appeared to have no modern counterpart and therefore must now be extinct, Ray replied, "Why, it is possible that many sorts of Shell-Fish may be lodged so deep in the Seas, or on Rocks so remote from the Shores, that they never come to Sight" (Ray: 149-50). Ray's views continued to be accepted by most scientists throughout the 18th century. To many it appeared absurd that God would create a species incapable of surviving indefinitely. It offended that strange genius Thomas Jefferson (1743-1826) and his feeling for the "economy of nature." It was such, he declared, "that no instance can be produced of her having permitted any one race of her animals to become extinct" (Jefferson 1904: III, 427).

Linked with such an immutable view of the earth's flora and fauna was an equally simple and comprehensive geology, seen in its most developed form in the work of Werner. His geology, according to Lyell, was "little other than a subordinate department of mineralogy" (Lyell 1834: I, 5). Werner actually wrote very little and traveled even less. Most of his life was spent as student and teacher in the Freiberg School of Mines in which, as an inspiring teacher, he exercised his considerable influence over students who came from all over Europe to learn from him. His basic position, although much modified later by himself and his students, was expressed in his *Kurze Klassification* (1787).[14] In it he saw the entire history of the earth in terms of a very simple scheme. In the beginning the earth was completely enveloped by the waters of the primeval ocean. In this ocean, in suspension, was found all the materials of the earth's crust. As the waters receded – just where the waters went was a problem that the Neptunists, as Werner's followers were called, never satisfactorily answered – various chemical and physical forces operated to form the earth's strata as they now appear.

The first to "crystallize" from the ocean were the primitive rocks – the granites, gneisses, schists, and so on, that lacked all trace of organic remains. On these were deposited limestones, sandstones, and chalks of the secondary strata. These were universal formations. The remaining and more recent alluvial and volcanic deposits, Werner allowed, varied from region to region. Additions were later made to the scheme, and within each basic division of the rocks further subdivisions continued to be made in ever-increasing number.

The Wernerian system could work only as long as geologists were mineralogists and concerned only with their own locality. Lyell made the obvious objection that Werner "merely explained a small portion of Germany, and conceived, and persuaded others to believe, that the whole surface of our planet, and all the mountain chains in the world, were made after the model of his own province" (Lyell 1834: I, 82). The more extensive travels of geologists like Lyell showed this was not the case. Also, new techniques, largely due to Smith, destroyed the mineralogical basis of Werner's theory. It ruled out that rocks distinguished by markedly different fossil content could present "the same extenal characters, and often the same internal composition" (Lyell 1834: I, 132).[15]

Such objections destroyed the geological aspect of the traditional view. Other 18th-century discoveries undermined the view of Ray and Jefferson and prepared the way for the catastrophism of Cuvier.

Catastrophism

Although it was plausible to maintain, with Ray, that apparently extinct shellfish could well exist somewhere in the unexplored oceans, the view became less plausible when applied to the remains of massive quadrupeds. Odd "giant" bones could be found in most parts of the world. When

Cortez arrived in Mexico in 1519, one of the first things he came across in Tlascala was a "leg-bone . . . thick and the height of an ordinary sized man, and that was a leg-bone from hip to knee" (Diaz: 181). Or, again, Otto von Guericke (1602-1686), one of the first to demonstrate the power of the atmosphere with his "Magdeburg hemispheres," reported the discovery of giant tusks and bones in the limestone quarries of Quedlinburg in the Harz mountains. He dismissed them as unicorn remains. A more common response, particularly if tusks were found, was to accept them as the remains of the elephants Hannibal had taken over the Alps in the third century B.C. If they were found in areas outside Hannibal's route, as with elephant bones excavated in London in 1715, they could be explained away as the remains of an invading Roman army that had been equipped with "heavy" cavalry.

More difficult for the traditionalists to explain was the discovery, in 1739, of a collection of large bones near present-day Louisville, Kentucky. We owe this find to de Longueuil, an officer in the French army, who was traveling to Louisiana to pacify the Chickasaws. Nearby, at Big Bone Lick, enormous quantities of bones, tusks, and teeth of elephants and other large quadrupeds lay scatterd about. Eventually some of these bones found their way to scholars like Benjamin Franklin (1706-1790) in the east and to main centers of learning in Europe.

The finds were, in Franklin's words, "extremely curious on a number of accounts." First was their very presence on a continent in which no related living forms had been found and which the Roman legions had not reached. The size of these fossils was also puzzling; estimates put them at six times larger than the modern elephant! The main problem lay, as with much else in paleontology, with the teeth found with the bones and tusks. The very distinctive teeth of a modern elephant consist of a number of transverse plates held together by a bony tissue called cementum, a maximum of ten such plates found in the African species and about twenty for the Indian variety. The teeth found at Big Bone Lick, in contrast, had raised pairs of smooth bumps. For this reason Cuvier, on the basis of their supposed resemblance to nipples, named the owner of such teeth *Mastadon,* literally "breast-tooth."

Today, we would readily accept the idea that what de Longueuil had found were the remains of a vanished species, distinct from anything known to exist, so willing are we to accept the idea of extinction. But in the 18th century, alternative explanations sprang more naturally. The first serious response came from Jean Daubenton, a colleague of the great Buffon, who flatly denied that the teeth and tusks belonged to the same animal. He argued that the teeth belonged to a hippopotamus. As the northern latitudes cooled, he conjectured, elephants and hippopotamuses that had once roamed over Kentucky moved on to more congenial climes. This explanation was compatible with Buffon's view of the history of the earth. Consequently, the remains were evidence not of extinction, but of migration.

In contrast, Jefferson believed the bones and teeth belonged to the same creature. Since it was too bulky to be the modern elephant, he posited instead the existence of a comparable American form he misleadingly called "mammoth."[16] Elephants were creatures of the tropics; to Jefferson the animals of America appeared more like the great quadrupeds reported to have been found in the wastes of Siberia. It was wrong, Jefferson declared, to suppose they were extinct. America was too large and unexplored to permit such a rash conclusion. In support of this view he searched the traditions of American Indians for references to such creatures and, as President, instructed Lewis and Clark in 1804 to look for live mammoths beyond the Mississippi. The best they could do was to provide him with yet further bones from Big Bone Lick. The bones reached him in 1808, during his second term, and for some weeks they were carefully examined in the East Room of the White House.

There was more than paleontology involved in this. It was a curious belief of several European savants that the New World was inherently inferior to the Old World. It was supposed to have emerged from the Flood later than other parts of the globe, and its consequent development had been inhibited by an inferior climate contaminated by humid and noxious vapors. Buffon, for example, was convinced that the fauna of America contained fewer species, all of which were smaller and less developed than their European counterparts. Minds less acute than Buffon's further exacerbated the issue. Abbé Raynal (1713-96), for one, expressed astonishment in 1770 "that America has not produced one good poet, one able mathematician, one man of genius in a single art or a single science" (Jefferson 1955: 64). As if, Jefferson politely replied in his *Notes on the State of Virginia,* Raynal's own France had been so rich with mathematicians and men of genius before their contact with Rome. Nature was not, he insisted, a "Cis- or Trans-Atlantic partisan."[17]

Jefferson published tables on the comparative weights and sizes of European and American beasts showing a complicated relationship. While living in Paris in the 1780s, he went to the trouble of importing the skin, antlers, and bones of an American moose to justify his claim that a European deer could walk under its belly. He was eager to find an American "mammoth" to outrank surviving Old World forms.

Cuvier finally demonstrated the facts of extinction in 1799. He examined all known elephant remains, New and Old World alike, and worked out their affinities in a clear manner. Whereas Linnaeus had recognized only one species of elephant, *Elephans maximus,* which he placed in an ill-assorted class of Bruta (together with sloths, anteaters, and pangolins), Cuvier recognized four. Two of these, the Indian and African elephant, were still extant; mammoths and mastodons he declared extinct. Cuvier placed them, with tapirs, rhinoceroses, and hippopotamuses, in a class he named Pachyderms. Of the mammoth and mastodon, he asserted that it was absurd to suppose that such creatures could continue to exist unknown.

Cuvier's work became acceptable because he had an ability to place it in a wider context: "Let me add to these two examples . . . whose originals are not known the crocodile . . . of Maestricht . . . [which] differs from the crocodile of the Nile and from that of the Ganges . . . , the fossil rhinoceri of Siberia, which I shall prove not to resemble either those of Africa or those of the Indies . . . , the petrified horns of a species of the deer genus which are not those of the elk, nor the reindeer, nor of any known species" (Greene: 108). It might be reasonable to imagine one or maybe two large quadrupeds hiding, unknown to man, in an obscure corner of the globe. But as the number of such species increased and as none of them had been found alive, they must be presumed to be extinct. It was possible to lose one species, but not a dozen.

One final problem needed to be resolved before the facts of extinction could be confidently accepted. Could not such fossils simply be varieties of existing species? Cuvier dismissed the argument on the grounds that traces of modification should be detectable, but "since the bowels of the earth have not preserved monuments of this strange genealogy, we have a right to conclude, that the ancient and now extinct species were as permanent in their form and characters as those which exist at present" (Cuvier 1817: 115).

Given the facts of extinction, the question of cause inevitably arose. "Why so many remains of unknown animals?" Cuvier asked in 1799, and promptly answered: "They have belonged to creatures of a world anterior to ours, to creatures destroyed by some revolution of our globe; beings whose place those which exist today have filled, perhaps to be themselves destroyed and replaced by others some day." What caused the revolutions?

The answer came from the work done by Cuvier and Brogniart on the geology of the Paris basin. Eleven strata were distinguished, ranging from the underlying chalk with its remains of reptiles, tortoises, and lizards, to the surface alluvial deposits with their remains of elephants, oxen, and other mammals. In between, dramatic changes were displayed in the constitution of the strata. A rock formation rich in freshwater shellfish would be succeeded by one containing marine specimens only to be replaced by a second strata of freshwater shellfish. It is difficult to look at the strata of the Paris Tertiary period without concluding that there had been a series of inundations from both the sea and a freshwater source. Given also that many of the fossils found in the strata were of extinct forms, it also seemed to follow that the flooding had been no gradual incursion but a deluge of sufficient force to wipe out whole species and genera. These views were expressed in *Discours sur les revolutions de la surface du globe* (1811), an introduction to Cuvier's most significant work, *Recherches sur les ossamens fossiles*, known in Britain through R. Jameson's influential edition, *Essay on the Theory of the Earth* (1817).[18]

Whatever use Cuvier's revolutions or catastrophes served for others, his account makes clear that his theories came from reading the rocks and not the Bible; his views were geological, not theological. Cuvier characterized them as universal, for "every part of the globe bears the impress of these terrible events" (Cuvier 1817: 17). Flooding was not the only cause of extinction, for if some species could find themselves stranded by the sudden elevation of the ocean floor, others could be destroyed by a sudden lowering of the temperature. Catastrophes could be sudden, in proof of which he quoted the example of the carcasses of large quadrupeds found in Siberia with much of their skin, hair, and flesh intact. "If they had not been frozen as soon as killed they must quickly have decomposed by putrefaction." But, Cuvier added – according to some reports – "the flesh was still in such high preservation that it was eaten by dogs" (Cuvier 1817: 16).

Concerning his violent, rapid, and frequent catastrophes, Cuvier felt it necessary to prove a further point, one that was both central and contentious. While it was possible to explain past political events "by an acquaintance with the patterns and intriques of the present day," the same could not be done for "physical history." For here, Cuvier argued, "the thread of operations is . . . broken, the march of nature is changed, and none of the agents that she now employs were sufficient for the production of her ancient works" (Cuvier 1817: 24). Lyell accepted much of Cuvier's work, including a good deal of his catastrophism. On this last point, however, Lyell was uncompromising; he refused to accept any position, however plausible it might be on other grounds, that might commit him to it. The issue is a deep and recurrent one in the history of science.

It is a constant temptation to the scientist to explain the phenomena presented to him in terms of forces or events beyond his observational range. While some scientists are easily seduced, others present a quite rigorous chastity. There is consequently a constant tension in science between those who feel compelled to introduce new explanatory forces and events and the purists who wish to maintain the original integrity of the discipline. Much of the impact of the scientific revolution was produced in this manner by the firm rejection of a variety of explanations that had been acceptable to previous generations of scientists. In this way, substantial forms, final causes, sympathies, and powers were all declared "occult properties" or "arcane principles" and banished from the domain of science. For some in the 17th and 18th centuries, the opposition was carried further and directed against such ideas as indivisible atoms and the force of gravity on the grounds that they were equally occult and fictitious.[19] In this tradition, too, is the uniformitarianism of Lyell, which is apparent uncompromisingly in the full title of his classic work, *Principles of Geology, Being an Attempt to Explain the Former Changes of the Earth's Surface, By Reference to Causes Now in Operation.*

Uniformitarianism

Against the catastrophism of Cuvier, Lyell insisted that "all former changes of the organic and inorganic creation are referrible [sic] to one interrupted succession of physical events, governed by the laws now in operation" (Lyell 1830: 144). Lyell's insistence that only causes now in operation were geologically acceptable was soon labelled by William Whewell (1794-1866), Master of Trinity College, Cambridge and an influential figure in Victorian science, in his review of *Principles* as the doctrine of uniformitarianism. In broad outline, the dispute between Lyell and Cuvier was far from new and had been heard in fields other than geology. Nor did the issue find its ultimate resolution with Lyell. Although his seemed to be the victory and uniformitarianism became the accepted orthodoxy for more than a century, scientists in recent years have begun to question some of the assumptions behind Lyellian geology. Singularities are heard of once more in cosmology while cataclysms are being spoken of again in paleontology.

In detail, Lyell's uniformitarianism involved three distinct claims. He first insisted that only the *actual* forces we see in operation today be permitted to account for geological change. Such forces as the erosion of rocks by rivers and weathering by wind were all that geology needed. As such forces exercised their power only slowly Lyell was therefore committed to a *gradualist* view of geology. This did not of course mean that he denied any role to more violent and sudden events like floods, earthquakes and volcanoes. All he demanded was that such events should not be allowed to violate his actualist assumptions and be endowed with greater power and frequency than they now possess. Within these limits he was willing to accept any number of continental inundations even though none had so far been observed. He even indicated how one could happen: the Great Lakes of America are 600 ft. above sea level but 1,200 ft. deep in parts. All that would be needed would be the "opening of fissures as have accompanied earthquakes since the commencement of the present century" and with the "breach of the barriers" the result would be the laying waste of "a considerable part of the American continent" (Lyell 1834: I, 128-30). In this way, Lyell claimed, unwitnessed events were not ruled out by his uniformitarianism, nor were events with an impact of continental proportions.

Much of *Principles* is in fact concerned with arguments of this kind. Wherever geologists had difficulty understanding some feature of the earth's topography in terms of current forces, Lyell felt an obligation to fill the gap. Consequently, Book II of *Principles* is devoted exclusively to the role of "aqueous" and "igneous" causes in shaping the surface of the earth. In an unending, almost cyclic, sequence, Lyell described a world in which "the aqueous agents are incessantly labouring to reduce the inequalities of the earth's surface to a level, while the igneous, on the other hand are equally active in restoring the uneveness of the external crust" (Lyell 1830: I, 167). To support this view, Lyell described the transportation of silt in the valleys of the Po and Mississippi, the volcanic outpourings from the Icelandic volcano, Skapter Jokul, in 1738, the elevation of the Chilean coast by three feet for a hundred miles as a result of an earthquake, and a host of other significant processes culled from his own investigations and his impressive knowledge of the literature.

But for such a scheme to be workable, one more element was needed – time. Consider once more the coast of Chile. If one shock could raise the coast an average of three feet, then two thousand shocks of equal force could produce a mountain chain one hundred miles long and six thousand feet high. Even if we assume one such shock a century, it would still take 200,000 years to produce one relatively small mountain chain. One compelling factor in favor of catastrophism was the assumption that the earth was not old enough to have allowed for periods of this magnitude. Just as it would be difficult to account for the construction of the Great Pyramid in a day, so, too, Lyell commented, it would be hard to see how the coastline of Chile could have emerged in less than a century or two. Given such short periods, it is inevitable that exceptional forces must be assumed.

Yet to earlier generations of geologists, periods as long as 200,000 years were out of the question. The earth's age was measured by 17th-century scholars according to Biblical chronology which could be made to yield no more than about 6,000 years. To many, like the notorious James Usher (1581-1656), archbishop of Armagh, even such a figure was too high. In its place he proposed in *Annales veteri testamenti* (1654) the year 4004 as the date of creation. This date was inserted into the margins of the Authorized Version in 1701 where it remained until the 19th century. Usher has unfairly gained a reputation of exceptional naivety, yet his views were shared by dozens of contemporary scholars. In 1738, A. des Vignolles in *Chronologie de l'histoire sainte* examined as many as 200 computations of the earth's age. Although there was much disagreement, he found only a slight variation, the range falling between 3483 B.C. and 6984 B.C. Some estimates were extraordinarily precise. John Lightfoot (1602-75), a vice-chancellor of Cambridge University, claimed "man was created by the Trinity on the twenty-third of October, 4004 B.C., at nine o'clock in the morning" (White 1960: I,256) while the German astronomer Hevelius (1611-87) insisted that the earth was created at six in the evening on 20 October 3963 B.C.

Ignoring the bogus precision, the broad agreement on a figure of no more than 6,000 years for the age of the earth made some form of catastrophism inevitable. It was only necessary to examine the still visible ruts left throughout Europe by Roman chariots. If 2,000 years' wind and rain had failed to remove such superficial features then how, the catastrophists asked, could similar forces, operating over 6,000 years at most, have gouged out the mighty Alpine valleys and innumerable Scandinavian lakes and fjords? To break the hold of catastrophism it was essential first to extend considerably the earth's age. Doubts against Usher's time scale began, in fact, to emerge during the 18th century. Their source was twofold.

One arose from considerations of the thermal properties of the earth. Newton had noted in *Principia* (Book III, Prop. xli) (Newton 1966: 522) that "a globe of red-hot iron equal to our earth, that is, about 40,000,000 feet in diameter, would scarcely cool in an equal number of days, or in about 50,000 years." Against such an uncongenial conclusion, Newton speculated on "latent causes" that might have hastened the cooling process, and he recommended that the process be "investigated by experiments." Performed by Buffon a century later the results, published in *Les époques de la nature* (1780), confirmed Newton's initial calculations and offered a figure of more than 70,000 years for the earth's age.

The other main source came from geologists' realization that certain formations could not have been formed in the short time available. For example, Jean Guettard (1715-86), traveling through the Auvergne district of France, observed in 1751 a profusion of hexagonal basalt rocks typical of those he had seen around Vesuvius. Could the rocks of the Auvergne also have been produced by volcanic activity? If so, where, Guettard asked, were the volcanoes? Could the conical peaks he saw in the area be extinct volcanoes? Why, then, wondered geologists, were the ancient histories of southern France silent on such a dramatic subject? One possibility was that the existence of active volcanoes predated the 6,000 or so years supposed to have elapsed since the creation of the earth.

The jump from such queries to the audacious conclusion that the earth's age might well be *indefinite* seems to have been made first by James Hutton (1726-1797) in his *Theory of the Earth* (1788). Lyell was much influenced by Hutton, considered to be the founder of uniformitarianism, and particularly by his claim, "In the economy of the world I can find no trace of a beginning, nor prospect of an end."

Given ample time, there was no reason why Lyell's aqueous and igneous causes could not produce or obliterate a mass of any magnitude. This did not mean that all geological processes must necessarily require millions of years for their completion. Some could take place almost before the observer's eyes and, indeed, one of the most striking examples of *Principles* almost did. On his visit to Puzzuoli in southern Italy in 1828, Lyell studied the remains of the Temple of Jupiter Serapis and noted: "This celebrated monument of antiquity affords, in itself alone, unequivocal evidence that the relative level of land and sea has changed twice at Puzzuoli since the Christian era; and each movement, both of elevation and subsidence, has exceeded twenty feet" (Lyell 1834: II, 262). The remains of the temple consisted of little more than three pillars, 42 feet high, standing by the sea. Their first ten feet were quite smooth as was their last twelve; the middle area, however, was extensively and deeply pierced by a distinctive bivalve, which indicated "a long continued abode." During one period, then, the pillars must have been submerged by some 20 feet, which was possible only if the coast had sunk such a distance. The lower twelve feet of the pillar showed no signs of attack from the bivalve because it had been "covered up and protected by a strata of tuff and the rubbish of the building." When sometime later the coastline was elevated to its initial position, the pillars would show their present distinctive and informative appearance. He later linked the elevation of the temple to an earthquake in the region in 1538. Lyell was so struck by the temple that he used his own elegant sketches of the site to serve as frontispiece to *Principles.*

There was, in addition to Lyell's actualism and gradualism, a third element in his uniformitarianism that was a quite unnecessary intrusion, and it was to cause him great difficulty and place him in an almost isolated position among scientific circles. If similar causes always operated in exactly the same manner, then it seemed obvious to Lyell that they must produce similar effects. Consequently, the earth must throughout its history have appeared always the same. Suppose, for example, we had two pictures of the earth painted at wildly different periods. Then, according to Lyell, there should be no way in which we could distinguish between them. This view, familiar to us through its use in modern cosmology,[20] sees the earth or the universe as occupying a steady state. Changes in such systems cannot be progressive or linear; where allowed, changes are seen as either cyclic or compensatory. Thus Lyell believed that while some forces were at work wearing away the surface of the earth, others were elevating the land to maintain its basic shape.

Three areas in particular suggested to most others that the earth's history showed signs of progressive development. Much of *Principles* and of Lyell's life were devoted to these areas and led him to defend some rather strange positions. Specifically, these troublesome areas involved evidence produced by rocks for the progressive development of life forms, the general cooling of the earth's climate, and the facts of extinction. They will be dealt with separately.

Progressivism

Those with even a minimal acquaintance with geology are aware that rocks record the steady succession of complicated life forms. The Age of Fishes was followed by the Age of Reptiles, which in turn made way for the Age of Mammals; by now, this is as widely accepted as the view that the earth is round. Therefore, it may be something of a shock to learn that in this area Lyell was a "non-progressivist" until converted late in his long life.

The progressivism apparent in rocks was already clear when Lyell wrote *Principles.* It had been described by Sir Humphry Davy in *Consolations in Travel* (1830) and quoted by Lyell. Davy had pointed out that in the oldest and deepest strata "forms even of vegetable life are rare." Above them we find "shells and vegetable remains," and so on through "the bones of fish and viviparous reptiles...the remains of birds...quadrupeds...and extinct species..., and it is only in the loose and slightly consolidated strata of gravel and sands...that the remains of animals, such as now people the globe, are found" (Lyell 1834: I, 215-16) – a description as acceptable today as it was 150 years ago.

Lyell did not deny the facts produced by Davy and by geological research. Instead, he replied (as have all since – scientists and cranks alike – who disagree with the paleontological record) that it was misleading and incomplete. It is one of the ironies of 19th-century science that it was Lyell and Darwin who refused to accept the record of the rocks and argued that they presented only partial and fragmentary evidence. On the other hand, catastrophists and theologians accepted nature's evidence without qualification.[21]

However, it was not enough simply to assert that the paleontological record was misleading. Lyell also had to explain just why the evidence should be so fragmentary. Fossils were missing from primary rocks not because none had ever been there, he declared, but because they had been unrecognizably transformed by the metamorphic process they had undergone. But what of later rocks? Why did they display such a profusion of fish and crustacea and a total absence of birds and mammals? Lyell replied to these questions by replying that most fossils are found in marine strata because they are best preserved there. What would happen to a bird or quadruped washed into the sea? To begin with, such cases are rare. Still rarer is the "contingency of such a floating body not being devoured by sharks or other predaceous fish." If it should survive such threats, "is it not contrary to all calculations of chances that we should hit upon the exact spot–that rare point in the bed of an ancient ocean, where the precious relic was entombed?" (Lyell 1834: I, 220-23).

Lyell argued repeatedly that the absence of fossil forms from a strata did not prove that the species did not exist during the period associated with the strata. He never tired of restating this point. Thus, in 1851, in his presidential address to the Geological Society, he referred to the results of some dredgings off the coast of Britain. Although a number of marine invertebrates and fish had been found, no cetacea had turned up. Once more Lyell hammered home his point: "If reliance could be placed on negative evidence we might deduce from such facts that no cetacea existed in the sea" (Wilson 1970: xxxvi).

Lyell also had a second argument: the presence of supposedly later forms of earlier strata. "The occurrence of one individual of the higher classes of mammalia . . . in the ancient strata is as fatal to the theory of successive development as if several hundreds had been discovered" (Lyell 1834: 224). Lyell devoured the literature for any mention of mammal remains found in the Age of Reptiles or of reptile remains found in the Age of Fishes. Inevitably, like anyone who searches long enough, he found what he was looking for. Each new edition of *Principles* contained new supportive evidence. His most impressive example was the 1814 discovery of the jaw of a small mammal in the Stonefield slate of Oxfordshire, a Secondary rock formation that progressivists had not expected to contain mammalian remains. None other than Cuvier identified the jaw as belonging to an opposum which was named *Didelphis.* Lyell could scarcely contain his excitement and triumphantly wrote his father in late 1827: "So much for . . . the theories of a gradual progress to perfection! There was everything but Man even so far back as the Oolite." The Oolite refers to certain rocks of the upper Jurassic in Britain and Europe or, in other words, Secondary or Mesozoic rocks (Wilson 1972: 182). For 150 years intensive research has failed to place the existence of mammals beyond the upper Triassic, a later part of the Secondary period.

Lyell was more impressed by *Didelphis* than were the progressivists. They were not committed to any particular time for the emergence of mammals, only to the view that they emerged late, sometime after the Paleozoic. It was not a matter of fundamental importance to them whether this event occurred in the early or late Mesozoic.

With many of his other examples, Lyell was often unlucky, accepting as established proof what turned out to be ill-founded identifications. The ninth edition of *Principles* (1853) incorporated a report of an intermediate between a frog and a lizard, discovered in Primary rocks, the Devonian epoch. The rocks were later shown to have been erroneously identified; they were, in fact, Tertiary. Other examples cited by Lyell suffered similar indignities. Nevertheless, for forty years, despite mounting evidence to the contrary and motivated only by his theoretical and narrow uniformitarianism, Lyell persisted in rejecting progressivism. He may be thus compared to Einstein who for the last thirty years of his life rejected quantum theory, despite overwhelming acceptance by the entire community of physicists. To stand so alone requires a colossal degree of self-confidence.

But unlike Einstein, Lyell did concede finally. His renunciation was made public in the tenth edition of *Principles* in which he wrote: "We have been firmly led by palaeontological researches to the conclusion that the invertebrate animals flourished before the vertebrate, and that in the latter class fish, reptiles, birds and mammalia made their appearance in chronological order analogous to that in which they would be arranged zoologically according to an advancing scale of perfection in their organisation" (Lyell 1867/68: I, Chap. 9). When it is realized that the tenth edition as the first edition of *Principles* to appear after 1859, the year of publication of Darwin's *Origins of Species,* the source of Lyell's conversion seems obvious.

Climate

Despite geological evidence showing that Britain enjoyed a much warmer climate during much of the Paleozoic and Mesozoic eras, Lyell refused to accept that the earth started hot and simply cooled down; to his way of thinking, this was merely another form of progressivism.

Believing in cyclic rather than progressive temperature changes, he proposed a connection between climate and the distribution of land and sea. The effect of sea moderates temperature extremes; the effect of land depends on the latitude. In high latitudes, it tends to trap ice and consequently lower temperatures; in lower latitudes, it radiates heat and thus warms the prevailing winds. As the pattern of land and sea is constantly modified by the action of the "aqueous" and "igneous" forces, temperatures are subject to a corresponding modification. By these arguments, Lyell provided himself with a sufficient number of variables to allow almost any type of climate to be produced anywhere, at any time, within a uniformitarian framework. As Lyell himself put it in a letter to Mantell in early 1830, he had at his disposal "a receipt for growing tree-ferns at the pole. . .pines at the equator" (Wilson 1972: 286). He also believed that climatic changes accounted for the extinction of species.

Extinction and Creation

Lyell was convinced of the truth of the following three proposition: First, there is a "strict dependence of each species of animal and plant on certain physical conditions;" second, these conditions are in "a state of continual fluctuation;" and third, transmutation of species is untenable and their variability is strictly limited (Lyell 1834: III, Chaps. 2 and 3).

From these assumptions, it followed that if species could not change to meet the changing physical conditions, then they must "in the course of ages, become extinct one after the other" (Lyell 1834: III, 91). But as a uniformitarian and an upholder of the steady state, Lyell could not allow a progressive disappearance of species without providing for their replacement.[22] Denying the possibility of transmutation, the only remaining alternative was that of creation. But if species are being created to compensate for those becoming extinct, why is it that these remarkable events are never noticed? Why, for example, has no new quadruped ever been reported?

Once more Lyell found himself forced to develop an argument to explain why something predicted by uniformitarianism never seemed to be observed. He began by arguing that the development of species was so rare that it was not surprising it had never been seen. He then offered some rather arbitrary calculations as evidence. Assume that, excluding microscopic life, there were one to two million species in existence and that each year one species became extinct and one was created. Assume also that this figure covers all species, aquatic as well as terrestial, and applies to the whole world. Adjust the estimate to European mammals and, Lyell concluded, "One mammifer might disappear in 40,000 years" (Lyell 1834: III, 100). Having convinced himself of the rarity of extinction and creation, Lyell made no attempt to answer the question of how and why new species appeared. Instead he merely asserted his conviction that "geological monuments alone are capable of leading to the discovery of ulterior truths," and moved on to other matters.

Nor had his thoughts much changed when he worked on the definitive tenth edition of *Principles.* Under pressure from Darwin to endorse the theory of evolution, Lyell did go so far as to add a new chapter on "Theories on the Nature of Species and Darwin on Natural Selection" in which he gave a sympathetic account of Darwin's views and concluded that Darwin "had explained how new races and species might be formed by Natural

Selection but also by showing that, if we assume this principle, much light is thrown on many very distinct and . . . unconnected classes of phenomena" (Lyell 1867/68: Chap. 35). And there the matter rested.[23] The term "evolution" does not occur in the index and "transmutation" is included only to be dismissed in its Lamarckian form. This was hardly the response Darwin was looking for.

As a matter of fact, Lyell's commitment to Darwinism was never total. Lyell's stumbling block was *Homo sapiens*. From very early in his geological career Lyell observed that acceptance of the idea of transmutation of species would inevitably lead to the displacement of man from his privileged position in nature. Evolution was not something that could be used at the discretion or whim of the naturalist; it was unlikely to provide an ancestry for man just as readily as it would for the horse. To avoid this dilemma, Lyell – too honest and perceptive to be able simply to deny Darwinism – took the weaker option. He dallied in offering Darwin's support and then yielded only the barest minimum.

The Tertiary

Principles also contains the first detailed description of the epochs of the Tertiary. Even in its use today, in its organization, structure, and terminology, the Tertiary is Lyell's creation. The Primary and Secondary periods were worked out over a number of decades by several scholars; Lyell is the only scientist to have charted by himself the geology of a whole period.

The great Primary, Secondary, and Tertiary divisions were somewhat vaguely established by a number of naturalists in the mid-18th century. One was Giovanni Arduino (1714-95) who, in 1760, classified rocks into three main types: Primary, or ore-bearing crystalline rocks; Secondary, or stratified rocks with marine fossils; and Tertiary, or rocks that covered alluvial and volcanic deposits. At this time, there was little agreement in either terminology or content among the various proposed tripartitie divisions. This was a time for the forging of broad distinctions; the detailed protocols could be worked out later. Omalius d'Halloy (1783-1875) made a crucial advance in 1822, when he identified the boundary between the Secondary and Tertiary with the chalk beds surrounding the Paris basin. For him, these constituted the Cretaceous epoch and marked the top of the Secondary rocks. Above the Cretaceous, d'Halloy recognized the Mastozoic, or mammal-rich levels, and the Pyroid rocks of igneous origin. Neither term survived Lyell's work.

Lyell's own account of the Tertiary was based on an impressive familiarity with the published literature and first-hand acquaintance with the main sites of France and Italy (obtained during his 1828-29 journey). His starting point was Cuvier and Brogniart's work on the Paris basin. As the basin rested directly on the chalk of the Cretaceous, it clearly defined the base of the Tertiary. The first sign that this was of more than local interest came in 1814 from a paper by Thomas Webster (1773-1844) entitled "On the Freshwater Formation of the Isle of Wight" that appeared in the *Transactions* of the Geological Society. Webster was an architect by training who served as Clerk of Works at the Royal Institution of London and had actually built its well-known lecture theatre. His main interest lay in geology, and after holding various posts with the Geological Society, he became in 1841 the first professor of geology at University College, London. In an earlier work on the strata of the Isle of Wight, he noted that a distinctive sequence of fossil-bearing rocks rested on the chalk, and though they differed mineralogically from those in Paris, the fossil content of both sites was virtually identical. This was the first identification of Tertiary rocks in Britain.

In the same year that Webster's paper appeared (1814), Giovanni Brocchi (1772-1826), professor of botany at Brescia, identified in *Conchiologia fossile subapennina* another Tertiary formation in a range of low hills flanking the Apennines and known a the subapennine beds. While the fossil content differed markedly from those of Paris and the Isle of Wight, it nonetheless resembled that described by James Parkinson, in 1822, in a study of the Suffolk crag. And when Jules Desnoyers (1800-80) published an account of the Tertiary of the Loire Valley in 1825, the link between the Paris basin and the Subapennines became clearer. The fossil content differed from both, but the uppermost part of the Paris strata could be traced continuously to the bottom of the Loire strata. In addition, fossils similar to those found in the Loire were also found in the beds *underlying* the rocks of the Subapennines. It was a neat and impressive fit, and its showed very clearly the power of the fossil record to identify and relate strata from widely separated sites.

Fossils, though not actually discovered by Lyell, were examined by him in his travels through southern Italy. In Sicily, he collected a large number of Tertiary shells which he took to Naples for identification There he was informed that virtually all of his collection were shells readily found in the Mediterranean. Since the fossils of the Paris, Loire, and Subapennine rocks were largely those of extinct shellfish, Lyell realized that he had identified a fourth and more recent Tertiary strata.

Lyell's Tertiary researches reached a climax during his visit in 1829 to Paris on returning to London from Sicily. He was introduced to Gerard Deshayes (1797-1875), a Paris physician, who possessed a superb collection of Tertiary shells. Centuries before, Galileo had declared that the great book of nature was written in the language of mathematics; for the Tertiary, Lyell was to find that nature delivered its message in the idiom of fossil mollusks. Deshayes proved considerably helpful in deciphering this new language. Trained as a physician, he indulged his geological passion by writing, teaching, and collecting, and was rewarded in 1869 with a chair at the Musée d'histoire naturelle.

Deshayes's particularly value to Lyell lay in his ability to identify shells from any site, and to indicate whether or not they belonged to extinct species. For this was to be the crucial factor in dating and ordering the strata: the greater the proportion of extinct shells found in any strata, the older it was. The great advantage of this principle was that it allowed two sites with no common fossils to be compared. For example, if site A contained only extinct fossil shellfish while site B contained five percent, then clearly site A was older. With Deshayes's aid, Lyell began a careful analysis of the fossil shellfish from the main sites of Europe. Their results are summarized in Table 12:1.

Table 12:1

Site	Number of species examined	Number found extinct	Percentage extinct
Paris	1,238	1,196	96
Isle of Wight			
Loire	1,021	845	82
Subapennines	569	331	82
Sicily and Naples	226	10	4

The impressive results confirmed Lyell's intuition that he was dealing with four quite distinct geological epochs, and he broadly grouped them according to percentage of extinct shellfish. So important was this evidence that Lyell included in the first edition of *Principles* Deshayes's tables of more than 3,000 species of Tertiary shellfish. Runing to eighty pages, it was dropped from later editions on grounds of expense.

Lyell was vulnerable to the charge that his subdivisions of the Tertiary had been taken from Deshayes. As a consequence, he was eager to point out that the idea occurred to him in 1828, long before he had met Deshayes. *Principles* contained this explanatory footnote: "I had conceived this four-fold division in 1828 and found on my return to Paris in February 1829 that. . . M. Deshayes had deduced from a comparision of the fossil shells in his collection the conclusion that three Tertiary periods might be established, the third or most modern of which comprehended the two last of my intended divisions" (Lyell 1834: III, 302). Lyell did not actually reveal the basis of his 1828 classification, but since Deshayes seems never to have objected to Lyell's claim to originality, it would be just as well to believe him.

There remained the important task of naming the four subdivisions. In 1831, Lyell approached Whewell on the issue, proposing to him the terms Asynchronous, Eosynchronous, Meiosynchronous, and Pleiosynchronous. Whewell objected that the termination "-synchronous" was "long, harsh and inappropriate," proposing instead the equally unattractive Aneous, Eoneous, Meioneous, and Pleioneous. In a postscript, Whewell made the further suggestion that the Greek root *kainos* should be used instead of *neous*—both terms mean something like new or recent—and he came up with Acene, Eocene, Miocene, and Pliocene. Lyell went along with this suggestion; but he had the sense to drop Whewell's Acene, fearful of it being mistaken for a mispelling of *acne*, and replaced it with a splitting of the Pliocene into a newer and older part. Details of the Tertiary as it was known to Lyell, together with a modern estimate of its dates, are listed in Table 12:2.

Table 12:2

Epoch	Site	First Described	Percentage Extinct	Dates (Million years ago)
Eocene	Paris	Cuvier 1811	90+	54-38
	Isle of Wight	Webster 1814		
Miocene	Loire	Desnoyers 1825	About 80	26-7
Older Pliocene	Subapennines	Brocchi 1814	40-65	7-2
Newer Pliocene	Sicily	Lyell 1829	less than 10	
	Naples			

A certain amount of readjustment and filling in was soon called for. Lyell himself, dissatisfied with the cumbersome division of the Pliocene, in 1839 changed Newer Pliocene to Pleistocene. In 1854, Heinrich von Beyrich introduced the term Oligocene to fill the gap between the Miocene and Eocene, while the sequence was completed with the later ad-

Table 12:3

Era	Period	Epoch	Millions of Years Ago
Cenozoic (Phillips 1841):	*Quaternary* (Desnoyers 1829)	Recent	
		Pleistocene (Lyell 1839)	2.5
	Tertiary (Arduino 1760)	Pliocene (Lyell 1833)	7
		Miocene (Lyell 1833)	26
		Oligocene (von Beyrich 1854)	38
		Eocene (Lyell 1833)	54
		Paleocene (Schimper 1874)	64
Mesozoic (Phillips 1841):	*Cretaceous* (d'Halloy 1822)		136
	Jurassic (von Buch 1839)		190
	Triassic (von Alberti 1834)		225
Paleozoic (Phillips 1841):	*Permian* (Murchison 1841)		280
	Carboniferous (Kirwan 1799)		345
	Devonian (Murchison and Sedgwick 1837)		395
	Silurian (Murchison 1835)		430
	Ordovician (Lapworth 1879)		500
	Cambrian (Sedgwick 1836)		570

Note: In America, it is customary to split the *Carboniferous* into a Lower and Upper, known as the Pennsylvanian and the Mississippian, terms which were introduced by A. Winchell (1870) and H.S. Williams (1891) respectively.

dition by W. Schimper of the earliest period, the Paleocene (later shown to stretch from about 64-54 million years ago).

All these ages are modern estimates. Lyell's own estimates – and indeed, those of 19th-century geologists in general – differed widely from today's figures and from each other. Before the discovery of radioactivity by Henri Becquerel (1852-1908), no accurate and reliable calculations of the earth's age existed. Most published estimates relied on unscientific measurements of the thickness of the earth's strata and on presumed rates of deposition. An examination of nineteen such estimates made during the period 1860-1909 produced ages ranging from 3 million years to 1.5 billion (Eicher: 15). Lyell's method of calculating the earth's age was based on his assumption that it took about 20 million years for a complete change of species of fossil mollusks. Since he thought there had been twelve such changes, this gave him a working figure of 240 million years as the span of life on earth. A figure twice as high would be nearer the truth, but his was of the right order of magnitude.

Table 12:3 provides a summary of geological epochs, their discoverers and modern age estimates.

Reception

Volume I of *Principles* was reviewed by Whewell in *British Critic* (1831), by A. Sedgwick (1785-1873), professor of geology at Cambridge, in a presidential address to the Geological Society in 1831, and by Lyell's friend, George Scrope (1797-1876), in *Quarterly Review* (1830). Few took issue with Lyell on specifically geological topics; in this respect, it was acknowledged by all as a work of considerable authority. On the other hand, many found it difficult to agree with the assumptions underlying Lyell's geology. The most perceptive criticism came from Whewell.

Critics were not particularly inspired to leap to the defense of catastrophism, but they found the actualism of Lyell difficult to accept. There may or may not have been past catastrophes that shaped earth's topography; they could be neither defended nor rejected. Nevertheless, they were quick to point out that Lyell appeared determined to exclude the very possibility of such events. Could we not, Whewell mused in a review of the second volume of *Principles,* simply be "ignorant of some of the most important of the agents which, since the beginning of time, have been in action"? For our knowledge is limited in time and space, he continued, and would it not be strange if from it we could have derived "all the laws and causes by which the natural history of the globe, viewed on the largest scale, is influenced" (Whewell 1832: 126). Elsewhere, Whewell raised the objection that new forces had in the past been introduced into science, had been unknown to one generation of scientists, introduced by a second, and accepted as orthodox by a third. At some point in the 17th century, all actual forces were rejected as inadequate to account for the movements of the heavens, and universal gravitation was introduced. This, surely, was a violation of Lyell's principles.

Whewell also seized upon a more specific weakness in Lyell's argument, noting that he was noticeably vague in his arguments on the reasons for the extinction and creation of species. Whewell believed, as indeed did many others of the period, that the creation of species was nothing but "a distinct manifestation of creative power, transcending the known laws of nature: and it appears to us, that geology has thus lighted a new lamp along the path of natural theology" (Whewell 1831: 194). It was left to Darwin to provide an answer to this real problem.

Others readily accepted Lyellian geology in its pre-Darwinian form, among them the poet Tennyson. Thomas Jefferson, it will be recalled, had written of an economy of nature so perfect that "no one instance can be produced for her having permitted any one race of her animals to become extinct" (Jefferson 1955: 63). The Victorian spirit was bleaker than the Enlightenment one, partly as a result of Lyell's picture of a universe in which impersonal climatic changes led regularly and inevitably to the destruction of a whole species. Those living in the middle of the 19th century were the first to cope with the idea of wholesale carnage. The phrase, "nature, red in tooth and claw," comes from Tennyson's *In Memoriam.* A pupil of Whewell, Tennyson knew Lyell's *Principles* well, and picking up on his themes of the extinction of species and an apparent lack of purpose, he was moved to ask (*In Memoriam,* sections 55-56)

> Are God and Nature then at strife,
> That Nature lends such evil dreams?
> So careful of the type she seems,
> So careless of the single life;
>
> . . . .
>
> "So careful of the type?" but no.
> From scarped cliff and quarried stone
> She cries, "A thousand types are gone:
> I care for nothing, all shall go".

The Significance of "Principles"

The significance of *Principles* in establishing the structure of the Tertiary period and in laying down the basic uniformitarian methodology for geology, despite the exessive lengths to which Lyell took it, has already been discussed. But this by no means exhausts the role it played in the development of 19th-century science. One may cite, for example, Darwin's dedication of the second edition of his *Journal of A Voyage Around the Wordl* (1845) to Lyell (see the quote at the head of this chapter). He had taken the first volume with him on the *Beagle;* it was a gift from his Cambridge tutor, J. C. Henslow (1796-1861), a botanist who, by the way, advised Darwin to "take no notice of the theories." The other two volumes reached Darwin at stops in Montevideo and Valparaiso. In his autobiography, Darwin recorded the impact of the work on him. "I had brought with me the first volume of Lyell's *Principles*...which I studied attentively; and this book was of the highest service to me in many ways. The very first place which I examined, namely St. Jago in the Cape Verde islands, showed me clearly the wonderful superiority of Lyell's manner of treating geology, compared with that of any author" (Darwin 1974: 44).

Darwin's intellectual debt to Lyell, though indirect, was still considerable. Lyell gave him a workable geology in which he could place his evolutionary theories. He also gave hint of the possibility of evolution itself. For Lyell, the "monuments of geology" – the mountains, valleys, rivers, oceans, and plains – all were the result of forces observable *now*. A pattern of argument had been set up to deal with a wide range of phenomena scattered over all known time and space. Darwin took this framework and applied it, not to the strata of the earth, but to its organic species. Just how these emerged, developed, were distributed, and become extinct, had to be explained by the operation of identifiable processes that were discernible in the present. Darwin had no more use than Lyell for catastrophes and cataclysms to explain the distribution of organic forms, either in the present or in the fossil record. Consequently, Darwin seized upon such actualist factors as an animal's tendency to vary, its struggle for food, and its competition to breed, in order to explain the origin of species; factors Lyell should also have appreciated if he had not been so obsessed by his steady-state view of the earth. Lyell could well have written *Origin of Species* some thirty or so years before Darwin. These points are not new; they were probably first made by T. H. Huxley in an obituary of Darwin when he wrote, "It is hardly too much to say that Darwin's greatest work is the outcome of the unflinching application of Biology of the leading idea and method applied in the *Principles*" (Huxley 1893: 168).

Traces of Lyell's insights can be found in numerous other fields throughout the 19th century. To document even a fraction of them requires a thick volume of its own, but to see just how far Lyellian actualism extended, consider the following from Edward Tylor (1832-1917), one of the founders of modern social anthropology and the first to held the anthropology chair at Oxford (1895-1909). Writing in *Primitive Culture* (1871), he began on a clearly Lyellian note: "If any one holds that human thought and action were worked out in primeval times according to laws essentially other than those of the modern world, it is for him to prove by valid evidence this anomalous state of things; otherwise, the doctrine of permanent principle will hold good, as in astronomy and geology" (Tylor: I, 33).

Publishing History

The first indication that Lyell intended to write a book on geology is found in an entry in his notebook dated March 1827, listing possible chapter headings. Shortly afterwards, in a letter to John Murray, the publisher, he wrote of his surprise that "the sale of Conybeare's *Introd. to Geology* has been. . . above 3000 copies and going on well." Ever keen to earn money, Lyell added, "I am vexed that I did not set to work 2 yrs. ago and shall lose no time now" (Wilson: 170). The first pages were, Wilson judges, probably written in December 1827, but as the period between May 1828 and February 1829 was spent in continuous travel through France and Italy, it is unlikely that much progress was made. Murray wrote to Lyell on 3 December 1829, agreeing to publish the work: "I will print 1500 copies of your work in 2 vols. 8vo to sell for 30/- at my own cost and risque, and give you Four Hundred Guineas for permission to do so.... I will tell you candidly that this is taking the whole risque of a first publication myself and securing to you one half the profit. After this edition shall be sold, the Copyright being entirely your own, we can bargain again for subsequent editions" (Wilson 1972: 270). Lyell sent his acceptance to Murray the next day. Before his death in 1875, twelve editions appeared.

First edition, 1830-33: Three volumes in four.

Volume I: With a frontispiece of the Temple of Jupiter Serapis, Volume I was published after much rewriting in July 1830 in two parts. Fifteen hundred copies were printed, each selling for 15 shillings. Lyell received 200 guineas. Book 1 begins with a long historical introduction (Chaps. 2-4) which is followed by an important discussion of the theoretical errors supposedly made by traditional geologists. The remainder of Book 1 is concerned with overthrowing any "assumed discordance of the ancient and existing causes of change" and includes Lyell's discussion of climatic change. It is in the latter part of Book 1 that Lyell's views on uniformitarianism are developed and defended. Book 2 is concerned exclusively with the "aqueous" and "igneous" causes of geological change.

Volume II: With a frontispiece of the Valley of Bove, Etna, Volume II was published in January 1832 at a price of 12 shillings in an edition of 1,500 copies. Lyell expected to receive a final 200 guinea payment, but Murray tried to persuade him that, in response to the lower price of the second volume, he was due only 150 guineas. Lyell would have none of it, and demanded a sum of 160 guineas, "being in the same proportion as the 200 guineas for...the first volume" (Wilson 1972: 341). As Lyell's calculation was correct, Murray (after a certain amount of bluster) conceded and paid him the full amount. Volume II consists entirely of Book 3 and begins with a long and critical account of Lamarck's theories. The remainder deals with the geographical distribution of species, their extinction and creation, and concludes with a number of chapters on fossils.

Volume III: Beginning with a frontispiece of the extinct volcanoes of Olot, Spain, Volume III was published in May 1833 in a larger edition of 2,500 copies, for which Lyell received £370. It consists entirely of Book 4 and is the volume most likely to be recognized by the modern student as a geological work. It is dedicated to Murchison and contained Lyell's detailed account of the geology of the European Tertiary. In a lengthy appendix, Deshayes's table of fossil shells is found.

Second edition, 1832-33: Two volumes.

Volume I was published in May 1832, and the second in January 1833. Lyell attempted to correct "positive errors" only, although for typographical reasons, the edition is some 80 pages longer than the first. He received £210 from Murray for the edition. One curiosity of this edition is that it was completed prior to the publication of Volume III of the first edition. In fact, Volume III can be assigned to either the first or second editions since it completed either set.

Third edition, 1834: Four volumes.

This is the first complete edition. Sold at the cheaper price of 24 shillings, it brought Lyell a further £255. Murray was reluctant to publish it so early as he still had several hundred copies of Volume III of the first edition, and he felt that few could be willing to buy an odd volume once the new, complete, four-volume set was announced. Murray asked Lyell to stand the loss, and Lyell finally agreed to compensate Murray for £150, about half the cost of the unsold books. The third and all subsequent editions lack Deshayes's table. This edition incorporated a fair amount of new material, particularly a discussion of earthquakes and volcanoes which is found in Volume II (pages 275-95).

Fourth edition, 1835: Four volumes.

Essentially a reprint of the third edition, with some new material, Lyell received £255. The price held at 24 shillings. Lyell incorporated some material of Sedgwick on metamorphic rocks. By this time Lyell suspected what the future had in store for him, writing to Sedgwick, "Indeed, I sometimes think I am in danger of becoming a perpetual editor to myself, rather than of starting anything new" (Wilson 1972: 412).

Fifth edition, 1837: Three volumes.

Still priced at 24 shillings and published in an edition of 2,000 copies, this is the last complete edition of the "original" *Principles.*

Sixth edition, 1840: Three volumes.

In 1838, Lyell published his *Elements of Geology* which contained most of the technical material that had originally appeared as Volume III of the first edition and later as Volume IV of the third. This new work, too, was regularly revised and reissued, first in 1841 in a two-volume set. A third edition, published in 1851, was retitled *A Manual of Elementary Geology,* but later editions, appearing in 1852, 1855, and 1865 (the sixth and last), contained the original title. But by 1865, *Elements* had lost its original rationale, having grown into a "large and expensive work." Urged to "bring the book back again to a size more approaching the original, so that it might be within the reach of the ordinary student" (Lyell 1885: 7), Lyell produced the more elementary *The Student's Elements of Geology* (1871). It was revised and reissued in 1874, shortly before Lyell's death. Revised by others, it continued to be printed as late as 1898.

An effect of these new publications meant that Lyell no longer needed a Volume IV of *Principles.* Consequently, it was dropped from the sixth edition, and Lyell transfered a fair amount of material to Book 1, to which he added four new chapters. No further substantial revisions of *Principles* was made until the tenth edition.

(The seventh edition was published in 1847 in a one-volume large format size; the eighth and ninth editions appeared in 1850 and 1853 respectively.)

Tenth edition, 1867/68: Two volumes.

This was the last major revision undertaken by Lyell and thus it contains the fullest expression of his mature views. Lyell added much new material, but the edition is noteworthy for Lyell's belated acceptance of the progressive nature of fossil rocks and grudging acceptance of Darwin's views.

(The eleventh and twelfth editions appeared in 1872 and 1875 in a two-volume format.)

American editions

The first American edition was published in 1837, in Philadelphia, by J. Kay Junior and Brother. Although taken from the four-volume fifth edition, it was published in only two volumes. A three-volume edition of the sixth was published in Boston in 1842 by Hilliard, Gray and Co. The main American publisher thereafter was D. Appleton of New York. Between 1853 and 1868 no fewer than twelve editions of the one-volume ninth edition appeared while between 1872 and 1892 eleven editions of the two-volume eleventh edition were issued.

Modern editions

As far as I am aware, no modern editions exist. The early editions of Lyell are expensive, and the first, second, and third editions are not likely to sell for less than £500. Even the tenth and eleventh editions cost £50 or more. This means that *Principles* is available only in large reference libraries. Later editions of the more accessible *Student's Elements* can still be found in second-hand bookshops. However, this textbook lacks Lyell's discussion of uniformitarianism. Since the three-volume sets of the early editions run to some 1,400 pages, it is not likely that the book will be reissued before 2030 in celebration of its bicentenary.

Translations

No translation into any language seems to have been published.

Further Reading

Principles is expensive and difficult to find. It remains a very readable work if, perhaps, too long for modern tastes. Extracts found in the usual anthologies are too brief to give anything but the barest glimpse of Lyell's style and argument. This is a real deprivation and should induce an enterprising publisher to produce an abridged Lyell in the way that many inexpensive one-volume editions of Gibbon have been published. There is a superb biography of Lyell of Wilson (1972); unfortunately, it treats his life only to 1841; a second biography by Sir Edward Baily (1962), though more complete, is of little value. Happily, geology has been well-treated in recent years by historians, and there are a number of excellent studies available: M. Rudwick, *The Meaning of Fossils* (1972); C.G. Gillispie, *Genesis and Geology* (1959); J.C. Greene, *The Death of Adam* (1959); M. Davies, *Earth in Decay* (1969); Toulmin and Goodfield, *The Discovery of Time* (1965); R. Porter, *The Making of British Geology* (1977). Two earlier works are still of considerable value: F.D. Adams, *The Birth and Development of Geological Sciences* (1954), and A. Geikie, *The Founders of Geology* (1962); both have been issued in paperback by Dover Books. There is also a useful volume of readings: K. Mather and S. Mason, *A Source Book in Geology, 1400-1900* (1967).

Notes

1. He joined the Navy in 1756 as an able-bodied seaman, and in 1766 was appointed purser on H.M.S. *Romney.* His fortune was probably made after 1770 when he gained, while still in the navy, the lucrative contract to supply Royal Naval ships stationed off the American coast. He retired in 1779.

2. She was the daughter of Leonard Horner (1785-1864), one of the founders of the University of London and its first warden.

3. Lyell began to study for the bar on leaving Oxford in 1819; he qualified in 1822, having been delayed by an eye infection, a recurrent complaint, which made reading difficult but proved no obstacle to wandering around geological sites. It seemed to return again in 1822, and it was not until 1825 that he actually practiced in court as a member of the western circuit. By 1827 he had had enough of the law.

4. This was increased to £500 a year. His wife brought a dowry of £4,000, worth a further £120 a year. After 1830, Lyell must have earned at least £100 or £200 a year from his writing, making his yearly income around £800. It may be mentioned that in the 1820s, an income of £500 could easily support a cook at £16 a year and four or five maids at 14 guineas each. Horses were more expensive, costing £65 a year to keep, and it was generally reckoned that a man needed £1,000 a year before he could afford a four-wheeled carriage. In 1832, the Lyells moved to a house at 16 Hart Street, off Bloomsbury Square, London, for which they paid £89 a year in rent. By no means as rich as Darwin, Lyell nonetheless lived comfortably and, without children, could devote himself to science with no real financial problems.

5. The reception given to Lyell's lectures at King's may be seen as yet one more sign of the extraordinary appeal of geology at this time. *Principles* had just been published, and Lyell found himself lecturing to classes of almost 300. He was pleased with his reception, quoting one judgement that they were "the best ever given of Geology for 200 years in Europe;" his own view that he was "doing an immense deal of good to science" was no less favorable (Wilson 1972: 353-56).

6. If volumes instead of books are counted, including Lyell's two travel books – *Travels in America* (1845, 2 vols.) and *A Second Visit to the United States of America* (1849, 2 vols.) – then from 1830 to 1875, Lyell published forty-six volumes, about one a year. According to the Royal Society catalogue, he also published seventy-six scientific papers.

7. The Lyell family, descended from his brother's side, still live at Kinnordy. The Third Baron of Kinnordy, also named Charles Lyell, was born in 1939; in 1979, he was appointed a government whip in the first Thatcher administration.

8. "The accurate surveys to which I have devoted my life... have enabled me to prove that there is a great deal of regularity in the position and thickness of these strata... and that each stratum is also possessed of properties peculiar to itself, has the same... extraneous or organised fossils throughout its course" (Mather and Mason: 202).

9. It was introduced in 1834, independently, by de Blainville and von Waldheim. The term must have caught on quickly because, in 1847, a Palaeontological Society was founded in Britain to publish details of undescribed fossils.

10. James Parkinson (1755-1824) was the same figure who, in *Essay of Shaking Palsy* (1817), first described the disease since associated with his name.

11. The story of the discovery of the archosaurs and subsequent work on them has been told many times in recent years. The most interesting and fullest account is found in A.H. Desmond, *The Hot-Blooded Dinosaurs* (1975).

12. The figures are by no means accurate. *Iguanadon* was given a small horn on its nose that was, in fact, a spike placed on its thumb. Its stance as a quadruped is equally wrong as the creature was a biped supported by enormous hind legs.

13. This office was held by the formidable "Boss" Tweed (1823-78), soon to be exposed for corruption. Tweed's reasons for not accepting Hawkin's work are not known. Mobsters later broke into Hawkin's studio and destroyed all models, casts, and molds.

14. Werner's reputation has declined. Here is Geikie on *Kurze Klassification:* "It would be difficult to cite from any other modern scientific treatise... a larger number of dogmatic assertions of which almost every one is contradicted by the most elementary fact of observation" (Geikie: 216).

15. Equally, rocks could be found at one site in a single series while elsewhere the same rocks had been assigned to widely different periods. Thus Lyell and Murchison visited Cantal in 1828 and there in a "tertiary lacustrine formation" they found rocks corresponding in Britain to the new red sandstone of the Devonian. The point was not lost on them (Wilson 1972: 202).

16. News of the mammoth reached 18th-century naturalists through fossil remains collected by agents of Peter the Great while they were surveying the wealth of Siberia. A virtually complete carcass of a woolly mammoth was discovered there by Michael Adams in 1806. To the modern systematist, the mastodon and mammoth belong to different families, distinguished primarily by their teeth: the Mastodontidae, all extinct, and the Elephantidae, including not only the mammoths but the two surviving species of the Indian and African elephant.

17. The argument is an all purpose one and had been used in a different context by the philosopher David Hume (1711-1776) who believed in the natural inferiority of "negroes" on the grounds that they, too, had failed to develop the arts and sciences (Hume 1898: III, 252). What the argument ignores is the extent to which such intellectual advances as the differential calculus, logarithms, sines, the alphabet, a numerical place system, and axiomatic geometry, were made by a small group of people, often an individual, on perhaps one occasion. Diophantine equations were imported by the Sorbonne, Oxford, and Cambridge, just as surely as they were taken to the New World and the shores of Africa.

18. Despite the work of Lyell, catastrophism is still with us and it is still defended with basically Cuvierian arguments. Thus, 65 million years ago, the end of the Mesozoic saw the disappearance of many reptiles, including all fifteen families of dinosaur, marine invertebrates, and certain primitive plants. Paleontologists now seem tempted to explain such a major discontinuity in terms of catastrophes

grander than any ever contemplated by Cuvier: a supernova explosion, or encounters with asteroids or comets. The reason for seeking such unique causes was the discovery in 1979 of unusually large traces of the element iridium in late Cretaceous rocks – precisely the period of the mass extinction – which, we are told, can only have an extra-terrestrial origin. Up-to-date details of this neo-catastrophism can be found in Dale Russell: "The Mass Extinction of the Late Mesozoic," *Scientific American,* January 1982.

19. Thus, Leibniz, on the suggestion that bodies could attract each other by means of an "invisible and intangible" force of gravity, responded "'Tis a chimerical thing, a scholastic occult quality" (H. G. Alexander, *The Leibniz-Clarke Correspondence,* 1956: 94). While later in the century, the physicist and mathematician D'Alembert (1717-83) refused to consider "motive causes": all he would allow into his mechanics was the fact that a body crossed a cetain space and took a certain time to do so.

20. Modern cosmologists accept the cosmological principle which states that the universe looks the same in all directions and that there is no preferred position or direction. A more extreme form of this principle emerged in the 1940s from such astronomers as Fred Hoyle. Known as the "perfect cosmological principle," it states that the universe looks the same, not just from wherever it is viewed, but from *whenever* it is viewed. Thus, the universe today appears to a modern astronomer as it would have appeared billions of years ago. This is the "steady state hypothesis." One difficulty presented by the principle is how to deal with the accepted fact that the universe is expanding with a consequence that in its early history it must have been significantly denser than it is now. Not, Hoyle argued in 1948, if matter is being continuously created at just the right rate to compensate for the expansion. Lyell would find it necessary to make equally grand assumptions to defend his own steady-state views.

21. Nor has the matter changed much. Creationists still insist that there is no evidence for evolution in the fossil record; the title of one of their best selling texts is *Evolution: The Fossils Say No!* by Duane T. Gish. Evolutionists are still concerned with what Stephen Jay Gould has called the "trade secret of paleontology," namely, the "extreme rarity of transitional forms." The latest explanation, first proposed by Gould and Niles Eldredge in 1972 and known as the "punctuated equilibrium model," attempts to make sense of this rarity. Speciation is not a gradual, continuous affair, as Darwin assumed, but occurs in rapid spurts punctuated by vast periods of tranquility. Not surprisingly, periods of stability leave no transitional forms in the fossil record, and in this way "the fossil record is a faithful rendering of what evolutionary theory predicts, not a pitiful vestige of a once bountiful tale" (Gould: 184).

22. In the long run, Lyell was even prepared to consider the return of dinosaurs. If the appropriate climatic conditions returned, Lyell predicted in a lyrical passage, "the huge iguanadon might reappear in the woods, and the ichthyosaur in the sea, while the pterodactyl might flit again through the umbrageous groves of tree-fern" (Gould: 134).

23. Unwilling to commit himself unreservedly to Darwin, Lyell was nonetheless quick to point out the important role he had played in preparing Darwin's way. He told one German scholar, "I had certainly prepared the way in this country, in six editions of my work, . . . for the reception of Darwin's gradual and insensible evolution of species" (Lyell 1881: II, 436).

CHAPTER 13

On the Origin of Species
(1859)

by Charles Darwin

We know that this animal, the tallest of mammals, dwells in the interior of Africa, in places where the soil, almost always arid and without herbage, obliges it to browse on trees and to strain itself continuously to reach them. This habit sustained for long, has had the result in all members of its race that the forelegs have grown longer than the hind legs and that its neck has become so stretched, that the giraffe, without standing on its hind legs, lifts its head to a height of six meters.

–Lamarck, (1809)

Neither did the giraffe acquire its long neck by desiring to reach the foliage of the more lofty shrubs, and constantly stretching its neck for the purpose, but because any varieties which occurred...with a longer neck than usual *at once secured a fresh range of pasture over the same ground as their shorter-necked companions, and on the first scarcity of food were thereby enabled to outlive them.*

–A. R. Wallace (1858)

A deer with a neck that was longer by half
Than the rest of his family's (try not to laugh),
By stretching and stretching, became a Giraffe,
Which nobody can deny.

–Popular song (1861)

So under nature with the nascent giraffe the individuals which were the highest browsers, and were able during dearths to reach even an inch or two above the others, will often have been preserved.

–Charles Darwin (1872)

The Neck of the Giraffe or Where Darwin Went Wrong

–Book by Francis Hitching (1982)

In 1982 the world commemorated the hundredth anniversary of Charles Darwin's death. At about the same time the Arkansas creationist trial was taking place, and the media discovered a new evolutionary theory, the doctrine of punctuated equilibrium proposed by Harvard paleontologist

Stephen Jay Gould and Niles Eldredge of the American Museum of Natural History.[1] For some time it was impossible to pick up a magazine without finding a discussion of an aspect of Darwin's life or thought. A major television series on the voyage of the *Beagle* was screened, a new popular biography by Peter Brent was published, Irving Wallace wrote a novel, *The Origin,* about Darwin, Dr. Gould could be seen on the cover of *Newsweek*, and leading publishers began to issue introductions to the life and thought of Darwin. In a more serious vein, scholars continued to explore the minutiae of Darwin's notebooks and to prepare the long awaited edition of his collected correspondence. So far, no *Journal of Darwin Studies* has yet been announced, but an 834-page concordance has been published listing "every word (excluding prepositions)" to appear in the first edition of *Origin of Species.* Yet despite such intense media and scholarly focus, it is still not widely appreciated that evolution and Darwinism are not synonymous. While Darwin and Darwin alone demonstrated the reality of the evolutionary process, many biologists became active supporters of evolution without adhering to the theory advanced in *Origin of Species.* At times it even appeared as if Darwin himself would be one such biologist.

Life and Background

Darwin was born in 1809, the son of a prosperous country doctor. He was supposed to follow in his father's profession, but the brutalities of life as an Edinburgh medical student so appalled him that he fled to Cambridge where he read theology. His interest in natural history sufficiently impressed the Cambridge botanist, John Henslow (1796-1861), to recommend him for the post of naturalist on the *Beagle*, a Royal Navy ship under the command of Capt. Robert Fitzroy (1805-65), which was setting out in 1831 to complete a survey of the South American coast.[2] For five years, Darwin sailed in and around South America, traveled extensively on the mainland, visited the Galapagos Islands, and eventually circumnavigated the globe.

It was a remarkable voyage. When the young twenty-two-year old Darwin joined the *Beagle*, he showed no signs of being anything other than a competent naturalist. Nor, when he returned in 1836, was there any indication that he had by then formulated his theory of evolution. He took with him Volume 1 of Lyell's *Principles*, a work which initially turned him into a Lyellian. It was during the voyage, however, that Darwin made numerous observations that on later reflection poorly fit either with the catastrophism of Cuvier or the steady-state views of Lyell. The *Journal* (1845) of the voyage is in part a record of these carefully expressed puzzles.[3] Thus, during his five-week visit in 1835 to the Galapagos, a group of volcanic islands some 600 miles west of South America, he noted that each of the islands possessed its own highly specialized species of birds, tortoises and much else. "One is astonished at the amount of creative force. . . displayed on these small, barren, rocky islands," Darwin declared. He had never dreamed "that islands, about fifty or sixty miles apart, and most of them in sight of each other, formed of precisely the same rocks, placed under a quite similar climate, rising to a nearly equal height, would have been differently tenanted" (Darwin 1845: Chap. xviii). This is just one of numerous problems Darwin recorded in his *Journal* and which continued to exercise his mind on his return to England.

Much of the next ten years were spent writing his account of the voyage.[4] Yet while working on his geological reports, he began to perceive in outline a solution to his problems which, in turn, led him to perceive a way to tackle the outstanding biological puzzle of the century: the origin of species. To this end Darwin began "on true Baconian principles" to collect any relevant facts on "the variation of animals and plants under domestication and nature" (Darwin: 1974:71). His results were recorded in a number of notebooks, the first of which was opened in July 1837 ("Notebook B"). He continued to record his observations during the period 1837-39 in a further three notebooks, "C," "D" and "E," known collectively as the "transmutation notebooks." Having independent means, Darwin was under no financial pressure to undertake any other work or to meet any particular deadline, and from 1837 onwards, he spent the rest of his life working out the details of his theory, publishing it, defending it, while at the same time pursuing his passion for natural history.

He married his cousin Emma Wedgewood in 1839. They settled in 1842 at Down in Kent which remained their home for the rest of their lives. Indifferent to the pull of London or other attractions, Darwin devoted himself to his writing, his research, his health, and his growing family. Of ten children, three died in early childhood.[5] Information about them can be found in Darwin's *Autobiography* (1974) while many more details about his children and household may be found in Gwen Raverat's *Period Piece* (1960).[6]

On his death in 1881, Darwin was buried in Westminster Abbey where his tomb now lies overshadowed by the hideous monument to Newton.[7] Although knighthoods had been awarded to his colleagues Lyell and Hooker, no such honor was offered to Darwin.[8]

The Writing of "Origin of Species"

In 1859 Darwin published *On the Origin of Species by Means of Natural Selection or the Preservation of Favoured Races in the Struggle for Life*. To scholars of the early 18th-century, the problem of the origin of species was unappreciated. Species were held to be permanent. Given that the universe was widely considered to be no more than 6,000 years old, it was not unreasonable to suppose that the flora and fauna visible to Buffon were the same as those described by Aristotle. Moreover, little was known of the paleontological record, and few naturalists had any first-hand experience of nature beyond the wildlife of their own country.

The explosive growth in geological knowledge in the early years of the 19th-century destroyed this picture. Evidence of massive and regular extinctions together with the apparent "re-appearance" of similar life forms began to be presented by geologists. It was not only great dinosaurs that had disappeared from the face of the earth but also large numbers of crustacea, fish, and insects. Unlike dinosaurs, these latter groups had been replaced by related but distinct species. The process had happened not once but a dozen or so times in the Paris region alone. It was not, perhaps, too difficult to find explanations for the extinction of species, but the regular and widespread appearance of new species of all kinds was more difficult to understand.

For those who found the idea of regular creations *ex nihilo* too bland, the only possible alternative was to find a continuity between the extinct and the new species and to declare that the latter had evolved from the former. Many took this course, including Darwin's grandfather Erasmus, in his *Zoonomia* (1794-96).[9] Yet, on reflection, the endorsement of evolution rather than creation proved to be no real advance. To say that species *A* evolved into species *B* is no more or less explanatory than to say that species *B* was created after species *A* had become extinct. Talk of evolution would only become fruitful when a plausible mechanism was proposed by which one species could evolve into another. Of the many who had sought to develop an evolutionary theory, all, before Darwin, failed to provide a plausible and comprehensive description of precisely how evolution took place.[10] The process Darwin proposed was natural selection and it is therefore on his development of a theory of selection, not a theory of evolution, that his genius and originality rest.

Darwin's first insight into the mechanism of evolution was later recorded in his *Autobiography*: "I happened to read for amusement 'Malthus on Population', and being well prepared to appreciate the struggle for existence which everywhere goes on from long continued observation of the habits of animals and plants, it at once struck me that under these circumstances favourable variations would tend to be preserved and unfavourable one destroyed. The result of this would be the formation of new species" (Darwin 1974:71).[11] Malthus had shown that while natural populations had the power to increase in a geometrical ratio, food resources could increase only in an arithmetical ratio. From this insight Darwin immediately saw there must be in nature competition for scarce resources, selection of those best fitted to capture those resources, with the important consequence that this process could lead to the formation of *new* species. Darwin began to work out details of his new theory. By June 1842, he had prepared a "brief abstract of my theory in pencil of 35 pages;" this was enlarged "during the summer of 1844 into one of 230 pages." Known as the 1842 *Sketch* and the 1844 *Essay* respectively they remained unpublished until 1909. It would seem natural at this point for Darwin to turn an *Essay* of 230 pages into a publishable treatise presenting what he knew to be an original and revolutionary theory.

Yet at this point, Darwin did something very strange: he devoted eight years of his professional life to the taxonomy of barnacles. Results of his labors were published in four volumes between 1851 and 1854. It remains the standard work on the subject, yet even Darwin later doubted whether it had been worth the time (Darwin 1974:70). The consequence of such prolonged labors was to delay publication of the work on species which appeared fifteen years after the 1844 *Essay*. The immediate question to be asked is: Why did Darwin, who possessed the solution to the greatest biological problem of modern times, delay its publication for so long?[12]

We can quickly dismiss the idea that Darwin had no wish to publish his work. As early as 1844 he had instructed his wife in the event of death to spend £400 on the publication of the *Essay*. In 1854 he returned to the same theme, noting on the back of a letter, "Hooker by far the best man to edit my species volume."

Nor more satisfactory is the proposal of Darwin's friend and defender, T.H. Huxley (1825-95), that Darwin "never did a wiser thing than when he devoted himself to the years of patient toil which the Cirripede-book cost him." Darwin, Huxley argued, lacked certain essential biological skills. Seeing the "necessity of giving himself such training. . . he did not shirk the labour of obtaining it." Hence, the need to study "Anatomy and Development and their relation to Taxonomy," a need satisfied by eight years work on barnacles *(Cirrepedia)* (Darwin 1958:166-7). The difficulty about Huxley's answer is that it makes Darwin sound to much like a self-conscious athlete, completing some self-imposed complicated training schedule, too much like Huxley himself, in fact. Nor is it a reason Darwin ever offered.

It has also been proposed that it was the poor state of Darwin's health that was responsible for his fifteen-year delay.[13] His son Francis noted that during the period 1842-54 "my father suffered perhaps more from ill-health than at any other period of his life." (Darwin 1958:167). Whatever the origin of his complaint, it is certain that for much of the time Darwin was either too ill or too busy seeking treatment to work. A recent study by Ralph Colp (1977) reveals some of the extraordinary lengths to which Darwin went in pursuit of a cure. In 1849, he spent four months in Malvern undergoing treatment at the hydropathic establishment of Dr. J.M. Gully. "I feel sure the water-cure is no quackery," he wrote to his cousin, R. Fox, also informing him later, "I consider the sickness to be absolutely cured" (Darwin 1958:170). To maintain his improved health, he had a small bathhouse built at Down where he continued the treatment.

Despite his optimism, Darwin was soon seeking further help. First, from a clairvoyant, a woman with a supposed power to see "the insides of people" and diagnose their ills. Prudently, Darwin asked her to tell him the number of a banknote in an envelope. The clairvoyant scorned the challenge yet went on to diagnose his complaint as centered in the stomach and lungs.

In 1851, Darwin found something new: electric chains.[14] Made from alternate brass and zinc wires, they were moistened with vinegar to produce a mild current and wrapped around the patient's waist and neck (Colp:46). Other therapies were tried over the years,[15] yet however much time his illness consumed, Darwin managed to get through more work than most healthy scientists. If his health was no bar to the completion of a thousand-page monograph on barnacles, then how could it have prevented the preparation of the *Essay* for publication?

Another reason offered for Darwin's procrastination is that he feared the ridicule and controversy his work would inevitably inspire. He had seen the reception accorded *Vestiges of Creation* (1844) by Robert Chambers and knew well his friends' contempt for Lamarck.[16] Consequently, the theory goes, Darwin kept his manuscript to himself. This makes no more sense than earlier proposals, for Darwin quite happily shared his ideas with associates and friends like Lyell, Hooker, Henslow and Huxley. He maintained an extensive scientific correspondence and as early as 1844 wrote to Hooker: "At last gleams of light have come, and I am almost convinced. . . that species are not (it is like confessing a murder) immutable." The same candor and horror at his own temerity can be found in many of his letters. In 1856, for example, he was still going through his confessional motions, this time with the American botanist Asa Gray: "As an honest man I must tell you that I have come to the heterodox conclusion that there are no such things as independently created species—that species are only strongly defined varieties. I know this will make you despise me." None of his friends ever advised him to keep his views secret; if anything, many, like Lyell, pressed him to publish.

But Darwin felt no pressure to publish, certainly not financial pressure. Unlike Huxley and A.R. Wallace, who had to struggle thoughout their careers to maintain a position in science, Darwin was a wealthy man and free to work at what he willed, at whatever pace he chose. In early life he received from his father an annual income of £400 per year and on his father's death in 1848 he inherited £40,000, a vast sum in those days. From his books Darwin earned a further £10,000. His income was thus so substantial that he was able to invest sizeable sums (as much as £5,000 in 1873) each year. On his death Darwin's estate was valued at £282,000. With such an income and without any ambition for academic preferment, Darwin could indulge his passions whether they be field studies, the taxonomy of barnacles or simply adding to and reorganizing his manuscript.

One danger remained. Others might be more willing to publish their results and so anticipate Darwin's claim to have solved the problem of the origin of species. It is the date of publication which counts in science, not discovery. If someone had rushed into print with a fully worked out theory of evolution, Darwin's work of many years would have been devalued overnight. Darwin claimed that he cared little whether men attributed originality to him. His friends nevertheless worried for him and pressed him to publish a preliminary account of his views. Lyell, for one, nagged Darwin on the issue, but after an initial if reluctant agreement, Darwin rapidly found the idea too distasteful to continue. In a letter to Hooker in 1856 he expressed his fear that all he could quickly prepare would be "a *very thin* and little volume, giving a sketch of my views and difficulties. . . *a resumé*, without exact references, of an unpublished work." But this, he insisted, "is dreadfully unphilosophical," the kind of work "I should sneer at anyone else doing." Later in the year he reported his failure to Lyell: "I have found it quite impossible to publish any preliminary essay or sketch;" he had rather returned to his "big book" at which he was "working very steadily" (Darwin 1958:191-92). Left to himself, it was clear that Darwin's "big book" would require several more years before it was completed.[17]

Yet, Darwin was not left to himself. On the morning of 18 June 1858 he received a letter from the then little-known bug hunter A. R. Wallace (1823-1913), at that moment at work in New Guinea.[18] "I never saw a more striking co-incidence," Darwin wrote to Lyell on the same day, "if Wallace had my MS. sketch written out in 1842, he could not have made a better short abstract! Even his terms now stand as heads of my chapters. . . . So all my originality, whatever it may amount to, will be smashed."

Darwin had probably first heard of Wallace in 1855 when the latter published a paper, "On the Law which Has Regulated the Introduction of New Species," in *Annals and Magazine of Natural History.* Lyell recommended Darwin to read it. It seemed clear to Lyell that Wallace was thinking very much on Darwinian lines and it could well have been this paper that led Lyell to urge his friend to publish some account of his theory. Sometime later, in 1857, Darwin and Wallace corresponded. "We have thought much alike," Darwin told Wallace, and pointedly added that he had been working on the problem for twenty years. In a further letter Darwin repeated his claim that the book was "half written" (Darwin 1958: 193-4).

A less naive man than Wallace would have recognized the territorial markers Darwin was carefully laying around his intellectual property and either moved to another problem or, at least, directed his letters elsewhere. Nevertheless, Wallace persisted and when he, too, finally solved the problem of the origin of species, he immediately set out his theory in a brief paper, "On the Tendency of Varieties to Depart Indefinitely from the Original Type," and trustingly sent it to Darwin.[19]

This put Darwin in a difficult position. The paper did say, although briefly, what Darwin wanted to say about the origin of species. If he published his own work after receiving Wallace's paper he would be accused, at best, of stealing Wallace's priority or, at worst, of plagiarizing his ideas. The other alternative was to publish Wallace's paper, in which case he would lose his own claim to scientific fame. Twenty years' work would have been wasted.[20]

In recent years, the scientifically famous have often received a bad press. They have been presented as more interested in advancing their careers and winning medals than pursuing truth, in resorting to almost any lie and deception to establish their claim to disputed discoveries. Few scientists of any stature have escaped such debunkings; Darwin is no exception. Until recently the behavior of Darwin and his friends over the Wallace affair was widely recognized as exemplary. After all, Wallace was an unknown figure, 8,000 miles away, unlikely to return to Europe for another three years, and remarkably trusting. If the powerful figures of British science had decided to cook the books, leaving Wallace out altogether, they could easily have done so. Newton would have done no less.

What actually happened was that Darwin, Lyell and Hooker arranged for credit to be shared through a joint publication presented to the Linnean Society on the evening of 1 July 1858. Lyell and Hooker presented Wallace's 1858 paper, together with extracts from Darwin's 1844 *Essay* and a letter he had written to Asa Gray on 5 September 1857. A covering letter from Lyell and Hooker gave a brief account of the circumstances surrounding the publication. Later commentators were impressed. Darwin behaved with "magnanimity," Irvine declared (1955: 63), or "admirably," according to Moorehead (1971: 261). Chancellor (1973: 154), even more impressed, saw it as "one of the noblest episodes in the history of science."

Opposed to this unusual consensus, there recently appeared a dissenting and harsher judgment from Arnold Brackman.[21] A one-time journalist, Brackman's professional skepticism was aroused by the sight of the establishment patting itself on the back and telling the world how well they behaved. A detailed examination of the documents persuaded Brackman otherwise. Darwin's behavior he viewed as "devious" and "unethical" (Brackman: 65); the affair itself was "the greatest conspiracy in the annals of science" (Brackman: 22).[22]

Brackman found his suspicions confirmed when he sought to examine the relevant correspondence. Much of it was missing. No letter from Wallace to Darwin before the Linnean Society publication had survived, nor had the letters of Lyell and Hooker to Darwin from the crucial period 18 June 1858 to 1 July 1858 when the three friends decided how to respond to Wallace's paper. What has survived, however, is Darwin's side of the correspondence. Inevitably, as the story has been told for the last century, it rests solely on the evidence of Darwin's letters, first published in 1887. A key point, as it has been repeatedly told, involves Darwin's intention, on receiving Wallace's paper, to publish it and to abandon his own claim to priority; and only under pressure from Lyell and Hooker did Darwin reluctantly agree to the appropriateness of a joint publication. This is the version the eighty-nine-year-old Hooker remembered in 1906: "After writing to Sir Charles Lyell, Mr. Darwin informed me of Mr. Wallace's letter and its enclosure. . . , explicitly announcing his resolve to abandon all claim to priority for his own sketch. I could not but protest against such a course. . . . I further suggested the simultaneous publication of the two and offered – should he agree to such a compromise – to write Mr. Wallace fully informing him of the *motives* of the course adopted" (Brackman: 67). There was also more contemporary evidence. Lyell and Hooker had noted at the Linnean Society meeting that Darwin had been "strongly inclined" to withhold his contributions in favor of Wallace.

Various comments from Darwin have also added to the legend. In his *Autobiography* he reported that he was "at first very unwilling" to allow his work to be published with Wallace's (Darwin 1974: 72). In a letter to Hooker, dated 13 July 1858, he went further and claimed that he thought his contribution would be "only an appendix to Wallace's paper" (Darwin 1958: 202). The impression is thus conveyed that almost against his will, and perhaps even without his knowledge, Darwin's friends dragged from him some early writings on natural selection and arranged for their joint publication with Wallace's 1858 paper. A careful reading of the surviving correspondence suggests that Darwin was not the villain painted by Brackman; nor was he quite so selfless as the traditional story maintains.

There is no doubt that Darwin felt deeply the possible loss of his priority. "I always thought it very possible that I might be forestalled, but I fancied that I had a grand enough soul not to care; but I found myself mistaken and punished," he wrote to Hooker on 13 July 1858 (Darwin 1958: 201). Four letters have survived from Darwin, three to Lyell (18, 25 and 26 June), and one to Hooker (29 June), from the time he received Wallace's paper and the joint publication in the Linnean Society. No replies have survived (Darwin 1958: 196-99).

In all of them he stressed how much he cared about the loss of his priority. It was a dreadful week for Darwin. His fourteen-year-old daughter Etty was seriously ill with diphtheria; his eighteen-month-old son Charles died on June 28.[23] "I am quite prostrated," he wrote to Hooker on the following day. Despite his "prostration," he was still able to sort out and send to Hooker the papers he needed for the Linnean Society. "It is miserable in me to care at all about priority," he confessed to Hooker in the same letter. Yet care he undoubtedly did.

His first reaction on receiving Wallace's letter had been to console himself that although his priority had been lost, "my book will not be deteriorated . . . as all the labour consists in the application of the theory." By 25 June this was clearly not sufficient. A week of thought had convinced him that "there is nothing in Wallace's sketch which is not written out much fuller in my sketch, copied out in 1844." Consequently, "he would be glad now to publish a sketch of my general views, in about a dozen pages or so." But no sooner having said this than once more he raised the question, would not such behavior be "base and paltry"? The third letter to Lyell was briefer but much the same. Again, there is a clear statement that it would be improper to publish. Yet, once made, it is followed by the lament: "it seems hard on me that I should be thus compelled to lose my priority of many years' standing."

Reading the correspondence 125 years later, incomplete though it may be, the impression is conveyed that Darwin did all he could to extract from Lyell an admission that it would be proper to publish. Thus, although it is certainly true that Darwin expressed an unwillingness to publish, he also proclaimed an eagerness. And while it may be technically true that the decision to publish came not from Darwin but from Hooker and Lyell, it is also true that the case for publishing was given to them by Darwin.

If any complaint can be held against Darwin it is that he allowed others to gain a distorted impression of his role in the affair.[24] Wallace certainly raised no objection. Indeed, he was somewhat apologetic for having intruded at all. "I felt that they had given me more honour and credit than I deserved, by putting my sudden intuition. . . on the same level with the prolonged labours of Darwin, who had. . . worked continuously. . . in order that he might be able to present the theory. . . with such a body of systematized facts and arguments as would almost compel conviction," Wallace wrote half a century later in his autobiography (Wallace: 193).

The fault lay, and still lies, as Wallace saw, with the system. As long as an absurd convention is followed that awards equal priority to someone who had developed and worked out in detail a complex theory with someone who merely sketched its outlines, such problems and dilemmas will continue to trouble the conscience of scrupulous scientists.[25]

After this fright, it remained for Darwin to write his book. His *Autobiography* records that he began to prepare a volume "on transmutation of species" in September 1858. His diary, however, noted, "July 20 to Aug. 12. . . began Abstract of Species book. . . Sept. 16, Recommenced Abstract" (Darwin 1958: 204). The work was initially described by Darwin as an "Abstract" because the bulk of it was abstracted from his notebooks and early drafts. By 25 January 1859 he wrote to Wallace that he had reached the penultimate chapter. Shortly afterwards the publisher John Murray agreed, without seeing the manuscript, to publish the book. Darwin sent him the first three chapters on 5 April 1859.

Darwin wished to call his work *An Abstract of an Essay on the Origin of Species and Varieties Through Natural Selection,* hoping that the word *Abstract* would cover his failure to provide complete references.[26] Murray had no faith in a title which began *An Abstract of an Essay* and refused to carry it. Nor was he impressed by the text. He thought it too technical for most readers. The opinion of Whitwell Elwin, editor of *Quarterly Review,* was sought. Showing a monumental lack of insight, Elwin advised Darwin to forget his *Abstract* and write a book on pigeons. "Everybody," he told Darwin, "was interested in pigeons. The book would be reviewed in every journal in the kingdom and would soon be on every library table" (Himmelfarb: 208). Darwin politely declined. Another reader reported to Murray that the text was "probably beyond the comprehension of any scientific man then living." Despite such gloomy advice, Murray persisted with his plans to publish and actually raised his initial print order of 500 to 1,250. Even Darwin thought this was excessive. "Rather too large an edition, but I hope he will not lose," he told Lyell on 30 September (Darwin 1958: 216).

On 11 September Darwin corrected his last proof. Publication was set for 24 November 1859, by which date the whole issue had already been taken up by the trade at Murray's annual fair. Darwin had gone to Ilkley, Yorkshire, where he tested yet one more hydropathic establishment.

"Origin of Species"

Origin of Species is a very readable work, perhaps the only one discussed in this book which can be read by most people with pleasure and understanding from beginning to end. The theorems of Euclid and the systems of Linnaeus are acquired tastes; Lyell's *Principles* is too vast a work for most readers; Vesalius is available only to readers of Latin, while the intricacies of Ptolemy, Copernicus and Newton are likely to remain closed forever to all but the specialist. It is significant that of the twelve classics discussed in this book only *Origin of Species* can be readily purchased in inexpensive, standard editions.

The first edition is divided into fourteen chapters. It begins with a brief introduction in which Darwin anxiously made a number of points. It is clear that he wanted no confusion over Wallace's role. Consequently, the first chapter pointedly refers to the year 1837 as the time he first began to consider the problems dealt with in the book and to 1844 as the time he recorded his conclusions. He also could not resist, against Murray, referring to the work as "this Abstract," a remark that gave him a chance to apologize for the absence of references and authorities. In many ways this limitation was a great blessing. An examintaion of Darwin's other major works, *The Variation of Animals and Plants* (1867) and *The Descent of Man* (1871), shows that Darwin could never resist a footnote, making his works bloated and often indigestible.

In terms of structure, excluding introductions and recapitulations, *Origins of Species* divides naturally into three parts. The first five chapters present the theory, the following four deal with major difficulties facing the theory, and the final four indicate ways in which the theory can be deployed to throw light on a range of geological, geograhpical, morphological, embryological and anatomical problems.

The kind of problems Darwin set out to tackle are best illustrated not from *Origin of Species* but from the 1844 *Essay*, written before his answers were fully worked out. Why, to take one example, are there three distinct species of rhinoceroses separately inhabiting the nearby sites of Java, Sumatra, and the Moluccas? Why do they differ markedly from the African rhino? Why do they closely resemble the "ancient wooly rhinoceros of Siberia"? Why do their "short necks . . . contain the same number of vertebrae with the giraffe"? Why are their thick legs "built on the same plan . . . of the wing of the bat"? Why do their embryo's arteries "run and branch as in a fish, to carry the blood to gills which do not exist"? We do not, Darwin concluded, attribute the particular details of a planetary orbit to specific, individual pushes and pulls of the Creator but to the "intervention of the secondary and appointed law of gravity." His implication was clear. We should not ascribe the rhinoceros' structure and distribution to specific acts of the Creator, but rather seek the secondary laws of development through which He operated.

In *Origin of Species* the sought-for laws are approached through a study of variation. The first two chapters deal with variation under domestication and in a state of nature respectively. By "variation" Darwin could mean a number of things. Sometimes it referred to what he called "individual differences" or the "many slight differences . . . such as are known frequently to appear in the offspring from the same parents" (Darwin 1968: 102). But it was also used to refer, as in the first chapter, to the variation seen between, for example, different varieties of pigeons or dogs. Why was there, he asked, so much diversity in pigeons? This was a field Darwin knew well; he had bred them himself, mixed with fanciers, and he had a good command of the literature.

He first dismissed the view that there was no real variety in pigeons, that the carrier, the short-faced tumbler, the pouter, turbit, runt, barb, and trumpeter, among others, all breed true because they are descended from distinct aboriginal stocks. The alternative, that it had been possible to get "seven or eight supposed species of pigeons to breed freely under domestication,' Darwin rejected as implausible.[27] They all derived from a common ancestor, the rock pigeon (*Columba liva*). How, then, did the variation arise? Some was no doubt due to the "direct action of the external conditions of life, and some little to habit." But, Darwin added, "He would be a bold man who would account by such agencies for the differences of a dray and a race horse, a greyhound and a bloodhound, a carrier and a tumbler pigeon." It was rather "man's power of accumulative selection: nature gives successive variations; man adds them up in certain directions useful to him" (chap. 1). The above process, Darwin insisted, was not a hypothetical mechanism but could be massively documented from the records of breeders. He went on to illustrate his thesis with examples ranging from King Charles' spaniel to Leicester sheep, from the pages of Genesis to the opinions of rabbit fanciers.

In chapter two, the shortest in the book, Darwin sought to establish that comparable variation can also be found in nature. The evidence presented is almost entirely botanical.

In the following two chapters, Darwin introduced the basic concepts and terminology with which his name has become associated: "struggle for existence" and "natural selection."[28] He began with the familiar observation that organic species increase their numbers so inordinately that in the absence of constraints they would swamp the earth.[29] Even a slow breeder like the elephant would from a single pair have produced in five hundred years a population of fifteen million living offspring. Apart from such obvious constraints as limited food supplies, Darwin showed that the limiting factors could be specific and complex. Heartsease (*Viola tricolor*), for example, depends for its survival on the number of cats in a given neighborhood: cats eat mice, mice destroy nests of bees, bees are indispensable in the fertilization of heartsease. Hence, where cat populations are large, fewer mice mean more bees and an abundance of heartsease (chap. 3).

So far, Darwin had simply prepared the ground. At the beginning of chapter four, he introduced the principle on which everything else would depend and which gave to *Origin of Species* its depth and originality: "many more individuals are born than can possibly survive. . . . Individuals having any advantage, however slight, over others, would have the best chance of surviving and of procreating their kind. . . . Any variation in the least degree injurious would be rigidly destroyed. This preservation of favourable variations and the rejection of injurious variations, I call Natural Selection." So far Darwin had merely shown how natural selection would preserve the fittest; it remained to show how it could operate to produce new species, how "the lesser differences between varieties become augmented into the greater difference between species" (Darwin 1968: 130-31). Here was the vital point on which biologists were awaiting enlightenment and upon which skeptics needed to be convinced.

Not surprisingly, Darwin found the question difficult.[30] So difficult, in fact, that he could only offer examples taken from the irrelevant process of domestic selection; for natural selection, he was reduced to illustrating the process in a diagram. He began with a large species *A*, ranging over a wide territory. Let *A*, a species of bird perhaps, produce two minor varieties, the long-beaked M_1 and the short-beaked A_1. For a thousand generations the darker forms of the long-beaked M_1 variety could be preferred, while the lighter forms of the short-beaked A_1 variety could be selected. By now very distinct varieties have been produced. Allow the process to continue with the new properties being preserved while, at each stage, a further variation like eye color or length of tail feather is added to the A_1 and M_1 varieties. By the time A_{10} and M_{10} appear after 10,000 generations they could well be new species. The process was presented in the form of the branching tree diagram seen so commonly in works of biology since the days of Darwin. For his model to work, Darwin needed a number of favorable conditions: the initial variability, conditions which would select one property rather than another, the tendency for such properties to be preserved in breeding, and time. Not all the conditions would prove easy to establish; some of them would preoccupy Darwin for his lifetime.

Darwin presented the final part of his theory, the laws of variation, in chapter thirteen. To account for the origin of variation he suggested a number of factors of which the effects of external conditions, the use and disuse of parts and correlated variation were particularly emphasized.[31] It was, however, a topic which would trouble Darwin throughout the various editions of *Origins of Species* and about which he candidly admitted, "our ignorance . . . is profound."

Having presented his theory in the first five chapters, Darwin proceeded in the following four to raise a number of formidable objections. He was acutely aware of the difficulties facing the theory, difficulties that became more troublesome with the passage of time and some of which remain today. One of the most important was the problem of the so-called missing links. Why is it, Darwin asked, "if species had descended from other species by insensibly fine gradations, do we not everywhere see innumerable missing forms?" Secondly, he raised the problem of organs of extreme perfection. He put the objection with some force: "To suppose that the eye, with all its inimitable contrivances for adjusting the focus to different distances, for admitting different amounts of light, and for the correction of spherical and chromatic aberration, could have been formed by natural selection, seems, I freely confess, absurd in the highest possible degree" (Darwin 1968: 217). A third major problem, and one much discussed in recent years by sociobiologists, concerned the problem of instinct (found in chap. 7). How, for example, asked Darwin, could natural selection produce worker ants with behavior and instincts so different from the normal male and female ants? Why not suppose, mused Darwin, "that all its characters had been slowly acquired through natural selection . . . by an individual having been born with some slight profitable modification of structure, this being inherited by its offspring, which again varied and were again selected, and so onwards." Such a solution was impossible for worker ants are sterile and consequently, "could never have transmitted successively acquired modifications of structure or instinct to its progeny" (Darwin 1968: 257-58).

There was never any question of Darwin presenting simple and conclusive answers to what were very real problems.[32] All he could hope to do most of the time was to minimize their force by showing they were theoretical difficulties and not undisputed refutations. He was further able to offer pointers to where possible solutions might lie, to indicate the form a future answer might take, rather than delivering the details of the solution immediately.

More direct was the final third of the book, chapters ten through thirteen, in which Darwin demonstrated that the principles of evolution could be applied to, and were not contradicted by, the main areas of biology. Chapter ten thus demonstrated that there was no conflict between evolution and the geological record, and chapter eleven made comparable claims for the geographical distribution of species. The concluding thirteenth chapter argued that the facts of comparative anatomy, taxonomy, and embryology were consistent with his theory.

Darwin was not content to show the mere consistency of this theory with biology as he knew it. In addition, he attempted on page after page to demonstrate its indispensability by presenting a vast range of facts which on any other hypothesis appeared arbitrary and inexplicable. Why, he asked, "should the species which are supposed to have been created in the Galapagos Archipelago, and nowhere else, bear so plain a stamp of affinity to those created in America?" (Darwin 1968: 386). Or, why, "on the theory of creation", are there nowhere any frogs, toads and newts on oceanic islands? Why do remote islands have bats but no terrestial mammals? (Darwin 1968: 382-83).[33] Moving to another topic, comparative morphology, Darwin found many more facts that were "inexplicable . . . on the ordinary view of creation." The skull of a mammal was made from "numerous and . . . extraordinary shaped pieces of bone" so that they could yield in the act of parturition. He asked why the same construction could be seen in the skull of birds who had merely to crack an egg (Darwin 1968: 417). Finally, turning to embryology, Darwin asked why are the embryos "of distinct animals within the same class. . . often strikingly similar," so similar in fact that on one occasion "Agassiz. . . having forgotten to ticket the embryo of some vertebrate animal. . . cannot now tell whether it be that of a mammal, bird or reptile" (Darwin 1968: 419). Multiply the above examples a dozen-fold and the flavor of Darwin's last four chapters of *Origin of Species* can be savored.

Although some of the problems raised were undoubtedly troublesome, Darwin could afford to treat them lightly. If he found it difficult to account for the development of the vertebrate eye and a host of other problems, then so, too, did all other biologists no matter what their theoretical position. There were other objections which could not be so complacently tolerated since they were directed specifically against the coherence of *Origin of Species.* With these objections in mind, Darwin prepared five more editions of the work during the following thirteen years.

The Defense of "Origin of Species"

Darwin's intellectual development can be traced in successive editions of *Origin of Species.* In each Darwin attempted to cope with criticisms made against him and to deal with what he saw as major difficulties in his theory. So extensive were the revisions he introduced that only about half of the 1859 first edition can be found unchanged in the 1872 final sixth edition.

Not much need be said of the second edition. Darwin began revising the first edition the day it appeared and referred to the second edition as "little more than a reprint." The third edition of 1861, however, was some thirty-five pages longer. It is notable because it contains Darwin's first explicit views on what earlier biologists termed sport or monsters and now known as mutations. In the first edition he had merely commented: "By a monstrosity. . . is meant some considerable deviation of structure in one part, either injurious to or not useful to the species, and not generally propagated" (Darwin 1968: 101). In the third edition Darwin spelled out the reasons why such forms could be safely ignored. It was extremely "improbable that any part should have been suddenly produced perfect." But even if it had been, it would still be a singleton and, assuming it to be fertile, would be forced to cross with normal forms and "their abnormal character would almost inevitably be lost."[34] He concluded by noting that he had never seen, even after diligent search, "cases of monstrosities resembling normal structures in nearly allied forms." Pigs, he pointed out, were occasionally born with a sort of proboscis. If any wild species of pigs were ever to be found with a trunk, then it would be reasonable to suppose they derived from such an original sport. But, Darwin insisted, no such case was known.

The point is important in that it removed from Darwin's evolutionary scheme any kind of monster or sport. But if natural selection did not operate on these forms to develop new species, what was its raw material? Throughout all editions of *Origin of Species* Darwin strove to show how natural selection could work on what he termed in 1859 "infinitesimally small inherited modifications." By the fourth edition they had become "small modifications," but despite minor changes in terminology Darwin remained committed to the essential point that evolution operated through the selection of individual differences and not by the control of sports. As early as 1860 Darwin was warned by the sympathetic Huxley that his commitment to individual differences was too strong: "Nature does make jumps now and then, and a recognition of the fact is of no small importance in disposing of many minor objections to the doctrine of transmutation" (Huxley: 77). There appeared in 1866 a publication that transformed the minor objections of Huxley into major difficulties that Darwin would spend the rest of his life pondering.

The challenge came from Fleeming Jenkin (1833-85), professor of engineering at Glasgow University, in a review of *Origin of Species* in *North British Review* (46 [1867]: 277-318). Darwin's theory required, Jenkin began, "that there shall be no limit to the possible differences between descendants and their progenitors" (Hull: 305). But, he declared, "there appears to be a limit to their variation in any one direction. This limit is shown by the fact that new points are at first rapidly gained, but afterwards more slowly, while finally no further perceptible change can be effected"(Hull: 311). In fact, "a given animal or plant appears to be contained. . . within a sphere of variation" (Hull: 308). It was possible to push individuals closer and closer to the perimeter, but never beyond it. Natural selection could consequently "improve hares as hares, and weasels as weasels" but "the origin of species requires not the gradual improvement of animals retaining the same habits and structure, but such modifications of those habits and structure, as will actually lead to the appearance of new organs" (Hull: 313). Darwin's evidence, whether taken from domestic or natural selection, Jenkin believed, concerned the improvement of species and not their creation.

The response that given sufficient time improvements alone could produce new species was unsatisfactory. It supposed that improvement was "constant or erratic" which would in "untold time. . . lead to untold distance"; in actual fact the "rate of deviation steadily diminishes till it reaches an imperceptible amount." Resorting to a physical analogy, Jenkin concluded "we are as much entitled to assume a limit to the possible deviation as we are to the progress of a cannon-ball from a knowledge of the law of diminution in its speed" (Hull: 308). The same argument was also presented by Hooker. He noted in 1869 the results of some breeding experiments with wheat. Rapid improvements were initially noticed when certain specific varieties were selected but, after a few generations, the rate of improvement declined until, finally, no further enhancement of the selected line could be gained. It had reached, supposedly, the limit of its variation. Darwin replied to Hooker as he did to Jenkin: "I am not at all surprised that . . . some varieties of wheat could not be improved in certain desirable qualities as quickly as at first. All experience shows this with animals; but it would, I think, be rash to assume . . . that a little more improvement could not be got in the course of a century, and theoretically very improbably that after a few thousands [of years'] rest there would not be a start in the same line of variation" (Vorzimmer: 155). The invitation to judge the merits of his theory in a few thousand years must have seemed weak even to Darwin. If natural selection was ever to be established it would have to be demonstrated by its ability to point out how species could be created here and now and not in some distant future.

At this point Darwin may have been tempted to fall back on the notion that natural selection might operate on sports. Precisely why this was so unattractive an option was also pointed out by Jenkin. For the attributes of a sport to be preserved, it was first necessary to reproduce. But theories of heredity in the mid-19th century, fifty years before Mendel's work, assumed a blending of the parent's attributes. "Suppose a white man to have been wrecked on an island inhabited by negroes" and suppose him, Jenkin argued with the racial assumptions of his time, "to possess the physical strength, energy, and ability of a dominant white race," then though he may end up king, "a highly-favoured white cannot blanch a nation of negroes" (Hull: 315-16). The same result would be seen if a sport, however adaptive, was produced in a natural population. The only way to preserve its attributes would be for two identical sports to appear and breed together, an event too infrequent for Darwin.

The objection was pushed further. Francis Bowen, from Harvard, in a review of the first edition of *Origin of Species,* argued that the same strictures would apply to individual differences: "Variations, if slight, are seldom transmitted by inheritance. . . . The very act of crossing the varieties tends, by splitting the difference, to diminish the distance between them. Under domestication, indeed, the varieties will be kept apart; but in the wild state, Nature has no means of preventing them for pairing. They will interbreed . . . and will thereby kill out instead of multiplying the variations" (Vorzimmer: 111-12). Darwin's theory worked by preserving and augmenting small differences. But if these differences or variations were not inherited, they could never be selected and marshalled together to form new species.

Darwin's problem was to show how he could consistently maintain the following three propositions he had committed himself to: (a) evolution operates through the action of natural selection; (b) natural selection operates on individual differences; (c) individual differences are blended rather than inherited. Because of the haste in which some of the earlier editions appeared, it was not really until the fourth edition of *Origin of Species* that Darwin began to introduce an answer to his problems.[35] Variations, he admitted, if rare would indeed be swamped out in a natural population. The solution would obviously be to make them so plentiful that their sufficient numbers alone would counterbalance the effects of blending. Here the advantages of insisting that natural selection operates on "many slight differences" that are present in any population become apparent. It would be absurd to propose that identical sports are likely to be distributed in a natural population in any number; yet, with individual differences, it is plausible to suppose that a large population will contain a fair number of specimens with the same "individual differences."

But this simply created a further problem: What was the source of this supposed convergence of individual differences? Darwin adopted quite unambiguously the principle of the inheritance of acquired characteristics, a principle as popular with today's biologists as perpetual-motion machines are with physicists.[36] Even in the first edition Darwin had insisted, "there can be little doubt that use in our domestic animals strengthens and enlarges certain parts, and disuse diminishes them; and that such modifications are inherited" (Darwin 1968: 175). In a later edition he warned that we should not be hasty in dismissing cases of the inheritance of "accidental mutilations" (Darwin 1958b: 133). According to modern orthodoxy, however, no amount of use or disuse of body parts can affect the germ cells which alone decide what is or what is not inherited. To quote the standard objection first proposed by St. George Mivart (1827-1900), if acquired characters can be inherited, why do Jews still need to circumcize their young men?[37]

To show how acquired characteristics could be inherited, Darwin constructed a new theory of heredity in *Variation of Animals and Plants* (1868). He supposed that body cells regularly gave off gemmules or pangenes that accumulated in the germ cells. The pangenes were affected by the condition of the somatic cells and in this way the effects of use and disuse could be inherited. This theory, known as the doctrine pf pangenesis, also ran into trouble. Darwin's cousin, Francis Galton (1822-1911), in 1872 transfused blood between two strikingly different breeds of rabbits. If pangenesis were true, he expected to see attributes of breed *A* turning up in the offspring of breed *B*, and of *B* in the attributes of *A*'s offspring. No such outcome was noticed.

Further attack came from Mivart in his *Genesis of Species* (1870).[38] Although there was little new, Mivart knew how to exploit the inconsistencies, omissions and obscurities that could be found after diligent search of the five published editions. There were also other works of Darwin to explore in the hope of finding some passage or other which would conflict with something said somewhere in one of the editions of *Origin of Species.* It must have been dispiriting for Darwin to read again that slight variations were not inherited and of little use in evolution. He dismissed Mivart as a "pettifogger" and an "old Bailey lawyer," and complained to Huxley, "I cannot possibly hunt through all my references for isolated points. . . . At present I feel sick of everything" (Darwin 1958: 292).

This may help to explain why there was no seventh edition of *Origin of Species.* Seven reprints of the sixth edition appeared in the remaining ten years of Darwin's life, yet a seventh edition remained unplanned. Darwin had lost little if any of his intellectual capacity or appetite for work. He published more in the 1870s than many scientists produce in their whole life: eight new books, the second editions of three other works, and much minor correcting and writing.[39] Most were substantial works based on fieldwork and experiment, no doubt carried out much earlier, but still requiring checking, organizing and writing.

Darwin's silence was not due to any lack of interest in biological problems. In letters of 1871-72, he explained his position. "I have resolved to waste no more time in reading reviews of my works or on evolution." It was now up to others. "There are so many good men fully as capable, perhaps more capable than myself, of carrying on our work." He even refrained, in his new works, from discussing evolution where it seemed to arise naturally. Wallace noticed the curious omission from his later work and raised the matter with Darwin over his *Insectivorous Plants* (1875): "You do not make any remarks on the origin of these extraordinary contrivances for capturing insects. Did you think they were too obvious? . . . . I feel sure they will be seized upon as inexplicable by Natural Selection, and your silence on the point will be held to show that you consider them so!" (Vorzimmer: 254).

The point is that Darwin had run out of ideas. To the objections of scholars like Mivart and queries from Galton, he could make no specific defense. The best he could offer was a general response that given sufficient time almost any kind of development was possible. This was too weak an argument to convince any critic. If it proved anything, it was just as capable of proving that evolution worked by inheriting acquired characteristics as it was in showing that natural selection was the prime cause of the origin of species. Darwin, thus, returned to the study of natural history; the kind he could practice in his garden at Down and which he had pursued for much of his life.

There remains a further objection to consider and one described by Darwin as "one of the gravest as yet advanced." The reference was to the calculations carried out by William Thomson, Lord Kelvin (1824-1907), on the age of earth and sun. Kelvin was a leading physicist and made important contributions throughout the discipline. On physical matters he spoke with great authority. In his *Energies of the Solar System* (1854) he first turned his attention to Lyell's uniformitarian geology, a theory he found unacceptable. Was Lyell's claim of the earth's great antiquity, Kelvin asked, physically sound?

Kelvin developed a number of independent arguments. One involved calculating the earth's age by using the known rate at which it was cooling. Given this information it was possible to infer how hot the earth must have been at any selected time in the past. His alarming conclusion was that "we are led to a limit of something like ten million years as the utmost we can give geologists for their speculations as to the history of the lowest orders of fossils" (Tait: 167). A second argument depended upon the age worked out for the sun since the earth could be no older. Here, too,

Kelvin's calculations demonstrated that the sun "cannot have supplied the earth. . . for more than fifteen or twenty million years" (Tait: 175).

Such figures contrasted sharply with Darwin's estimates in the first edition of *Origin of Species* that "the denudation of the Weald must have required about 306,662,400 years; or say three hundred million years" (Darwin 1968: 297).[40] For physicists like P. G. Tait (1831-1901), professor at Edinburgh, the issue was clear. "So much the worse for geology," he declared, adding with a confidence shared by Kelvin that it was "utterly impossible that more than ten or fifteen million years can be granted" (Tait: 167-68).

Geologists' response was various. Some, like Wallace in *Island Life* (1880), happily adjusted their biology to Kelvin's physics. Huxley, too, agreed in 1869 that "biology takes her time from geology." But, at the same time, in a presidential address to the Geological Society, he carefully noted that "pages of formulae will not get a definite result out of loose data" (Burchfield: 84).

In contrast to Wallace and Huxley, Darwin would have none of Kelvin's legislation. Although he dropped his calculations of the denudation of the Weald – the once-forested area of Kent, Surrey and Sussex – from the third edition of *Origin of Species,* and although unsophisticated in the field of mathematical physics, Darwin nonetheless was prepared to insist that Kelvin was wrong.[41] In this, his judgment – not faith – was sounder than the physicists'. He suspected them of overconfidence. Did we really know enough "of the constitution of the universe and of the interior of our globe to speculate with safety on its past duration?" he asked in the sixth edition (Darwin 1958a: 431). Kelvin's repeated pronouncements can only have confirmed Darwin's skepticism. His first calculations in 1854 suggested an age for the earth of only 32,000 years. In 1862 this had increased enormously to 400 million years but in 1868 was down to 100 million. Thereafter the trend continued downward: 50 million years in 1876, 20-50 million in 1881, and in 1897 his final estimate was put at 24 million years.

Kelvin always allowed that his calculations would break down if forces and operations unknown to the physics of his day were later to be discovered. Just such an event occurred in 1903 when Pierre Curie (1859-1906) reported that heat was constantly being released from radium salts.[42] Darwin had been vindicated. Before long physicists like R. B. Boltwood (1870-1927) were making radiometric estimates of the rate at which uranium decays into lead and coming up with ages of 340 million years for the Carboniferous period (Eichler: 18).

In 1904 Ernest Rutherford lectured in London on the influence that new work on radioactivity bore on estimates of the earth's age. Present in the audience was the eighty-year-old Kelvin. Rutherford later described the occasion (Andrade: 80):

> I came into the room, which was half dark, and presently spotted Lord Kelvin in the audience and realised that I was in for trouble at the last part of my speech dealing with the age of the earth, where my views conflicted with

> his. To my relief, Kelvin fell asleep, but as I came to the important point, I saw the old bird sit up, open an eye and cock a baleful glance at me! Then a sudden inspiration came, and I said Lord Kelvin had limited the age of the earth, *provided no new source was discovered.* That prophetic utterance refers to what we are now considering tonight, radium! Behold! the old boy beamed upon me.

Later Reception of "Origin of Species"

The idea that the community of biologists for the past century or more have uncritically accepted Darwin's theory of evolution without considering alternative views is nonsense. For much of the past century, Darwin's own account of evolution has been rejected by most biologists. His commitment to the importance of small "individual differences" meant that Darwin was looking at features like height, beak length, color of tail feather, features that vary continuously throughout a population. Many favored a different model. One such was William Bateson (1861-1926) who in the late 1880s investigated the distribution of shellfish in the lakes of Kazakastan, Russia. Although he was able to arrange the lakes in a continuous series of increasing salinity, he found nonetheless that specific shellfish characters were not graded accordingly. Variation in shellfish was seen to be discontinuous or discrete. The jumps Huxley had warned Darwin to be cautious about had been demonstrated in a natural population. Bateson published the results, with much troublesome data, in his *Materials for the Study of Variation* (1894).

Nor did the rediscovery of the work of G. Mendel (1822-84) help Darwin at this point; the factors inherited in Mendelian breeding experiments with peas (*Pisum sativum*) were discrete factors like smooth/wrinkled, yellow/green, not at all the factors Darwin had in mind.[43] So confused did the issue become that those who supported Darwin and continuous variation, like W. F. R. Weldon (1860-1906), the Oxford geneticist, ended up rejecting Mendel's theory. As the victory in this dispute went to Mendel, Darwin's work was somewhat eclipsed for a number of years. The first two decades of the century were exciting ones for geneticists with a new discipline to develop. Research was partisan and results tended to confirm expectations; data also tended to be ephemeral with one day's results contradicting the next.

In 1903 T. H. Morgan (1866-1945), one of the first geneticists at Columbia University, published *Evolution and Adaptation* in which he strongly denied that natural selection as described by Darwin could work. The objections were old but were made more convincing by his presentation of new experimental evidence. Natural selection, Morgan argued, could only operate as a negative factor; it could eliminate the unfit without much trouble, but there was no way that continuous variation could produce new species. Support came from Hugo de Vries (1848-1955) in *Mutätionstheorie* (1901). Working with the evening primrose (*Oenethera lamar-*

ckiana), de Vries demonstrated how two distinct strains which bred true when self-fertilized, on being crossed produced strains different enough from the parent strain to be considered new species. He concluded that "new elementary species arise suddenly without transitional forms" and termed them *mutations*. For de Vries, the role of natural selection was to choose between the new and the old species, a means of preserving existing species rather than creating new ones. All Darwin's objections to the role of sports in evolution now seemed misguided.

A second damaging finding came from the work of Wilhelm Johannsen (1857-1927), the man who in 1909 coined the word gene. He first isolated pure strains of beans which, when self-fertilized, bred true for such properties as seed weight and size. Selection applied to a natural population could produce any one of the pure strains but, when applied to any of the pure lines themselves, proved to be of limited value. If weight was selected for, it might show some mild success in the first few generatons, but selection would soon cease to yield any additional improvement. Further, if the selection pressure was removed, regression to the original weight would occur.

With such evidence mounting, traditional Darwinism began to look less and less convincing. Darwinians responded by rejecting much of the newly established Mendelian genetics. Weldon, for example, was particularly vociferous in claiming results that conflicted with the Mendelian ratios. Japanese waltzing mice and other exotic breeds were enlisted in support of Darwin. Later, a particularly bitter dispute erupted over the issue of racehorses, and whether the color chestnut was recessive to bay and brown. Not so, declared Weldon, in 1906, after devoting many hours to the twenty volumes of Weatherby's *General Stud Book* and emerging with a handful of counterexamples. "Errors of entry," replied Mendelians. Only the death of Weldon shortly thereafter put an end to this particular dispute. The fact is that early in the century the necessary experimental techniques had not been worked out, essential distinctions had not been made, knowledge of various strains had yet to be acquired, and new mathematical disciplines created and mastered. Without these foundations any breeding experiments were likely to be difficult to interpret at best and incomprehensible at worst. Add to this a factionalism so bitter that it led to the distortion if not falsification of experimental results, then it is easy to see how startling results could emerge from one school on one day only to be countered with equally startling but inconsistent results from a hostile school on the following day.

Looking back, it is surprising how quickly progress was made. Major schools were established, like Morgan's famous Drosophila group at Columbia University; concepts and techniques were purified and within a decade many acknowledged that it had been premature to dismiss natural selection so completely. One confusion behind much of the disagreement was avoided when in 1909 Johannsen distinguished between the genotype, the genetic constitution of an individual, and the phenotype or the characteristics shown by an individual. It was soon recognized that individuals with the same phenotype could possess different genotypes and, equally important, an apparently simple phenotype could be controlled by a quite complex genotype. The fact that an ear of corn, for example, was red did not mean that its color was controlled by the presence or absence of a single gene.

In 1909, moreover, this point was expanded by geneticists to show that simple properties could be the result of a complex interaction between several genes. A few independent factors, interacting, could produce several hundred different forms, each with a different genotype. With this insight the continuous variation of Darwin could be linked to the new Mendelian orthodoxy.[44] As a result of this and other work, T. H. Morgan, who in 1903 had been strongly opposed to Darwin, wrote in 1909 a popular article with the title "For Darwin." It took another twenty years for this early work to be fully developed. The man who first presented the grand synthesis of neo-Darwinism in all its complexity was R. A. Fisher (1890-1962) in *The Genetical Theory of Natural Selection* (1930), a title which would have puzzled Weldon and his Oxford colleagues a generation earlier. Fisher's work was followed in the 1930s and 1940s by the great texts of modern evolutionary theory such as Julian Huxley's *Evolution—A Modern Synthesis* (1940), G. Simpson's *Tempo and Mode in Evolution* (1942), T. Dobzhznsky's *Genetics and the Origin of Species* (1937) and E. Mayr's *Systematics and the Origin of Species* (1942). The dates reveal all; there are no comparable texts from the 1920s.

Any attempt to assess Darwin's impact on the intellectual life of the last one hundred and twenty-five years would need to cover every subject from anthropology to zoology. For while zoologists at the turn of the century were somewhat suspicious of Darwin, sociologists had begun to welcome his ideas enthusiastically. The most obvious sign of this acceptance was the appearance of the term "evolution" in such titles as *Mind in Evolution* (1901) and *Morals in Evolution* (1906) by L. T. Hobhouse (1864-1929), and *The Evolution of Culture* (1906) by A. H. Pitt-Rivers (1827-1900). Earlier, Karl Marx (1818-83) read *Origin of Species* and declared it "a basis in natural science for the class struggle in history." He proposed to Darwin that he be allowed to dedicate the English translation of *Das Kapital* to him. Darwin politely refused on the grounds that identification with a work so hostile to religion would distress some of his family (Himmelfarb: 316).

Frequently the reference to evolution or Darwin was little more than a bow to fashion. A. C. Haddon's *Evolution in Art* (1895), for instance, has nothing about evolution, but it contains much about the origin of certain patterns and their spread throughout a particular region. Excluding such almost ritual-like borrowings, it is plain why historians, sociologists, anthropologists, theologians and moralists have all been so attracted to *Origin of Species.* For centuries, students of the humanities have been concerned with two particular problems of understanding. How could the development and change witnessed in all societies be understood? And how could the variety of social institutions distributed throughout the world be explained? If anything, such problems became more pressing as knowledge of the world's contents and past accumulated in the great libraries and museums of 19th-century Europe and America. Facile talk of three ages or three stages, whether in Hesiod or Comte, was no more attractive than the meteorological theories of culture that had persisted in western thought from Hippocrates to the French Revolution.

Some, of course, merely used evolutionary theory to justify their own condition. Even J. D. Rockefeller described the growth of a large business as "a survival of the fittest. . . , merely the working-out of a law of nature and a law of God" (Himmelfarb: 346).[45] There were others, though, who made a genuine attempt to use Darwinian insights in order to understand better the nature of man and society. At its simplest this can be seen in attempts by ethnographers to comprehend the variety and distribution of customs, institutions and technologies in terms of the key Darwinian concept of adaptation. Why do Indians refuse to eat cows, Muslims refuse to eat pigs, Kwakiutl burn their property, and people of the New Hebrides await the return of John Frum, King of America? At first sight such beliefs and behavior to the outsider can seem odd, if not ludicrous, and at second sight tend to remain intractably puzzling. Yet in the hands of anthropologists like Marvin Harris in *Cows, Pigs, Wars and Witches* (1977), all are shown to be adaptive, serving to maintain the stability and integrity of the society they are found in.[46] Just as Darwin could show how the cuckoo's instinct to lay eggs in other birds' nests is somehow in her interest, equally, and using the same concepts, anthropologists can show how some persistent structures of society survive. It is difficult now to conceive how such institutions would appear without the illumination gained from Darwin's work.

Another problem in the humanities – how does knowledge grow – has been tackled by the philosopher Karl Popper, using Darwinian insights. The process, Popper says, closely resembles "what Darwin called 'natural selection'; that is, *the natural selection of hypotheses.*" The hypotheses we hold at any one moment "have shown their (comparative) fitness by surviving so far in their struggle for existence; a competitive struggle which eliminates those hypotheses which are unfit" (Popper: 261). Popper's views had been heard almost a century before from Huxley: "The struggle for existence holds as much in the intellectual as in the physical world. A theory is a species of thinking, and its right to exist is coextensive with its power of resisting extinction by its rivals" (Huxley: 229). In fact, between Huxley and Popper, many have resorted to the analogy, although its use as anything more than an analogy has yet to be demonstrated.

It remains, however, in the field of science that Darwin's main achievements lie. Here, it is impossible to estimate their full significance because the impact of Darwin's ideas on biology has yet to run full course. Several points stand out nevertheless. There is, first, the pervasiveness of the evolutionary process. In Darwin's hands it was not restricted to a part of the organic world; it applied also to man, his habits, instincts, and emotions as well as his dentition and limbs.

Origin of Species actually ignored the problem of man. Only in the last few pages did Darwin, speaking of future researches, add, "Light will be thrown on the origin of man and his history" (Darwin 1968: 458). Few, if any, before Darwin were prepared to question the great gulf which separated man from all other parts of nature. Even without his theological status as made in the image of God, there were still plenty of other reasons, like man's intellect, his moral sense, and even his anatomy, not just to make him different from all other creatures, but to place him on a separate plane of creation.[47]

The implications of Darwin's work were nonetheless clear to most readers. How could Darwin exclude man from the general evolutionary process except by an arbitrary and unconvincing fiat? Such was the charge made against Darwin at the famous encounter between Bishop Wilberforce, widely known as "Soapy Sam," and Huxley at the meeting of the British Association for the Advancement of Science in Oxford in 1860. "Was it through his grandfather or his grandmother that he claimed his descent from a monkey?", he publicly asked Huxley. Either would be preferable, Huxley replied, than to have as an ancestor "a man who used great gifts to obscure the truth" (Irvine: 3-7).

Huxley excepted, most of Darwin's colleagues – Lyell and Wallace for example – wished to exempt some aspect of man's nature from the general evolutionary process. Darwin's own detailed views were published in *The Descent of Man, and Selection in Relation to Sex.* The bulk of the work is concerned with sexual selection and only deals specifically with the descent of man in the first part. His conclusion on the status of man was unequivocal: "We thus learn that man is descended from a hairy, tailed, quadruped, probably arboreal in its habits, and an inhabitant of the Old World. . . , classed among the Quadrumana."

In the century following, more and more aspects of nature have been shown to fall within the domain of evolutionary biology. Much of this was implicit in *Origin of Species* but has by now been allowed to develop to such an extent that new subdisciplines have emerged. Ecology, ethology, molecular biology, cladistics and sociobiology are just some areas of biology which in the last quarter-century or so have established themselves as new disciplines. Uniting them is the discarded assumption that their subject matter was not amenable to scientific analysis. Either the areas were, like an ecological system, too complex to analyze, or, like the basic molecules of life, too inaccessible, reasons thought sufficient to exclude them from the central evolutionary debate. An examination of the special September 1978 issue of *Scientific American,* devoted entirely to evolution, with contributions on "Chemical evolution and the origin of life" by R. E. Dickerson, "The evolution of ecological systems," by R. May and "The evolution of behavior" by J. Maynard Smith, shows that such topics are now amongst the most intensively investigated in modern biology.

Such advances emphasize the essential contribution of *Origin of Species* to biology. Before Darwin, it was difficult to see what, for example, a physiologist, a paleontologist, and an ecologist had in common. The problems they tackled appeared distinct, while the languages adopted in various disciplines had little overlap. After Darwin, all such problems could be seen to have a common focus. The physiologist no longer demonstrated physico-chemical systems in living organisms, but revealed adaptations selected in the evolutionary process; the paleontologist uncovered the evolutionary history of extinct forms; while the ecologist analyzed adaptations of inter-connections observable in any biome. So solidly based is this unity by now that the emergence of new disciplines like molecular biology in the 1950s present no threat. Within a few years of its appearance, biologists were happily talking of *molecular evolution* and exhibiting the effect of natural selection on DNA molecules.

It should, however, be appreciated that it is the basic ideas of *Origin of Species* and not their specific application that have been incorporated into modern biology. Consequently, while many working scientists are happy to call themselves Darwinians and are committed supporters of evolutionary biology, they nonetheless frequently reject specific Darwinian ideas and modify others. One of the best-known such re-appraisals in recent years has come from Niles Eldredge and Stephen Gould with their punctuational view of the fossil record. Darwin's evolutionary model had been a gradualistic one with new species emerging by a long series of imperceptible changes from old ones. The fossil record should therefore be expected to show the many intermediate links between any two separated forms. Darwin asked in 1859 why "is not every geological formation and every stratum full of such intermediate links?" The explanation lay, he believed, in the "extreme imperfections of the geological record" (Darwin 1968: 292). A further century of exploration has found the geological record no more perfect than in Darwin's day. In the 1970s Gould and Eldredge reversed the argument by insisting the record was accurate and the model imperfect. Species, they proposed, underwent fairly rapid periods of change followed by long periods of stasis. The rocks were thus largely empty of intermediate forms because for long periods species were stable and there simply had been no intermediate forms. Lest there be any misunderstanding, Gould emphasized that "it is gradualism that we must reject, not Darwinism" (Gould: 182).

Others, like John Maynard Smith (1978), have sought not so much to revise Darwin but to take his central insights, endow them with greater precision, and thereafter use them to analyze problems of evolutionary theory with a depth and authority undreamt by Darwin. One problem he considered is how conventional fighting in animals evolved. Male fiddler crabs, for example, fight each other for possession of a burrow. Though they possess powerful claws they seldom seem to damage each other. Could such conventional behavior increase the "Darwinian fitness" of the animals adopting it? Why have not individuals emerged to take advantage of the ritual claw-waving of their opponent to tear them apart? Unhappy with the explanation that it was all done "for the good of the species," Maynard Smith hypothesized that natural selection would not produce a population of aggressive, as opposed to conventional, crabs (or "hawks" rather than "doves" in his terminology). He then turned to game theory to throw light on this and many other related problems, and formulated the concept of an evolutionary stable strategy: a strategy with the property "that if most members of a population adopt it, then no mutant strategy can invade the population." Opportunistic individuals could always gain a personal victory by adopting a hawkish strategy but "consistently playing hawk is not an evolutionary stable strategy" (Maynard Smith 1978: 98). Widespread adoption of the hawk's strategy would invite competition from mutant doves who would "reproduce more often than hawks." (Much of Maynard Smith's impressive work was made available to a wider public in *The Selfish Gene* [1976] by Richard Dawkins.)

Maynard Smith's problems are similar to those encountered in the new discipline of sociobiology. In *Origin of Species* Darwin had raised the problem of how natural selection could produce sterile castes among social insects and how altruism could evolve. An animal that lays down its life for a member of the herd can hardly be said to benefit itself. Such responses, together with sexual behavior and aggression, are examples of genetically determined social behavior. This means they must all have evolved. But how? Animals that sacrifice their lives for others are unlikely to have descendants to continue their genetic line while sterile insects, by definition, can have no offspring.

Yet, sterile castes have evolved at least a dozen times: some eleven times among ants, bees and wasps, the *Hymenoptera,* and just once outside the order, among termites (Wilson: 415). Darwin's solution to the problem was based, typically for him, on insights derived from the familiar facts of artificial selection. Selection, he noted, could be "applied to the family, as well as the individual." Consider, for example, how we select a particular quality of meat. The animal is slaughtered and never breeds but "the breeder goes with confidence to the same family." We could, if we wished, Darwin insisted, produce "oxen with extraordinarily long horns. . . by carefully watching which individual bulls and cows, when matched, produced oxen with the longest horns." In the same way "a slight modification of structure or instinct correlated with the sterile condition of certain members of the community, has been advantageous to the community: consequently the fertile males and females of the same community flourished, and transmitted to their fertile offspring a tendency to produce sterile members having the same modification" (Darwin 1968: 258-59). Suggestive though Darwin's proposals may be, they lack the precision and genetic detail now demanded from evolutionary biology. Such an answer was suggested in 1964 by W. D. Hamilton.[48]

Given work as original and powerful emerging from within the Darwinian tradition as that of Maynard Smith's and Hamilton's, it is clear to most scientists that evolutionary biology is still a rewarding and significant discipline. Excluding Lamarckians and Creationists few are prepared to challenge, in detail, the basic Darwinian model. There are, however, some. One of the best known of these is the Japanese population geneticist Motoo Kimura with his "neutral" theory of molecular evolution.[49]

There is still the occasional Lamarckian to be found within the confines of academic biology. One such was H. Graham Cannon (1897-1963), professor of zoology at Manchester (1931-63), author of two books defending Lamarck. His work found little support. If scientists have overwhelmingly preferred Darwin to Lamarck, dramatists and novelists have not always followed their lead. Samuel Butler (1835-1902), well known as the author of *Erewhon* (1872), published a number of polemical works defending Lamarckian positions against Darwinian attacks.[50] Butler sought to restore mind and purpose to the universe, factors he considered Darwin to have willfully discarded. In more recent times G. B. Shaw (1856-1950) in such plays as *Back to Methusalah* (1921) argued much as Butler had earlier done.

Today, the reputation of Darwin and *Origin of Species* has never been higher. Books continue to appear on Darwin at an alarming rate and the British Museum catalogue already contains three times more entries for Darwin than for Newton. Nothing, however, quite shows the esteem granted to Darwin than the use of his name. In this respect he is probably the most honored figure in science. Among those honors are dozens of plants with the epithet *darwinii*; five London streets; a Cambridge college; a medal of the Royal Society; towns in Australia, the United States and the Falkland Islands; a bay and a channel in Chile; a mountain and a glacier in Antarctica; and, appropriately enough, a volcano on the Galapagos.

Publishing History

Origin of Species has been one of the most successful books published in the last hundred years. It has sold steadily throughout the world, and even today it can be obtained in more than a dozen different English editions. Six editions were issued during Darwin's lifetime.

First edition: November 1859

Murray published 1,250 copies. Of these, he sent five to Stationer's Hall for copyright purposes, forty-one to reviewers, and twelve to Darwin. The remaining 1,192 copies were offered to the trade on November 22 and, by some accounts, immediately sold out. The publication date was November 24. The whole edition cost Murray £486 13*s* 4*d* to publish, including the £180 paid to Darwin. Selling at 15 shillings a copy, Murray's profit of £57 4*s* 2*d* was considerably less than Darwin's. At least twenty-four presentation copies have been identified, none in Darwin's hand since this task was performed by Murray's clerk. Darwin's own copy is in the University Library, Cambridge.[51] A first-edition copy today is unlikely to sell for less than £2,000.

Second edition: December 1859 (1860)

Murray issued a new edition of 3,000 copies on December 31 at the reduced price of 14 shillings. It was "little more than a reprint," according to Darwin, who earned £636 13*s* 5*d* and Murray £313 12*s* 5*d*.

Third edition: March 1861

A further 2,000 copies were published in March, earning Darwin £372. This was the first edition to contain the "Historical Sketch" ("the progress of opinion on the origin of species previously to the publication of this work"). It had in fact been published earlier in the first German translation in 1860. Additions added thirty-five pages to the text.

Fourth edition: December 1866

Fifteen hundred copies were published on December 15 at the price of 14 shillings. Darwin's profit was £250. It was fifty-two pages longer than the first edition.

Fifth edition: August 1869

Two thousand copies, priced at 15 shillings, were published, earning £315 for Darwin. It is in this edition that the phrase "survival of the fittest" first appeared.[52]

Sixth edition: 1872

This edition of 3,000 copies was published at the much reduced price of 7 shillings, 6 pence, and it was the last prepared by Darwin. It was reissued twice in 1872, in 1873, 1875, and as the "sixth corrected" in 1876. Two further reissues appeared in Darwin's lifetime in 1878 and 1880. As Darwin's final word on the subject, the "sixth corrected" forms the basis of most modern editions. Surprisingly, it is the first in which the word "evolution" was used by Darwin.[53] Before, he had written about adaptations that had "evolved." It also contains a new chapter seven, introduced in response to various criticisms, and a glossary, which was compiled by W. S. Dallas. As a result of the new chapter, entitled "Miscellaneous Objections to the Theory of Natural Selection," chapters seven through fourteen of the first five editions became chapters eight through fifteen in the sixth edition.

New editions continued to appear after Darwin's death. By the time the copyright of the first edition ran out, in 1901, Murray had published about 56,000 copies of *Origin of Species* in its original format. Murray's first cheap edition, published in 1900, sold 48,000 copies in its first two years. With such interest, Murray shrewdly issued a number of editions to suit almost every conceivable need. Within a few years, he had come out with a two-volume "library" edition (12 shillings), a "popular" (6 shillings), a "cheap with portrait" (2 shillings, 4 pence), and a "paper cover" edition for only one shilling.

Upon the expiration of Murray's copyright in 1901, the publisher soon faced competition from other publishers wishing to produce cheap editions. It appeared first in the Minerva Library (1901) and thereafter in series after series, some famous and still flourishing, others long forgotten. Among them: Grant Richard's World Classics (no. 11, 1902); Unit Library (no. 3, 1902); Rationalist Press Association (no. 11, 1903); Oxford's World Classics (no. 11, 1905); Hutchinson's Popular Classics (1906); Cassell's Popular Library (no. 73, 1909); Ward Lock's World Library (no. 5, 1910); Collin's Illustrated Pocket Classics (no. 149, 1910); Murray's Library (no. 163; 1910); Everyman's Library (no. 811, 1928); and Thinker's Library (no. 8, 1928). In more recent times, it was chosen to launch the Penguin Classics series in 1968, and since 1958 has been available as a Mentor Book. Up to 1975, according to Freeman (1977), 156 editions had appeared in English in Britain and the United States.

Translations

Between 1859 and 1975, translations of *Origin of Species* have been made in twenty-nine languages in 169 editions, amounting to 325 editions in all in the 116 years of the book's existence. Listed in chronological order by language and number of editions, they are as follows:

German 1860: 26
French 1862: 17
Russian 1864: 18
Dutch 1864: 5
Italian 1864: 12
Swedish 1869: 3
Danish 1872: 5
Polish 1873: 4
Hungarian 1873: 4
Spanish 1877: 25
Serbian 1878: 2
Japanese 1896: 15
Chinese 1903: 3
Czech 1914: 2
Latvian 1914: 2
Greek 1915: 2
Portuguese 1920: 3
Finnish 1928
Armenian 1936: 2
Ukranian 1936: 2
Bulgarian 1946: 2
Rumanian 1950: 2
Slovene 1951: 2
Korean 1957: 4
Flemish 1958
Lithuanian 1959
Hebrew 1960: 2
Hindi 1964
Turkish 1970

Table 13:1
Basic data on the editions of *The Origins of Species* 1859-72

Edition and date	Number of pages	Price	Number published	Darwin's profit	Sentences Dropped	Sentences Changed	Sentences Added	Additions
First 1859	502	15s.	1,250	£180	Contains 3,878 sentences			
Second 1860	502	14s.	3,000	£636	9	483	30	
Third 1861	538	14s.	2,000	£372	35	617	266	"Historical sketch"; Table of additions & corrections
Fourth 1866	593	14s.	1,500	£250	36	1,073	435	
Fifth 1869	596	15s.	2,000	£315	178	1,770	227	
Sixth 1872	458	7s.6d.	3,000	£210	63	1,669	571	Glossary; "Chapter 7"
TOTAL			12,750	£1,963				

Further Reading

Origin of Species is available in paperback in either its first edition (Darwin 1968) or its sixth edition (Darwin 1958a). Other editions are not so easy to find, but details from them can be found in the variorum text of Peckham (1959). Despite the multitude of books recently published to mark the 1859 and 1882 centenaries, no adequate biography of Darwin has yet appeared and his life is still best approached through his *Autobiography* (Darwin 1974) and the three-volume *Life and Letters* (1887), both of which are also available in a shortened version in a Dover Books paperback (Darwin 1958). The three current biographies: *Charles Darwin* (1963) by G. de Beer, Chancellor (1973) and Peter Brent (1981) are too sketchy, too superficial or too indifferent to illuminate the difficulties Darwin faced. More light is gained from such popular and less ambitious works as Irvine (1955) and Moorehead (1971). By way of commentary, Vorzimmer's study (1972) remains unsurpassed but should be read in conjunction with Dov Ospovat's *The Development of Darwin's Theory* (1982) and M. Ghiselin's *The Triumph of the Darwinian Method* (1969). More background can be found in the two splendid and complementary studies of Michael Ruse (1979 and 1982). Modern attitudes by biologists to Darwin can be followed in Maynard Smith (1966), Wilson (1975) and *Scientific American* (September 1978). Other accessible and relevant works of Darwin include his *Voyage of the Beagle* (1959), the early writings on evolution (Darwin and Wallace, 1958), and *The Descent of Man* (1982). At a more basic level there are Darwin's own notebooks which have been published in eleven parts by the British Museum while a single volume issue of the "theoretical notebooks (1836-44)" has been promised for the future. Finally, there are the two invaluable works of Freeman (1977 and 1978) providing between them the essential bibliographic and biographical data on Darwin's work and career.

Notes

1. In March 1981 the Arkansas legislature decreed "balanced treatment" had to be given in schools to the theory of evolution and "creation science." Suit was brought by the American Civil Liberties Union that such a course would violate the separation of church and state guaranteed by the First Amendment. The trial began in Little Rock on 7 December 1981, and on 5 January 1982 Federal Judge William Overton ruled that Act 590 was unconstitutional. Eyewitness accounts can be found in *New Scientist* (4 February 1982): "A philosopher at the monkey trial" by Michael Ruse, and in *Discover* (February 1982): "Judgement day for creationism" by James Gorman. Ruse has also written on the creationist case in *New Scientist* (27 May 1982) and in his book *Darwinism Defended* (1982). The creationist case is also treated in G. E. Parker's *Creation: The Facts of Life* (1980).

2. FitzRoy was a fundamentalist who observed with apprehension the trend of Darwin's thought throughout the voyage. Relations between the two were not always good: "We had several quarrels; for when out of temper he was utterly unreasonable." FitzRoy later was "very indignant" with Darwin for publishing so

"unorthodox a book." "His end," Darwin noted, "was a melancholy one, namely suicide" (Darwin 1974: 41-44).

3. This work was first published in 1839 as Vol. 3 of *Narrative of the Surveying Voyages of HMS 'Adventure' and 'Beagle' between 1826 and 1836* and was first issued separately in 1845 under the title *Journal of Researches into the Natural History and Geology of the Countries Visited during the voyage of H.M.S. "Beagle" round the World.* It is now more conveniently published under the title *The Voyage of the Beagle* and has formed the subject matter of an excellent study by Alan Moorehead, *Darwin and the Beagle* (1971).

4. Between 1836 and 1846 he published three geological works: *Coral Reefs* (1842), *Volcanic Islands* (1844), *Geological Observations on South America* (1846). He also edited the 585-page *The Zoology of the Voyage of H.M.S. Beagle* (1838-43), in five parts), and published his own 615-page *Journal of Researches* (1839 and 1845). Darwin also has left from this period many notebooks and several fairly complete works in manuscript.

5. The children were: William (1839-1914), a banker and his father's financial adviser; Anne (1841-51); Mary (1842); Henrietta (1843-1929), a noted eccentric; George (1845-1912), a mathematician and Fellow of the Royal Society (F.R.S.); Elizabeth (1847-1925); Francis (1848-1925),botanist, F.R.S. and editor of his father's *Life and Letters* (1887); Leonard (1850-1943), soldier and politician; Horace (1851-1928), maker of scientific instruments, F.R.S.; and Charles (1856-58).

6. The *Autobiography* was written mainly in 1876 with additions made in 1878 and 1881. It was first published by Francis Darwin (1887) in an incomplete form. Emma Darwin had been unwilling to allow some of her husband's frank comments on religion to be published. "The mystery of the beginning of all things," he had written, "is insoluble by us; and I for one must be content to remain an Agnostic" (Darwin 1974: 54). The first unexpurgated edition did not appear until 1958 when it was edited by Nora Barlow, a granddaughter of Charles Darwin.

7. He suffered a heart attack on 7 March, recovered and continued carrying out his experiments and working on two papers for the Linnean Society. A further attack on 18 April proved fatal: he died on the 19th. His last recorded words were: "I'm not in the least afraid to die." The family wished to bury him at Down but his friend John Lubbock (1834-1913), banker, scholar and politician, persuaded Emma and the authorities that Westminster Abbey would be more appropriate. Among the pall bearers were Hooker, Huxley, and Wallace.

8. Not so his children, three of whom were knighted: George (1905), Francis (1913) and Horace (1918).

9. In his *Darwin's Place in History* (1959), C. D. Darlington lists as precursors of Darwin the following: Hippocrates, Lucretius, Descartes, Maupertuis, Erasmus Darwin, Lamarck, William Wells, Patrick Mathew, and Robert Chambers. The list could be extended without difficulty.

10. The most famous, if not notorious, attempt to explain evolution before Darwin came from J. B. Lamarck (1744-1829), Curator of Invertebrates at the Musée d'histoire naturelle, Paris. In his *Philosophie zoologique* (1809) he proposed two laws. The first asserted that regular use of an organ strengthens and develops it while permanent disuse will lead to atrophy and loss. According to the second law such gains and losses are, under certain circumstances, "preserved by reproduction." Thus, by use, giraffes could stretch their necks and produce baby giraffes with slightly longer necks. Over many generations the process would produce the creature we see today. Opponents of Lamarckism strongly believed that what happens to the body, the somatic cells, cannot affect the reproductive or germ cells. The bulging muscles produced in a blacksmith's arms can have no possible impact on his germ cells and thereby on his progeny. The declaration of the fundamental independence of germ cells from somatic cells was made by August Weismann

(1834-1914) in *The Germ-Plasm* (1892). The removal of rats' tails over many generations led to no apparent decrease in the size of the tails of newly-born rats. The "continuity of the germ-plasm" thus became one of the major dogmas of modern biology, while the claim that acquired characteristics could be inherited was no longer seriously accepted.

11. Thomas Malthus (1766-1834), teacher of political economy at the East India Company's College at Haileybury, published *An Essay on the Principles of Population* in 1798. Darwin's *Notebook D* for 28 September 1838 has the entry: "Population is increas[ed] at geometrical ratio in *far shorter* time than 25 years – yet until the one sentence of Malthus no one clearly perceived the great check amongst men."

12. Copernicus, Harvey and Newton, as we have seen, delayed publication, for at least a decade, of their classic works. Although their reasons varied, it is more than a coincidence that four major works of western science were withheld from publication for so long. Nonetheless, it remains unclear why the supreme scientific genius behaves in this untypical manner. There are more modern examples of the trait.

13. Much has been written on the subject of Darwin's health. The fullest account is found in Colp (1977). Many explanations have been offered for his chronic ill health, ranging from hypochondria to arsenic poisoning and including Chaga's disease supposedly contracted in South America. From 1849-55 Darwin kept a *Diary of Health* in which he noted recurrent symptoms of flatulence, nausea, anxiety, palpitations, boils, headaches, together with the numerous remedies and medicines he adopted as regularly as he quickly dropped.

14. "16-10-1851: Electric chains to waist; 19-10-1851: ditto, neck" (Colp: 46).

15. One such remedy was the "ice treatment" adopted in 1865, on the advice of Dr. John Chapman, author of *Sea Sickness: Its Nature and Cure.* Apparently a bag of ice was applied to the spine three times a day for an hour and a half. Darwin tried it in June, but soon gave it up because, in matters of health as unpredictable as ever, he claimed it made his pulse race (Colp: 84).

16. This early best seller, four editions in the first year, was published anonymously. Its author, Robert Chambers (1802-71), was an Edinburgh publisher and writer. He argued that the geological record showed a development from simple to complex organisms. Just as in the inorganic world there was one law governing all matter, that of gravitation, so also was the organic world ruled by a single law of development. But when he came to consider how transmutation could take place, he was reduced to talking of "impulses" which in the course of generations could "modify organic structures in accordance with external circumstances." Darwin was most sympathetic to the work but, like many others, derived little insight from the idea of "impulses."

17. Would it have ever been published without outside pressure? Darwin, in a letter to Wallace in 1859, seemed to think not: "I owe indirectly much to you . . . for I almost think Lyell would have proved right, and I should never have completed my larger work" (Darwin 1958: 206).

18. Wallace began his working life as a surveyor but to indulge a passion for natural history made a living for many years as bug hunter. Prolonged trips to the Amazon and the Malay Archipelago brought him enormous numbers of specimens which he sold to dealers. It was profitable work. On returning from the Far East in 1862, he had saved enough to become financially independent. (Unwise investments in foreign railways and Welsh lead mines, and some costly legal adventures destroyed Wallace's financial security.) Despite his role in the development of the theory of evolution, Wallace was unable to obtain either university or museum employment. He supported himself by writing popular science works and eventually received a civil service pension of £200 a year.

19. Wallace was in Dorey, New Guinea, suffering from fever and a leg ulcer when

Darwin received his letter. It had been written in Ternate, a small island in the Moluccas, in February 1858 and posted on a Dutch mail boat in early March. At the time, Wallace once more was suffering from a fever and "every day during the cold and succeeding hot fits I had to lie down for several hours, during which time I had nothing to do but to think over any subjects then particularly interesting me" (Wallace: 190).

20. Why did he send it to Darwin in the first place? In a letter to the ornithologist A. Newton (1829-1907) in 1887, Wallace described his early relation with Darwin. He had seen him just once, in the British Museum, and knew of him through his agent as a customer for any "curious varieties." Wallace thought he had read *Voyage of the Beagle* at that time, but was not sure. As to the correspondence between them Wallace commented, "I must have heard from some notices in the *Athenaeum*. . . that he was studying varieties and species, and as I was continually thinking of the subject, I wrote to him giving some of my notions, and making some suggestions" (Darwin 1958: 200).

21. Brackman's is not the only voice to query Darwin's originality. Loren Eiseley in *Darwin and the Mysterious Mr X* (1979) argued the case for Edward Blyth (1810-73) as the source of Darwin's idea of natural selection, with the implication that Darwin was a plagiarist. Blyth spent the period 1841-62 in India as Curator of Vertebrates in the Calcutta Museum. He knew Darwin well and corresponded with him on questions of natural history. Before leaving for India he published in 1835 and 1837 two papers in *Magazine of Natural History* from which, Eiseley has claimed, Darwin derived his key idea of natural selection. Eiseley's views have not been widely accepted. See the 1979 review by Sydney Smith in *New Scientist.*

22. According to Brackman, when Darwin read Wallace's 1855 paper he had not yet solved the problem of the origin of species, and Wallace's paper gave him the solution. To establish priority, he wrote to Gray and, supposedly on the advice of Lyell, began writing his big book. Wallace's 1858 letter did not arrive at Down on June 18 but some days before. Darwin delayed announcing its arrival until he had had time to adjust his "big book." Later, he destroyed all the correspondence from Wallace, Lyell and Hooker received during the relevant period. Lyell and Hooker are also accused of a "devious, unethical ploy" by arranging at the Linnean Society for Darwin's paper to be read first. Brackman's case is persuasively argued and deserves a fuller answer. It falls down on three counts. No hard evidence is ever produced to prove the existence of a conspiracy; instead, the negative evidence he presents is the sort that depends on the absence of evidence disproving a conspiracy. Secondly, it is unclear that Darwin needed Wallace's two papers to complete his theory. Finally, Brackman overvalues the role of a particular idea in the development of a scientific theory at the expense of the application of the theory.

23. Etty, or Henrietta, developed into a hypochondriac of extraordinary dimensions. Advised on one occasion to breakfast in bed, Gwen Raverat reports "she never ever got up to breakfast again in all her life. . . . Ill health became her profession. . . and absorbing interest." At Down, "ill health was considered normal" (Raverat: 122-23). Upon the death of the young Charles, a mentally retarded child, Darwin later admitted he was thankful as well as sorrowful.

24. It is difficult to find a more disingenuous statement than Darwin's report to Wallace in January 1859: "I had absolutely nothing to do in leading Lyell and Hooker to what they thought a fair course of action" (Darwin 1958: 206).

25. Moreover, they can lead to confusion and injustice. An example is described in *Betrayers of Truth* (1982) by W. Broad and N. Wade. In 1978 Helena Rodbard submitted a paper to *New England Journal of Medicine* on "Insulin Receptor Abnormalities in Anorexia Nervosa." It was refereed by a Yale biochemist, P. Felig, who showed the paper to a subordinate, V. Soman, who was working on the same problem. Felig recommended that the paper be rejected; Soman used Rodbard's results to complete a paper of his own which he attempted to publish in *American Journal of*

Medicine. Rodbard, however, saw a copy of this paper, recognized her own work, reported the matter to the various editors, and accused Felig and Soman of plagiarism. After much delay and prolonged investigation, Soman resigned from Yale while Felig, innocent of fraud, remained at Yale. The issue was complicated by other factors, but parallels to the Darwin-Wallace affair are apparent.

26. The title *Abstract* indicated that the material was abstracted from the "big book" begun by Darwin on the advice of Lyell in 1855. Apart from haste, one reason *Origin of Species* carried few references was because Darwin assumed its publication would be soon followed by the publication of the *Abstract* carrying all such necessary information. Some of the material was used in other books but the full text was not published until 1975 when it appeared under the title *Charles Darwin's Natural Selection* (edited by R.C. Stauffer).

27. "For these several reasons, namely, the improbability...these supposed species being quite unknown in the wild state, and their becoming nowhere feral; these species having very abnormal characters...as compared with all other Columbidae, though so like in most other respects to the the rock pigeon" (Darwin 1968: 86-7).

28. Both of these phrases are found in *Origin of Species* as headings of chapters three and four respectively.

29. Darwin's reading of Malthus finds a parallel, oddly enough, in Wallace's own research. Sick with fever on the island of Ternate Wallace reported: "One day something brought to my recollection Malthus's 'Principles of Population'....I thought of his clear exposition of 'the positive checks to increase'....It then occurred to me that these causes...are continually acting in the case of animals also" (Wallace: 190).

30. Chapter four contains a section on "Illustrations of the Action of Natural Selection" which clearly reveal one source of Darwin's thought. Writing of the power of natural selection to guide "infinitesimally small inherited modifications" into the creation of new species, he anticipated the objection that nothing so small could have effects so large by referring to the work of Lyell in which "trifling and insignificant causes" produce dramatic effects like "the excavation of gigantic valleys" and the "lines of inland cliffs" (Darwin 1968: 142).

31. As examples of the application of the laws, Darwin noted "that animals of the same species have thicker and better fur the further north they live" (external conditions); "the eyes of moles...are rudimentary in size...probably due to gradual reduction from disuse"; "variations of structure arising in the young or larvae naturally tend to affect the structure of the mature animal" (correlated variation).

32. Just how a modern Darwinian deals with these problems can be seen in J. Maynard Smith's *The Theory of Evolution* (1966).

33. Darwin's concern with island ecology is very much at the center of current biological concern and the questions he posed are in the spirit of those tackled by E.O. Wilson and R. MacArthur in their *Theory of Island Biogeography* (1967).

34. Darwin's assumed the prevalent blending theory of his day. Before Mendel, it was widely held that a cross between two individuals would produce a blend of parental properties: a tall black crossed with a short white would thus produce a medium-sized brown.

35. He had considered, but rejected, earlier answers. They are analyzed by Vorzimmer, 97-111.

36. The doctrine was associated with the name of Lamarck, and short of being caught cheating, "Lamarckian" is as bad a charge as can be made against a modern biologist. Paul Kammerer (1880-1926) had the misfortune to be accused of both. The story of his work and ultimate suicide are told in Arthur Koestler's *The Case of the Midwife Toad* (1971). A more recent case for Lamarckian effects, this time in the immune system, has been made by a young Australian, Ted Steele. It has inevitably

led to the same controversy and suspicion that greeted Kammerer. Attempts to perform crucial experiments, difficult and time consuming, have so far led nowhere. The story can be followed in *New Scientist* (1981: 19 Feb.; 23 Apr.; 7 May; 21 May) and *Discover* (Aug. 1981).

37. To a traditional objection, Lamarckians offer a traditional reply: they are concerned with the response of organisms to conditions they meet in nature; surgical mutilations are a different matter.

38. St. George Jackson Mivart, son of the founder of Mivart Hotel, better known under its present name of Claridge's, was a well-known Catholic convert. He wrote a number of popular works on natural history. Although he accepted evolution, he rejected Darwin's account, finding its materialistic and atheistic tendencies conflicted with his religious views.

39. They are: *Expression of the Emotions in Man and Animals* (1872); *Descent of Man,* second edition (1874); *Insectivorous Plants* (1875); *Climbing Plants* (1875); *Variation in Animals and Plants,* second edition (1875); *The Effects of Cross and Self Fertilisation in the Vegetable Kingdom.* (1876); *On the Various Contrivances by which British and Foreign Orchids are Fertilised by Insects,* enlarged second edition (1877); *Different Forms of Flowers on Plants of the Same Species.* (1877); *Life of Erasmus Darwin* (1879); *Power of Movement in Plants* (with the assistance of Francis Darwin) (1880); *The Formation of Vegetable Mould Through the Action of Worms* (1881).

40. Darwin assumed that "the sea would eat into cliffs 500 feet in height at the rate of one inch in a century." In response to criticism from the geologists, he had qualified his claim in the second edition: "But perhaps it would be safer to allow two or three inches per century, and this would reduce the number of years to 100 or 150 million years." All references to the Weald were dropped from later editions.

41. In this he may have been supported by his son George, a one-time assistant to Kelvin. As a mathematical physicist, George Darwin was no doubt aware of how speculative many of the assumptions of his discipline had to be. At the British Association meeting in 1886 he declared: "At present our knowledge of a definite limit to geological time has so little precision that we should do wrong summarily to reject any theories which appear to demand longer periods than those which now appear allowable."

42. The phenomenon of radioactivity was unknown to physicists until 1896 when Henri Becquerel (1852-1908) noticed that uranium salts were capable of spontaneously fogging photographic plates.

43. Mendel, an Austrian monk at Brunn, published in 1866 a paper describing some breeding experiments. From his research he inferred the particulate nonblending theory of heredity so badly needed by Darwin. The paper, though read by competent scholars, was ignored until 1900 when his results were independently rediscovered by three widely separated workers. One of the essential discoveries of Mendel was that properties did not blend; they were either present or absent. Thus, when he crossed yellow and green peas the result was yellow peas. The green character had not been completely lost for when the hybrids were self fertilized, a constant proportion of the progeny was green. Mendel described, in this context, the green varieties as dominant and the yellow as recessive.

44. Genes are strung along pairs of homologous chromosomes. As the chromosomes segregate and recombine they do so in a variety of ways. Genes thus appear in the chromosomes as two variants (alleles). Assume that a particular species has only 1,000 genes. Segregation and recombination alone would permit as many as 2^{1000} different genetic variants, a figure in excess of the number of particles in the universe. Not all variants are possible or viable; there still remains sufficient variation for natural selection to work on.

45. Thus, William Sumner, the leading American social Darwinian, declared

"millionaires are the product of natural selection"; and German generals concluded that war was a "biological necessity" (see Darwin 1968: 45).

46. For an example, Harris (17-21) discusses the sanctity attached to cows in India. "India has 60 million farms, but only 80 million traction animals," he writes. "India's cattle annually excrete about 700 million tons of recoverable manure. Approximately half of this total is used as fertilizer, while most of the remainder is burned to provide heat for cooking." Without this supply India would need the thermal equivalent of 35 million tons of coal. So vital is the cow to the Indian economy that "the taboo on slaughter and beef eating may be as much a product of natural selection as the small bodies...of the zebu breeds."

47. Huxley and Sir Richard Owen (1804-92) quarrelled bitterly on the nature of the human skull and brain. Owen claimed that they could be distinguished from other primate skulls and brains, on purely anatomical grounds, insisting that the posterior lobe, the posterior cornu and the hippocampus could be found only in the human brain. The dispute was savagely pursued in the early 1860s in the pages of *The Athenaeum* and was parodied in Kingsley's *The Water Babies.* Far from absent, Huxley claimed, "they are the most marked cerebral characters common to man with the apes." Neither Huxley nor Owen had much idea of the role of the hippocampus.

48. Hamilton's solution was based in large part on the peculiar haplo-diploid genetic structure of the *Hymenoptera.* This means that while females develop from fertilized eggs and possess a double set of chromosomes (diploid), males develop from unfertilized eggs and consequently possess a single set of chromosomes (haploid). In a normal diploid population the co-efficient of relationship, or in more familiar language the average fraction of shared genes, between mother and daughter will be exactly the same as that between two full siblings. Both groups will, on average, have exactly half their genes in common. In a haplo-diploid species females will inherit half their mother's genes but both their haploid father's chromosomes. So, too, will all their sisters. Consequently, although sisters will share half their genes with their mothers, they will have three-quarters of their chromosomes in common, on average, with each other. It follows that if individuals strive to maximize the number of their genes present in future populations then they will be better served by caring for their sisters than breeding themselves. Hamilton's argument was first presented in "The Genetical Theory of Social Behaviour," *Journal of Theoretical Biology* 7 (1964): 1-32.

49. Kimura objects not to evolution itself but to the view that a mutant gene "increases in the population only by passing the stringent test of natural selection." Instead, he states, "most of the mutant genes that are detected only by the chemical techniques of molecular genetics are selectively neutral." Consequently, instead of the natural selection of genes Kimura speaks of evolutionary processes being caused by the "random drift" of mutant genes (Kimura: 94).

50. Butler's most important biological works are *Life and Habit* (1878), *Evolution: Old and New* (1879), *Unconscious Memory* (1880) and *Luck or Cunning* (1887).

51. That it still exists is a minor miracle, according to Darwin's son Francis. Few people showed less respect for books than Darwin. If one was too bulky to be read with comfort, he ripped it in two. If a chapter, illustration, reference or quotation in a book interested him, he tore it out. Those books that escaped dismemberment were likely to be scarred by Darwin's annotating pencil. His offprint collection in the Cambridge Botany Library, about 2,000 items, is embellished with more than 100,000 words of marginal comment.

52. The concept was introduced in the first four editions at the beginning of chapter three. In the fifth edition, Darwin added the comment, "But the expression often used by Mr. Herbert Spencer of the Survival of the Fittest is more accurate, and is sometimes equally convenient" (Darwin 1958a: 74).

53. The word was used by Darwin in his *Descent of Man* (1871). In *Origin of Species* it was first used in chapter eight on "Instinct." Speaking of the habit of *Molothorus,* an American bird similar in habit to the cuckoo, Darwin noted: "Mr. Hudson is a strong believer in evolution" (Darwin 1958a: 238).

Afterword

> As for what I have done as a poet, I take no pride in whatever. Excellent poets have lived at the same time with me, poets more excellent lived before me, and others will come after me. But that in my country I am the only person who knows the truth in the difficult science of colors – of that, I say, I am not a little proud, and here have I a consciousness of superiority to many.
>
> – Goethe

This exploration of the classics of science began with the query, "Are there classics of science?" Having discussed twelve at length, we are in a position to make a few generalizations. We have shown that the "classics" are those books that have been widely read and disseminated, more than has been appreciated, either in book form or as adaptations or plagiarized versions, in anthologies, commentaries, guides, and epitomes. Some have reached their audiences in oral form or been passed around from hand to hand. All the classics have exerted a profound impact on science and continue to do so today, whether as formulations of great scientific truths or as standards of scientific inquiry. We have shown that the classics of science have not only influenced scientific thought, but have also absorbed the interest of historians of ideas, literature, art, politics, and philosophy. Indeed, the lines between the history of science and the philosophy of science are often blurred – so close are the two disciplines. For the great scientific texts are known not for the faulty data or weak observational support on which they were often based, but rather on the originality of the ideas they presented, the avenues of thought they opened, the problems they posed. All acted to stimulate scientific and intellectual thought long after they first appeared.

It is widely accepted that scientific discoveries are seldom unique events. Significant problems have a way of engaging the attention of several scholars at any one time and, given the universality of the scientific method, several of them, most probably, will pursue similar lines of investigation. Consequently, nearly simultaneous announcements of identical results will often emerge from scientists working independently of each other. In the terminology of Robert Merton, who has been an influential writer on the subject (see Merton 1961), multiples are the norm in science; singletons are so rare as to be safely discounted. Examples are not hard to find, and some – Newton and Leibniz on calculus, Gauss and Bolyai on non-Euclidean geometry, Darwin and Wallace on natural selection – have been discussed in this book. The results can sometimes be startlingly close in content and timing. A famous 19th-century instance concerned the attempt of chemists to liquify atmospheric gases. Louis Cailletet (1832–1913) announced in Paris on 24 December 1877 his success in liquifying oxygen. Two days earlier, on 22 December, the secretary of the Académie des Sciences received a telegram from Raoul Pictet (1846–1929), a Swiss chemist, announcing identical success. If two days could threaten a scientist's claim to priority, how could Copernicus or Darwin have risked delaying publication of their work for decades? Or how do we explain the reluctance of such major figures as Thomas Hariot (1560–1621), Henry Cavendish (1731–1810) or Claude Shannon (b. 1916) to publish much of their work?—work often published later, piecemeal, by many lesser writers.

The reasons why scientists seek the earliest possible publication of their work can thus be well understood, and writers of classic works, such as Linnaeus, Vesalius, Lyell, and Galileo, showed no hesitation in publishing, speedily and without any hitch. These men were relatively young, with reputations to make, money to be earned, and positions to be gained: they used their work to advance careers, seek patronage, consolidate their scientific fame or simply to earn money. In contrast, Copernicus, Harvey, Newton and Darwin were under no such pressures. They were men either of independent means or had achieved patronage and position early in their lives. It is tempting to dismiss the subject of why they failed to publish early in their careers as a trivial matter. Such an answer fails to resolve the question of why, later on, they showed such determination and eagerness to publish.

Might there be a psychological element in the make-up of these scientific geniuses that inhibited them from bringing forth their work in public? They had less need to fear anticipation for they must have been aware of the unique character of their work. The more original the work, the greater the genius, the rarer the chances of duplication.

If we can speculate, part of the answer may lie in a not uncommon drive for perfection and completeness. Darwin, for example, was constantly on the lookout for more facts to support his theories. As the data from natural history were never likely to run out, Darwin—as he himself suspected—could easily have spent his life contentedly collecting material for a never-to-be-completed "big book." Newton exasperated his contemporaries by claiming that his work was "imperfect and confus'd," to which the seventy-nine-year-old John Wallis (1616–1703) commented, "modesty is a virtue, but too much diffidence . . . is a fault." And when Newton complained further that his work was unfinished, Wallis testily objected, "It may be so, (and perhaps never will be), but pray let us have what is; and while that is printing you may (if ever) perfect the rest." And in a modern example, the great philosopher Ludwig Wittgenstein (1889–1951) left behind a substantial body of work, much of it in publishable form, which was only made available after his death. He commented on one of his works in 1945, "I should have liked to produce a good book. This has not come about, but the time is past in which I could improve it." The work to which he was referring was his *Philosophical Investigations.*

Writers of classic works are not the only scholars constrained by feelings of inadequacy and drives for perfection. Pedestrian minds are also afflicted by such qualms. The two cases are quite different. The latter springs from a fussy and cautious personality, the former from the nature of the task pursued. How is a revolutionary work of science, claiming to overthrow the accepted view of the work, likely to appear to its author? It must, first, be tempting to rush into print and to collect the inevitable fame and prizes which go along with such achievement. But what if the work is flawed and rendered worthless by an elementary error? Many ordinary authors may feel this. What they cannot have is the sense of possession and power that authors of classics of science feel. As their works contain deep truths about the mechanisms of nature, known only to them, they may well be tempted only to forgo publication in order to enjoy the feeling of being the thinker who uniquely understands how nature really works. Publication would rob them of this power. Such psychological feelings may thus explain why authors of classics of science have sometimes found it difficult to publish their work.

There remains the question of the role played by scientific classics in the development of science. When attempts have been made to identify the basic currency of scientific thought, the units most commonly adopted have been theories and experiments. Consequently, historians and philosophers have long concentrated on identifying the origin, analyzing the structure, and tracing the development of such basic units as atomic theory, planetary theory, Hippocratic theory, relativity theory, and quantum theory, punctuated by equally detailed studies of such supposedly decisive events as the Michelson-Morley experiment, Newton's experiments on light and color, and Lavoisier's phlogiston experiments. With such a wealth of material at their disposal, some philosophers have taken as their primary task the elucidation of the precise logical relationship between theory and experiment.

Valuable as such approaches undoubtedly are, they are not without their limitations. Theories often become tidied up, neatly ordered logical structures in contrast to the messy, tentative, changeable entity not infrequently found outside the textbooks. To talk, for example, of Darwin's theory of evolution as if he held a single set of related propositions throughout his life is a commonly met distortion of the facts. Experiments, too, can be misread, and the event which appears so decisive to the historian may well have been largely ignored by the participants. Recently, for example, it has been demonstrated that the failure of the Michelson-Morley experiment to detect ether in 1887 was not, as was widely assumed, the dominant influence that led Einstein to formulate in 1905 his special theory of relativity.

A greater realism can be introduced into the discussion by concentrating less on idealized theories and experiments and more on the classic texts from which they are derived. Merely to know that there were six editions of *Origin of Species*, or that Newton spent much of his later life correcting and rewriting *Principia,* is to see the theories in a different light. At once they become more tentative, more changeable and altogether vaguer entities than the sets of timeless relations often presented in textbooks.

The classic is clearly a more objective entity than a theory. It is always a difficult task to pursue the spread of a theory over a period or throughout a region; it is less of a problem to trace the publishing history of a text. While responses to a theory may often be nebulous, reactions to a specific book can often be identified with great precision.

Nonetheless, such considerations are relatively minor ones. Since the seminal work of Kuhn (1966), it has become increasingly clear that scientific progress is made through a series of revolutionary advances. The idea that science developed in a Baconian manner by the careful collection of facts, by processes of accumulation, has long been discarded and replaced by an image of a decisive and sudden transformation of thought. Kuhn has spoken of periods of "normal" science during which there is widespread acceptance by the scientific community of an agreed orthodoxy, which he calls a "paradigm." It is these paradigms which are overthrown and replaced by new orthodoxies during a period of revolutionary science. What entities are these "paradigms" and what replaces them? They are normally spoken of as if they were highly successful theories, such as evolutionary theory, Copernican theory, or relativity theory. The term "theory," we suggest, is a confusing one, and it is perhaps better to identify, at least in certain areas, the emerging paradigms of a period of revolutionary science with more easily distinguishable classics of science.

Appendix: Editorial Details of the Classics of Science (Summarized)

ANTIQUITY

Author, Title	Date	First Printed Edition Latin	 Greek	 English	Standard Edition	No. of Editions	Translations
Hippocrates, *Corpus*	400-300 B.C.	1525 Rome F. Calvus	1526 Venice Asulanus	1849 London F. Adams	E. Littré (Paris) 1839-61 10 vols.	20	Arabic, Latin French, German, English, Italian, Swedish
Euclid, *Elements*	c. 300 B.C.	1482 Venice E. Ratdolt	1533 Basel Grynaeus	1570 London H. Billingsley	J. L. Heiberg (Leipzig) 1833-1916 5 vols.	Several hundred	20 languages
Ptolemy, *Almagest*	c. 150 A.D.	1515 Venice Gerard of Cremona	1538 Basel Grynaeus and Camerarius	1952 Chicago R. Taliaferro	J.L. Heiberg (Leipzig) 1898-1903 2 vols.	13	Arabic, Latin, German, Polish, French English

MODERN

Author, Title	Editions Prepared by Author	Other Editions	Opera omnia	First English Translation	Other Translations
Copernicus, *De revolutionibus*	1543, Nuremberg	19	Planned	1952 Chicago G. C. Wallis	Polish, German, Russian, English
Vesalius, *De fabrica*	1543, Basel 1555, Basel	5	1725 Leyden H. Boerhaave and B. Albinus 2 vols.	–	–
Galileo, *Siderius nuncius*	1610, Venice 1610, Frankfurt	16	1656 Bologna 2 vols.	1880 London E. S. Carlos	Italian, German, English
Harvey, *De motu cordis*	1628, Frankfurt	40	1766 London M. Akenside	1653 London	English, German, French, Danish, Russian
Newton, *Principia*	1687, London (ed. by E. Halley) 1713, London (ed. by R. Cotes) 1726, London (ed. by H. Pemberton	39	1779-85 London S. Horsley 5 vols.	1729 London A. Motte	English, French, German, Russian, Italian, Japanese, Swedish, Rumanian
Linnaeus, *Systema naturae*	1735, Leyden 1740, Halle, Stockholm 1744, Paris 1747, Halle 1748, Leipzig, Stockholm 1753, Stockholm 1756, Leyden 1758, Stockholm 1762, Leipzig 1766, Stockholm	3 facsim- iles	–	–	–

Author, Title	Editions Prepared by Author	Other Editions	Opera omnia	First English Translation	Other Translations
Dalton, *New System of Chemical Philosophy*	1808-27, Manchester 1842, Manchester (vol. 1, pt 1)	1 facsimile	–	–	–
Lyell, *Principles of Geology*	1830-33, London (3 vs.) 1832-33, London (3 vs.) 1834, London (4 vs.) 1835, London (4 vs.) 1837, London (4 vs.) 1840, London (3 vs.) 1847, London (1 v.) 1850, London (1 v.) 1853, London (1 v.) 1867-68, London (2 vs.) 1872, London (2 vols.) 1875, London (2 vols.)	–	–	–	–
Darwin, *Origin of Species*	1859, London 1860, London 1861, London 1866, London 1869, London 1872, London	320	–	–	170 editions in 28 languages

Bibliography

Alexander, H.G. 1956. *The Leibniz-Clarke Correspondence.* Manchester: Manchester University Press.

Andrade, E.N. da C. 1965. *Rutherford and the Nature of the Atom.* London: Heinemann.

Aristotle. 1956. *Metaphysics.* London: Everyman's Library.

Armstrong, Angus. 1957. *Copernicus, the Founder of Modern Astronomy.* New York: Thomas Yoseloff.

Aubrey, John. 1972. *Brief Lives.* Harmondsworth: Penguin Books.

Bailey, Sir Edward. 1962. *Charles Lyell.* London: Thomas Nelson & Sons.

Bagley, J.J. and Rowley, P.B. 1966. *A Documentary History of England.* Harmondsworth: Penguin Books.

Balfour, J.H. 1875. *A Manual of Botany.* London: A & C Black.

Ball, W.W. Rouse. 1893. *An Essay on Newton's "Principia".* London: Macmillan.

Barber, Lynn. 1980. *The Heyday of Natural History.* London: Jonathan Cape.

Barrett, P.H., Weinshank, D.J., and Gottleber, T.T. 1982. *Concordance to Darwin's Origin of Species, First Edition.* Ithaca: Cornell University Press.

Blackwelder, R.E. 1967. *Taxonomy.* New York: Wiley.

Blunt, Wilfrid, 1971. *The Compleat Naturalist.* London: Collins.

Boss, Valentin. 1972. *Newton and Russia.* Cambridge, MA: Harvard University Press.

Brackman, A.C. 1980. *A Delicate Arrangement.* New York: Times Books.

Brewster, Sir David. 1855. *Memoirs of the Life, Writings and Discoveries of Sir Isaac Newton* (2 vols.) Edinburgh: Thomas Constable and Co.

Broad, W. and Wade, N. 1983. *Betrayers of Truth.* London: Century Publishing Co.

Brock, A.J. 1929. *Greek Medicine.* London: J.M. Dent & Sons.

Buckland, William. 1835. On the Discovery of a New Species of Pterodactyl in the Lias at Lyme Regis. *Geological Transactions,* Series 2, 3: 217-22. London.

Burchfield, J.D. 1975. *Lord Kelvin and the Age of the Earth.* New York: Science History Publications.

Butler, E.M. 1949. *Ritual Magic.* Cambridge: Cambridge University Press.

Byrne, Oliver. 1847. *The First Six Books of the Elements.* London.

Cannon, H.G. 1958. *The Evolution of Living Things.* Manchester: Manchester University Press.

Capp, Bernard. 1979. *Astrology and the Popular Press.* London: Faber and Faber.

Cardwell, D.S.L. (ed.) 1968. *John Dalton and the Progress of Science.* Manchester: Manchester University Press.

Celsus. 1935. *De Medicina* (3 vols.) Cambridge, MA.: Loeb Classical Library.

Chadwick, J. and Mann, W. 1978. *Hippocratic Writings.* Harmondsworth: Penguin Books.

Chancellor, John. 1973. *Charles Darwin.* London: Weidenfeld and Nicolson.

Clark, G.N. 1964. *The Royal College of Physicians.* Vol. I. Oxford: Clarendon Press.

Clendening, Logan. (ed.) 1960. *A Source Book of Medical History.* New York: Dover Publications.

Cohen, I.B. (ed.) 1958. *Isaac Newton's Papers and Letters on Natural Philosophy.* Cambridge: Cambridge University Press.

1978. *Introduction to Newton's Principia.* Cambridge: Cambridge University Press.

1980. *The Newtonian Revolution.* Cambridge: Cambridge University Press.

1981. Newton's Discovery of Gravity. *Scientific American,* 244 (March): 122-33.

Colp, Ralph. 1977. *To Be an Invalid.* Chicago: University of Chicago Press.

Copernicus, Nicolaus. 1978. *On the Revolutions.* London: Macmillan.

Coxeter, Harold. 1961. *Introduction to Geometry.* New York: Wiley.

Crombie, A.C. 1964. *Augustine to Galileo.* London: Heinemann.

Crosland, M.P. 1962. *Historical Studies in the Language of Chemistry.* London: Heinemann.

1971. *The Science of Matter.* Harmondsworth: Penguin Books.

Cushing, Harvey. 1962. *A Bio-bibliography of Andreas Vesalius.* Hamden, CT: Archon Books.

Cuvier, Georges. 1799. *Mémoire sur les espèces d'éléphans vivantes et fossiles.* Paris.

1812. *Recherches sur les ossemens fossiles* (4 vols.) Paris.

1817. *Essay on the Theory of the Earth.* 3rd edition, Translated by R. Jameson. Edinburgh.

Cuvier, Georges and Brogniart, Alexandre. 1811. *Essai sur la géographie minéralogique des environs de Pairs.* Paris.
Darlinton, C.D. 1959. *Darwin's Place in History.* Oxford: Basil Blackwell.
Darwin, Charles. 1845. *Journal of a Voyage Around the World.* London: John Murray.
1871. *The Descent of Man* (2 vols.) London: John Murray.
1887. *Life and Letters* (3 vols.) Edited by Francis Darwin. London: John Murray.
1958. *The Autobiography of Charles Darwin and Selected Letters.* New York: Dover Publications.
[1872] 1958a. *The Origin of Species,* Reprint of 6th ed. New York: Mentor Book.
1959. *The Origin of Species.* Variorum text, edited by M. Peckham. Philadelphia: University of Pennsylvania Press.
[1859] 1968. *The Origin of Species.* Reprint of 1st edition. Harmondsworth: Penguin Books.
1974. *Autobiographies.* Edited by Gavin de Beer. London: Oxford University Press.
Darwin, Charles and Wallace, A.R. 1958. *Evolution by Natural Selection.* Cambridge: Cambridge University Press.
Darwin, Erasmus. 1791. *The Botanic Garden.* Part II: The Loves of the Plants. London.
Davy, Sir Humphry 1839-40. *The Collected Works* (9 vols.) London: Smith, Elder & Co.
D'Elia, P. M. 1961. *Galileo in China.* Cambridge, MA.: Harvard University Press.
De Santillana, Giorgio. 1961. *The Crime of Galileo.* Chicago: University of Chicago Press.
De Villamil, Richard. 1931. *Newton, the Man.* London: Gordon D. Knox.
Diaz, Bernal 1963. *The Conquest of New Spain.* Harmondsworth: Penguin Books.
Dobbs, B. J. T. 1975. *The Foundations of Newton's Alchemy.* Cambridge: Cambridge University Press.
Dodds, E. R. 1951. *The Greeks and the Irrational.* Berkley: University of California Press.
Drake, Stillman. 1978. *Galileo at Work.* Chicago: University of Chicago Press.
Dreyer, J. L. T. 1953. *A History of Astronomy from Thales to Kepler.* New York: Dover Publications.
Dupree, A. H. 1968. *Asa Gray.* New York: Atheneum.
Edelstein, E. and Edelstein, L. 1945. *Asclepius.* Baltimore: John Hopkins University Press.
Eicher, D. L. 1968. *Geologic Time.* London: Prentice/Hall International Inc.
Eiseley, Loren. 1979. *Darwin and the Mysterious Mr. X.* London: J. M. Dent and Sons.
Eliot, A. H. and Stern, B. 1979. *The Age of Enlightenment,* Vol. II. London: Ward Lock.
Evans-Pritchard, Edward. 1956, *Nuer Religion.* Oxford: Clarendon Press.
Eysenck, H.J. 1970. *The Inequality of Man.* London: Fontana.
1971. *Race, Intelligence and Education.* London: Temple Smith.
1975. *Crime and Personality.* London: Paladin.
Feyerabend, Paul. 1975. *Against Method.* London: New Left Books.
Freeman, R. B. 1977. *The Works of Charles Darwin.* Folkestone: Dawson.
1978. *Charles Darwin: A Companion.* Folkestone: Dawson.
Frisch, Otto. 1979. *What Little I Remember.* Cambridge: Cambridge University Press.
Galileo, Galilei. 1953. *Dialogue on the Great World Systems.* Chicago: University of Chicago Press.
1957. *Discoveries and Opinions of Galileo.* New York: Doubleday.
1960. *The Controversy on the Comets of 1618.* Philadelphia: University of Pennsylvania Press.
Garland, M. M. 1980. *Cambridge Before Darwin.* Cambridge: Cambridge University Press.
Geikie, Sir Archibald. 1962. *The Founders of Geology.* New York: Dover Publications.
Geymonat, Ludovico. 1969. *Galileo Galilei.* New York: McGraw-Hill.
Gingerich, Owen. 1973. Copernicus and Tycho. *Scientific American,* 229 (December): 86-101.
1980. The Great Copernicus Chase. *American Scholar,* 49 (Winter): 81-88.
Glacken, C. J. 1976. *From the Rhodian Shore.* Berkeley: University of California Press.
Glanville, S. R. K. 1942. *The Legacy of Egypt.* Oxford: Clarendon Press.
Gombrich, Ernst. 1960. *Art and Illusion.* London: Phaidon Press.
Gould, S. J. 1980. *Ever Since Darwin.* Harmondsworth: Penguin Books.
1980. *The Panda's Thumb.* New York: W. W. Norton & Co.
Grant, Edward. 1974. *A Source Book in Medieval Science.* Cambridge, MA: Harvard University Press.
Graubard, Mark. 1964. *Circulation and Respiration.* New York: Harcourt, Brace & World Inc.
Greenaway, Frank. 1966. *John Dalton and the Atom.* London: Heinemann.
Greenberg, M. J. 1980. *Euclidean and Non-Euclidean Geometries.* San Francisco: Freeman.
Greene, J. C. 1974. *The Death of Adam.* Ames: Iowa State University Press.

Hall, A. R. 1962. *The Scientific Revolution.* London: Longmans.
1980. *Philosophers at War.* Cambridge: Cambridge University Press.
Harris, Marvin. 1977. *Cows, Pigs, Wars and Witches.* London: Fontana Books.
Harrison, John. 1978. *The Library of Isaac Newton.* Cambridge: Cambridge University Press.
Hartley, Sir Harold. 1971. *Studies in the History of Chemistry.* Oxford: Clarendon Press.
Haskins, C.H. 1957. *The Renaissance of the Twelfth Century.* New York: Meridian Books.
1960. *Studies in the History of Medieval Science.* New York: Frederick Ungar.
Heath, T. L. 1921. *Greek Mathematics* (2 vols.) Oxford: Clarendon Press.
1932. *Greek Astronomy.* London: Dent.
1956. *Euclid's Elements* (3 vols.) New York: Dover Publications.
Heiberg: see Heath 1956.
Henrey, Blanche. 1975. *British Botanical and Horticultural Literature Before 1800* (2 vols.) London: Oxford University Press.
Henry, W. C. 1854. *Memoirs of the Life and Scientific Researches of John Dalton.* Manchester.
Hilbert, David. 1899. *Grundlagen der Geometrie.* Leipzig.
Hill, Christopher. 1964. William Harvey and the Idea of Monarchy. *Past and Present,* 27 (April): 54-72.
Hill, John. 1756. *The British Herbal.* London.
1759-75. *The Vegetable System* (26 vols.) London.
1760. *Flora Britanica.* London.
Himmelfarb, Gertrude. 1959. *Darwin and the Darwinian Revolution.* London: Chatto and Windus.
Hodgkin, Thomas. 1960. *Nigerian Perspectives.* London: Oxford University Press.
Hoffmann, Friedrich. 1971. *Fundamenta Medicinae.* London: Macdonald and Co.
Holmyard, E. J. 1931. *Makers of Chemistry.* Oxford: Clarendon Press.
Hurgronje, Snoek. 1970. *Mekka.* Leiden: J. Brill.
Hull, D. L. 1973. *Darwin and His Critics.* Cambridge MA: Harvard University Press.
Hume, David. 1878. *The Philosophical Works* (4 vols.) London.
Hurd, D. L. and Kipling, J. J. 1964. *The Origins and Growth of Physical Science,* Vol. I. Harmondsworth: Penguin Books.
Hutton, James. 1788. Theory of the Earth. *Transactions of the Royal Society of Edinburgh,* 1: 209-304.
Huxley, T. H. 1893. *Darwiniana.* London: Macmillan.
1900. *Life and Letters* (3 vols.) London: Macmillan.
Irvine, William. 1955. *Apes, Angels and Victorians.* London: Weidenfeld and Nicolson.
Jacob, Margaret. 1976. *The Newtonians and the English Revolution 1689-1720.* Ithaca: Cornell University Press.
Jefferson, Thomas. 1904. *The Works of Thomas Jefferson,* Vol. III. New York: Putnam's.
1955. *Notes on the State of Virginia.* Chapel Hill: University of North Carolina Press.
Johnson, F. R. 1958. Astronomical Textbooks in the 16th Century. In E. A. Underwood, (ed.), *Science, Medicine and History,* Vol. I., 285-303. Oxford: Clarendon Press.
Jones, B. Z. and Boyd, L. G. 1971. *The Harvard College Observatory.* Cambridge MA: Harvard University Press.
Jones, W. H. S. 1945. Hippocrates and the Corpus Hippocraticum. *Proceedings of the British Academy,* 31: 103-25.
Kastner, Joseph. 1978. *A World of Naturalists.* London: John Murray.
Keele, Kenneth. 1965. *William Harvey.* London: Nelson.
Kennedy, E. S. 1966. Late Medieval Planetary Theory. *Isis,* 57: 365-78.
Kermode, Frank. 1975. *The Classic.* London: Faber and Faber.
Kesten, Hermann. 1945. *Copernicus and His World.* London: Secker and Warburg.
Keynes, Sir Geoffrey. 1928. *A Bibliography of the Writings of William Harvey.* Cambridge: Cambridge University Press.
1947. The History of Blood Transfusion. *Penguin Science News* 3: 25-48. Harmondsworth.
1978. *The Life of William Harvey.* Oxford: Clarendon Press.
1981. *The Gates of Memory.* Oxford: Clarendon Press.
Kimura, Motoo. 1979. The Neutral Theory of Molecular Evolution. *Scientific American,* 241 (November): 94-104.
King, L. S. 1971. *A History of Medicine.* Harmondsworth: Penguin Books.
1982. *Medical Thinking.* Princeton: Princeton University Press.
Kirk, G. S. and Raven, J. E. 1966. *The Presocratic Philosophers.* Cambridge: Cambridge University Press.

Klebs, A. C. 1938. Incunabula Scientifica et Medica. *Osiris* 4: 1-359.
Knight, David. 1967. *Atoms and Elements.* London: Hutchinson.
Koestler, Arthur. 1964. *The Sleepwalkers.* Harmondsworth: Penguin Books.
Koyré, Alexandre. 1965. *Newtonian Studies.* London: Chapman and Hall.
. 1973. *The Astronomical Revolution.* Ithaca: Cornell University Press.
Kuhn, Thomas. 1957. *The Copernican Revolution.* Cambridge, MA: Harvard University Press.
1962. *The Structure of Scientific Revolutions.* Chicago: University of Chicago Press.
Lamarck, J. B. de.1914. *Zoological Philosophy.*
Leicester, H. M. and Klickstein, H. S. 1952. *A Source Book in Chemistry.* New York: McGraw-Hill.
Letwin, W. L. 1963. *Origins of Scientific Economics.* London: Methuen.
Levene, T. H. 1971. *Affinity and Matter.* Oxford: Clarendon Press.
Levi-Strauss, Claude. 1966. *The Savage Mind.* London: Weidenfeld and Nicolson.
Ley, Willy. 1963. *Watchers of the Skies.* London: Sidgwick and Jackson.
Linebaugh, Peter. 1977. The Tyburn Riots Against the Surgeons. In P. Linebaugh, J. G. Rule E. P. Thompson and C. Winslow, *Albion's Fatal Tree.* Harmondsworth: Penguin Books.
Linnaeus, Carl. 1938. *Critica Botanica.* Translated by A. Hort for the Ray Society from the 1st edition (1737). London.
[1758] 1956. *Systema Naturae.* Vol. I. Reprint of 10th edition. London: British Museum.
[1753] 1957-59. *Species Plantarum.* Reprint of first edition (2 vols.) London: Ray Society.
[1735] 1964. *Systema Naturae.* Reprint of 1st edition. Dutch Classics on History of Science; III. Nieuwkoop: B. de Graaf.
1973. Olund and Gotland Journey. *Biological Journal of the Linnean Society,* 5: 1-220.
1979. *Travels.* Edited by David Black. London: Elek.
Lyell, Sir Charles. 1830-33; 1834; 1867-8. *Principles of Geology.* 1st, 3rd and 10th editions. London: John Murray.
1885. *The Student's Elements of Geology.* 4th edition. London: John Murray.
Lyell, K. M. 1881. *Life and Letters* (2 vols.) London: John Murray.
Manuel, Frank. 1963. *Isaac Newton, Historian.* Cambridge MA: Harvard University Press.
1968. *A Portrait of Isaac Newton.* Cambridge MA: Harvard University Press.
1974. *The Religion of Isaac Newton.* Oxford: Clarendon Press.
Mather, K. F. and Mason, S. L. 1964. *A Source Book in Geology.* New York: Hafner Publishing Co.
Mayr, Ernst. 1969. *Principles of Systematic Zoology.* New York: McGraw-Hill.
Maynard Smith, John. 1966. *The Theory of Evolution.* Harmondsworth: Penguin Books.
1978. The Evolution of Behavior. *Scientific American,* 239 (September): 136-45.
McMullin, Ernan (ed.) 1967. *Galileo, Man of Science.* New York: Basic Books.
Medewar, Sir Peter. 1974. *The Hope of Progress.* London: Wildwood House.
Mendelssohn, Kurt. 1966. *The Quest for Absolute Zero.* London: Weidenfeld and Nicolson.
Merton, Robert. 1961. Singletons and Multiples in Scientific Discovery: A Chapter in the Sociology of Science. *Proceedings of the American Philosophical Society,* 105 (October): 470-86.
Midonick, Henrietta. 1968. *The Treasury of Mathematics,* Vol. I. Harmondsworth: Penguin Books.
Miller, Hugh. 1922. *The Old Red Sandstone.* London: Everyman's Library.
Mitchell, O. M. 1862. *The Planetary and Stellar Worlds.* Glasgow: William Collins.
Moorehead, Alan. 1971. *Darwin and the Beagle.* Harmondsworth: Penguin Books.
More, L. T. 1962. *Isaac Newton.* New York: Dover Publications.
Musson, A. and Robinson, E. 1969. *Science and Technology in the Industrial Revolution.* Manchester: Manchester University Press.
Nakayama, Shigeru. 1969. *A History of Japanese Astronomy.* Cambridge MA: Harvard University Press.
Needham, Joseph. 1954. *Science and Civilization in China,* Vol. I: Introductory Orientations. Cambridge: Cambridge University Press.
1959. *Science and Civilization in China,* Vol. III: Mathematics and the Sciences of the Heavens and the Earth. Cambridge: Cambridge University Press.
Neugebauer, Otto. 1969. *The Exact Sciences in Antiquity.* New York: Dover Publications.
1975. *A History of Ancient Mathematical Astronomy* (3 vols.) Berlin: Springer-Verlag.
Newton, Sir Isaac. 1959-77. *The Correspondence of Isaac Newton* (7 vols.) Cambridge: Cambridge University Press.
1966. *Principia* (2 vols.) Berkeley: University of California Press.

1967-1981. *The Mathematical Papers of Isaac Newton* (8 vols.) Edited by D. T. Whiteside. Cambridge: Cambridge University Press.
1972. *Principia.* The Third Edition (1726) with Variant Readings. Assembled and edited by A. Koyré and I. B. Cohen (2 vols.) Cambridge: Cambridge University Press.
Newton, Robert. 1977. *The Crime of Ptolemy.* Baltimore: John Hopkins University Press.
1977a. Claudius Ptolemy: Fraud. *Scientific American,* 237 (October): 79-81.
Nicholson, Marjorie. 1946. *Newton Demands the Muse.* Princeton: Princeton University Press.
Nicholson, R. N. 1962. *A Literary History of the Arabs.* Cambridge: Cambridge University Press.
O'Malley, C. D. 1964. *Andreas Vesalius of Brussels.* Berkeley: University of California Press.
O'Malley, C. D. and Saunders, J. B. de C. M. 1950. *The Illustrations from The Works of Andreas Vesalius of Brussels.* Cleveland, NY: The World Publishing Company.
Orr, M. A. 1956. *Dante and the Early Astronomers.* London: Wingate.
Osler, Sir William. 1929. *Bibliotheca Osleriana.* Oxford: Clarendon Press.
Pagel, Walter. 1967. *Harvey's Biological Ideas.* Basel and New York: Karger.
Pannekoek, Antoine. 1961. *A History of Astronomy.* London: Allen and Unwin.
Partington, J. R. 1962. *A History of Chemistry,* Vol. III. London: Macmillan.
Pedersen, Olaf. 1974. *A Survey of the Almagest.* Odense: Odense University Press.
Pedoe, Dan. 1970. *A Course of Geometry.* Cambridge: Cambridge University Press.
Peters, C. and Knobel, E. 1915. *Ptolemy's Catalogue of Stars.* Washington, DC: Carnegie Institution of Washington.
Phillips, E. D. 1973. *Greek Medicine.* London: Thames and Hudson.
Plato. 1941. *The Republic.* Edited by F. M. Cornford. Oxford: Clarendon Press.
Pliny. 1942-63. *Natural History* (10 vols.). Edited by W. H. S. Jones and D. E. Eichholz. Cambridge, MA: Harvard University Press.
Poincaré, Henri. 1952. *Science and Hypothesis.* New York: Dover Publications.
Popper, Sir Karl. 1972. *Objective Knowledge.* Oxford: Clarendon Press.
Poynting, J. H. and Thomson, J. J. 1904. *A Text-Book of Physics,* Vol. I: Heat. London: Charles Griffin and Company.
Proclus. 1970. *A Commentary on the First Book of Euclid's Elements.* Translated by G. R. Morrow. Princeton: Princeton University Press.
Rae, Isobel. 1964. *Knox, the Anatomist.* Edinburgh: Oliver and Boyd.
Raven, Charles. 1950. *John Ray.* Cambridge: Cambridge University Press.
Raverat, Gwen. 1960. *Period Piece.* London: Faber and Faber.
Ray, John. 1693. *Synopsis Methodica Animalium.* London.
[1724] 1973. *Synopsis Methodica Stirpium Britannicarum.* Reprint of the 3rd edition. London: Ray Society.
Richardson, E. G. 1929. *Sound.* London: Edward Arnold.
Romer, A. S. 1954. *Man and the Vertebrates* (2 vols.) Harmondsworth: Penguin Books.
Ronan, Colin. 1974. *Galileo.* London: Weidenfeld and Nicolson.
Rook, Arthur 1964. *Origins and Growth of Biology.* Harmondsworth: Penguin Books.
Roscoe, H. E. 1895. *John Dalton and the Rise of Modern Chemistry.* London: Cassell and Company.
Roscoe, H. E. and Harden, A. 1896. *A New View of the Origin of Dalton's Atomic Theory.* London: Macmillan.
Rosen, Edward. 1947. *The Naming of the Telescope.* New York: Henry Schumann.
1959. *Three Copernican Treatises.* New York: Dover Publications.
1965. *Kepler's Conversation with Galileo's Starry Messenger.* New York: Johnson Reprint Company.
Runes, D. D. (ed.) 1962. *Treasury of World Science.* Paterson, NJ: Littlefield Adams and Co.
Ruse, Michael. 1979. *The Darwinian Revolution.* Chicago: University of Chicago Press.
1982. *Darwin Defended.* Reading, MA: Addison-Wesley Publishing Company.
Russell, Bertrand. 1967. *Autobiography,* Vol. I: 1872-1914. London: Allen and Unwin.
Russell, C. A. 1968, Berzelius and the Development of the Atomic Theory. In D. S. L. Cardwell, (ed.), *John Dalton and the Progress of Science.* Manchester: Manchester University Press.
Russell, D. A. 1982. The Mass Extinction of the Late Mesozoic. *Scientific American,* 246 (January): 48-55.
Sarton, George. 1927-1947. *Introduction to the History of Science.* 3 vols. in 5. Baltimore: Johns Hopkins University Press.
1953. *A History of Science,* Vol. I: Ancient Science through the Golden Age of Greece. Cambridge MA: Harvard University Press.

1959. *A History of Science*, Vol. II: Hellenistic Science and Culture in the Last Three Centuries B.C. Cambridge MA: Harvard University Press.
1961. *Ancient and Medieval Science during the Renaissance.* New York: Barnes.
Schofield, R. E. 1970. *Mechanism and Materialism.* Princeton: Princeton University Press.
Scientific American. 1978. Evolution, Vol. 239 (September): 39-169.
Singer, Charles. 1956. *Galen on Anatomical Procedures.* Oxford: Clarendon Press.
1957. *A Short History of Anatomy and Physiology.* New York: Dover Publications.
Singer, Charles and Rabin, C. 1946. *A Prelude to Modern Science.* Cambridge: Cambridge University Press.
Smerdlow, N. 1979. Ptolemy on Trial. *American Scholar*, 48: 523-31.
Smith, William. 1815. *A Delineation of the Strata of England and Wales.* London.
1816. *Strata Identified with Organised Fossils.* London.
1817. *Stratigraphical System of Organised Fossils.* London.
Southern, R. W. 1953. *The Making of the Middle Ages.* London: Hutchinson University Library.
Sowerby, James. 1790-1814. *English Botany*, (36 vols.) London.
Stearns, W. T. 1957-9. Introduction. In Linnaeus 1957-9.
1971. Appendix. In Blunt 1971.
1973. Introduction. In Linnaeus 1973.
1979. Linnean Classification. In Linnaeus 1979.
1981. *The Natural History Museum.* London: Heinemann.
Sugimoto, M. and Swain, D. 1978. *Science and Culture in Traditional Japan 600-1854.* Cambridge, MA: M.I.T. Press.
Tait, P. G. 1876. *Lectures on Some Recent Advances in Physical Science.* London: Macmillan.
Targ, R. and Puthoff, H. 1977. *Mind-Reach.* London. Paladin.
Thackray, Arnold. 1968. Quantified Chemistry: the Newtonian Dream. In D. S. L. Cardwell, (ed.), *John Dalton and the Progress of Science.* Manchester: Manchester University Press.
1970. *Atoms and Powers.* Cambridge, MA: Harvard University Press.
1972. *John Dalton.* Cambridge, MA: Harvard University Press.
Thayer, H. S. 1953. *Newton's Philosophy of Nature.* New York: Hafner Publishing Co.
Thomas, Keith. 1973. *Religion and the Decline of Magic.* Harmondsworth: Penguin Books.
Thomas-Stanford, C. 1926. *Early Editions of Euclid's Elements.* London: Bibliographical Society Illustrated Monographs, No. 20.
Thorndike, L. T. 1923-58. *History of Magic and Experimental Science* (8 vols.) New York: Columbia University Press.
Todhunter, Isaac. 1933. *Euclid.* London: Everyman's Library.
Toulmin, S. and Goodfield, J. 1963. *The Fabric of the Heavens.* Harmondsworth: Penguin Books.
Turner, E. S. 1958. *Call the Doctor.* London: Michael Joseph.
Tylor, Sir Edward. 1929. *Primitive Culture* (2 vols.) London: John Murray.
Van der Waerden, B. 1963. *Science Awakening.* New York: Wiley.
Voltaire. 1980. *Letters on England.* Harmondsworth: Penguin Books.
Vorzimmer, Peter. 1972. *Charles Darwin: the Years of Controversy.* London: University of London Press.
Wallace, A. R. 1908. *My Life.* London: Chapman and Hall.
Wallis, P. and Wallis, R. 1977. *Newton and Newtoniana.* 1672-1975. Folkestone: Dawson.
Weil, E. 1943. The Publisher of Harvey's De Motu Cordis. *The Library*, 24: 142-64.
Westfall, R. S. 1980. *Never at Rest.* Cambridge: Cambridge University Press.
Whewell, William. 1831. Review of *Principles of Geology*, Vol. I. in *British Critic*, 17: 180-206.
1832. Review of *Principles of Geology*, Vol. II. in *Quarterly Review*, 47: 103-32.
White, A. D. 1960. *History of the Warfare of Science and Technology with Christendom* (2 vols.) New York: Dover Publications.
White, T. H. 1954, *The Bestiary: a Book of Beasts.* London: Jonathan Cape.
Wilks, Ivor. 1968. Islamic Learning in the Western Sudan. In Jack Goody, (ed.), *Literacy in Traditional Societies.* Cambridge: Cambridge University Press.
Wilson, L. G. 1970. *Journal on the Species Question.* New Haven: Yale University Press.
1972. *Charles Lyell, The Years to 1841.* New Haven: Yale University Press.
Wilson, E. O. 1975. *Sociobiology.* Cambridge MA: Harvard University Press.
Winslow, C-E. A. 1980. *The Conquest of Epidemic Disease.* Madison: University of Wisconsin Press.
Withering, William. 1801. *Botanical Arrangement*, (4 vols.) Fourth Edition. London.
Wittgenstein, Ludwig. 1953. *Philosophical Investigations.* Oxford: Basil Blackwell.

Index